A Beginners' Guide to Scanning Electron Microscopy

Anwar Ul-Hamid

A Beginners' Guide to Scanning Electron Microscopy

Anwar Ul-Hamid
Center for Engineering Research
King Fahd University of Petroleum & Minerals
Dhahran, Saudi Arabia

ISBN 978-3-030-07498-2   ISBN 978-3-319-98482-7 (eBook)
https://doi.org/10.1007/978-3-319-98482-7

© Springer Nature Switzerland AG 2018
Softcover re-print of the Hardcover 1st edition 2018
This work is subject to copyright. All rights are reserved by the Publisher, whether the whole or part of the material is concerned, specifically the rights of translation, reprinting, reuse of illustrations, recitation, broadcasting, reproduction on microfilms or in any other physical way, and transmission or information storage and retrieval, electronic adaptation, computer software, or by similar or dissimilar methodology now known or hereafter developed.
The use of general descriptive names, registered names, trademarks, service marks, etc. in this publication does not imply, even in the absence of a specific statement, that such names are exempt from the relevant protective laws and regulations and therefore free for general use.
The publisher, the authors, and the editors are safe to assume that the advice and information in this book are believed to be true and accurate at the date of publication. Neither the publisher nor the authors or the editors give a warranty, express or implied, with respect to the material contained herein or for any errors or omissions that may have been made. The publisher remains neutral with regard to jurisdictional claims in published maps and institutional affiliations.

This Springer imprint is published by the registered company Springer Nature Switzerland AG
The registered company address is: Gewerbestrasse 11, 6330 Cham, Switzerland

To my wife

To my wife

Preface

The ability of the scanning electron microscope (SEM) to characterize materials has increased tremendously since its inception on a commercial basis at Cambridge, United Kingdom, in 1965. The tremendous prospects offered by this invention have been consistently built upon, thanks to steady advances in instrumentation and computer technology in the past few decades. Presently, surface morphology of materials ranging from biological, polymers, alloys to minerals, ceramics, and corrosion deposits is routinely studied from micrometer to nanometer scale. The SEM has emerged as a vital, powerful, and versatile tool in the advancement of modern day nanotechnology by contributing to the area of characterization of nanostructured materials. Its ease of use, typically prompt sample preparation and straightforward image interpretation combined with high resolution and high depth of field as well as the ability to undertake microchemical and crystallographic analysis, has made it one of the most popular techniques used for characterization. Presently, the SEM is being used by professionals with a diverse technical background, such as life science, materials science, engineering, forensics, and mineralogy, in various sectors of the government, industry, and academia.

A significant number of in-depth and specialized accounts of the scanning electron microscopy are available to interested readers. This book is meant to serve as a concise and brief guide to the practice of scanning electron microscopy. In this treatment, the material has been developed with the goal of providing an easily understood text for those SEM users who have little or no background in this area. It provides a solid introduction to the subject for the uninitiated. The instrumentation and working and image interpretation have been explained in a succinct practical guide to the SEM. The aim is to provide all useful information regarding SEM operation, applications, and sample preparation to the readers without them having to go through extensive reference material. Essential theory of specimen-beam interaction and image formation is treated in a manner that can be effortlessly comprehended by the readers. The SEM technique is described in simple terms to help operators and users of the SEM to get the best imaging results possible for their materials of interest. The capabilities and limitations of the SEM are also described to enable students, engineers, and materials scientists to identify and apply this technique for their work.

Necessary background to the SEM is developed in Chap. 1. Primary and secondary components of the instrument are introduced in Chap. 2. Basic concepts of electron beam-specimen interaction and contrast formation are described in Chap. 3. Chapter 4 elaborates on the mechanisms of image formation in the SEM. The working of the SEM is introduced, and the factors affecting the quality of images are discussed. Specialized SEM techniques are described briefly in Chap. 5. Chapters 6 and 7 elaborate the characteristics of x-rays and principles of EDS/WDS microchemical analysis, respectively. Chapter 8 includes sample preparation techniques used for various classes of materials. Images, illustrations, and photographs are used to explain concepts, provide information, and aid in data interpretation. The effect of various imaging conditions on the quality of images is described to help users get the best results for their materials of interest. The book is structured in a way that can help a novice find necessary information quickly.

The support of the King Fahd University of Petroleum & Minerals (KFUPM), Dhahran, through project number BW161001 is gratefully acknowledged. I am utterly indebted to Mr. Abuduliken Bake for drawing with great skill almost all of the illustrations appearing in this book. In addition, he has taken a number of SEM images and photographs. His help has been instrumental in the timely completion of the manuscript. I am also grateful to personnel working in various organizations who permitted the use of relevant material. I especially thank Mr. Tan Teck Siong from JEOL Asia Pte Ltd. for providing me with a number of wonderful images. I also extend my appreciation to my colleagues at the Materials Characterization Laboratory, Center for Engineering Research, KFUPM, for their continued support.

In the end, I am grateful to have been blessed with family and friends who make life truly worthwhile.

King Fahd University of Petroleum &   Anwar Ul-Hamid
Minerals, Dhahran, Saudi Arabia

Contents

1 Introduction 1
   1.1 What Is the SEM 1
   1.2 Image Resolution in the SEM 1
   1.3 Image Formation in the SEM 4
   1.4 Information Obtained Using the SEM 4
   1.5 Strengths and Limitations of the SEM 8
   1.6 Brief History of the SEM Development 11
   References 14

2 Components of the SEM 15
   2.1 Electron Column 15
      2.1.1 Electron Gun 17
   2.2 Thermionic Emission Electron Guns 20
      2.2.1 Tungsten Filament Gun 22
      2.2.2 Lanthanum Hexaboride (LaB_6) Emitter 28
   2.3 Field Emission Electron Guns 30
      2.3.1 Working Principle 30
      2.3.2 Advantages/Drawbacks 31
      2.3.3 Cold Field Emitter (Cold FEG) 32
      2.3.4 Schottky Field Emitter 34
      2.3.5 Recent Advances 36
   2.4 Electromagnetic Lenses 37
      2.4.1 Condenser Lens 39
      2.4.2 Apertures 41
      2.4.3 Objective Lens 41
      2.4.4 Lens Aberrations 44
      2.4.5 Scan Coils 52
      2.4.6 Magnification 53
   2.5 Specimen Chamber 56
      2.5.1 Specimen Stage 56
      2.5.2 Infrared Camera 58
   2.6 Detectors 59
      2.6.1 Everhart-Thornley Detector 59

		2.6.2	Through-the-Lens (TTL) Detector	64

| | 2.6.3 | Backscattered Electron Detector | 66 |

| 2.7 | Miscellaneous Components | 71 |

	2.7.1	Computer Control System	71
	2.7.2	Vacuum System	72
	2.7.3	High-Voltage Power Supply (HT Tank)	74
	2.7.4	Water Chiller	74
	2.7.5	Heater	75
	2.7.6	Anti-vibration Platform	75

References 76

3 Contrast Formation in the SEM 77

- 3.1 Image Formation 77
  - 3.1.1 Digital Imaging 78
  - 3.1.2 Relationship Between Picture Element and Pixel ... 80
  - 3.1.3 Signal-to-Noise Ratio (SNR) 82
  - 3.1.4 Contrast Formation 85
- 3.2 Beam-Specimen Interaction 86
  - 3.2.1 Atom Model 86
  - 3.2.2 Elastic Scattering 87
  - 3.2.3 Inelastic Scattering 88
  - 3.2.4 Effect of Electron Scattering 89
  - 3.2.5 Interaction Volume 90
  - 3.2.6 Electron Range 93
- 3.3 Origin of Backscattered and Secondary Electrons .. 95
  - 3.3.1 Origin of Backscattered Electrons (BSE) .. 95
  - 3.3.2 Origin of Secondary Electrons (SE) 95
- 3.4 Types of Contrast 97
  - 3.4.1 Compositional or Atomic Number (Z) Contrast (Backscattered Electron Imaging) 97
  - 3.4.2 Topographic Contrast (Secondary Electron Imaging) ... 113

References 127

4 Imaging with the SEM 129

- 4.1 Resolution 129
  - 4.1.1 Criteria of Spatial Resolution Limit 131
  - 4.1.2 Imaging Parameters That Control the Spatial Resolution .. 134
  - 4.1.3 Guidelines for High-Resolution Imaging .. 138
  - 4.1.4 Factors that Limit Spatial Resolution ... 140
- 4.2 Depth of Field 141
- 4.3 Influence of Operational Parameters on SEM Images .. 146
  - 4.3.1 Effect of Accelerating Voltage (Beam Energy) .. 146
  - 4.3.2 Effect of Probe Current/Spot Size 147
  - 4.3.3 Effect of Working Distance 151
  - 4.3.4 Effect of Objective Aperture 154

Contents

		4.3.5	Effect of Specimen Tilt	157
		4.3.6	Effect of Incorrect Column Alignment	159
	4.4	Effects of Electron Beam on the Specimen Surface		160
		4.4.1	Specimen Charging	160
		4.4.2	Surface Contamination	166
		4.4.3	Beam Damage	167
	4.5	Influence of External Factors on SEM Imaging		168
		4.5.1	Electromagnetic Interference	169
		4.5.2	Floor Vibrations	169
		4.5.3	Poor Microscope Maintenance	170
	4.6	Summary of Operating Conditions and Their Effects		171
	4.7	SEM Operation		172
		4.7.1	Sample Handling	173
		4.7.2	Sample Insertion	174
		4.7.3	Image Acquisition	175
		4.7.4	Microscope Alignment	176
		4.7.5	Maintenance of the SEM	177
	4.8	Safety Requirements		178
		4.8.1	Radiation Safety	178
		4.8.2	Safe Handling of the SEM and Related Equipment	179
		4.8.3	Emergency	179
	References			180
5	**Specialized SEM Techniques**			181
	5.1	Imaging at Low Voltage		181
		5.1.1	Electron Energy Filtering	182
		5.1.2	Detector Technology	183
		5.1.3	Electron Beam Deceleration	186
		5.1.4	Recent Developments	187
		5.1.5	Applications	189
	5.2	Imaging at Low Vacuum		189
		5.2.1	Introduction	189
		5.2.2	Brief History	190
		5.2.3	Working Principle	190
		5.2.4	Detector for Low Vacuum Mode	192
		5.2.5	Gas Path Length	193
		5.2.6	Applications	196
		5.2.7	Latest Developments	197
	5.3	Focused Ion Beam (FIB)		197
		5.3.1	Introduction	197
		5.3.2	Instrumentation	202
		5.3.3	Ion-Solid Interactions	204
		5.3.4	Ion Imaging	205
	5.4	STEM-in-SEM		206
		5.4.1	Working Principle	206

		5.4.2	Advantages/Drawbacks	208
		5.4.3	Applications	208
	5.5	Electron Backscatter Diffraction (EBSD)		210
		5.5.1	Brief History	211
		5.5.2	Working Principle	211
		5.5.3	Experimental Setup	213
		5.5.4	Applications	215
	5.6	Electron Beam Lithography		218
		5.6.1	Introduction	218
		5.6.2	Experimental Set-Up	219
		5.6.3	Classification of E-beam Lithography Systems	220
		5.6.4	Working Principle	221
		5.6.5	Applications	222
	5.7	Electron Beam-Induced Deposition (EBID)		224
		5.7.1	Mechanism	224
		5.7.2	Advantages/Disadvantages of EBID	225
		5.7.3	Applications	225
	5.8	Cathodoluminescence		226
		5.8.1	Introduction	226
		5.8.2	Instrumentation	227
		5.8.3	Strengths and Limitations of SEM-CL	229
		5.8.4	Applications	230
	References			230
6	**Characteristics of X-Rays**			233
	6.1	Atom Model		233
	6.2	Production of X-Rays		234
		6.2.1	Characteristic X-Rays	234
		6.2.2	Continuous X-Rays	236
		6.2.3	Duane-Hunt Limit	239
		6.2.4	Kramer's Law	239
		6.2.5	Implication of Continuous X-Rays	240
	6.3	Orbital Transitions		242
		6.3.1	Nomenclature Used for Orbital Transition	242
		6.3.2	Energy of Orbital Transition	242
		6.3.3	Moseley's Law	244
		6.3.4	Critical Excitation Energy (Excitation Potential)	244
		6.3.5	Cross Section of Inner-Shell Ionization	246
		6.3.6	Overvoltage	247
	6.4	Properties of Emitted X-Rays		249
		6.4.1	Excited X-Ray Lines	249
		6.4.2	X-Ray Range	250
		6.4.3	X-Ray Spatial Resolution	251
		6.4.4	Depth Distribution Profile	253

		6.4.5	Relationship Between Depth Distribution φ(ρz) and Mass Depth (ρz)	254
		6.4.6	X-Ray Absorption (Mass Absorption Coefficient)	256
		6.4.7	Secondary X-Ray Fluorescence	261
	References			263
7	Microchemical Analysis in the SEM			265
	7.1	Energy Dispersive X-Ray Spectroscopy (EDS)		265
		7.1.1	Working Principle	267
		7.1.2	Advantages/Drawbacks of EDS Detector	272
	7.2	Qualitative EDS Analysis		272
		7.2.1	Selection of Beam Voltage and Current	273
		7.2.2	Peak Acquisition	273
		7.2.3	Peak Identification	273
		7.2.4	Peak to Background Ratios	275
		7.2.5	Background Correction	275
		7.2.6	Duration of EDS Analysis	275
		7.2.7	Dead Time	276
		7.2.8	Resolution of EDS Detector	276
		7.2.9	Overlapping Peaks	278
	7.3	Artifacts in EDS Analysis		278
		7.3.1	Peak Distortion	278
		7.3.2	Peak Broadening	278
		7.3.3	Escape Peaks	281
		7.3.4	Sum Peaks	282
		7.3.5	The Internal Fluorescence Peak	283
	7.4	Display of EDS Information		283
		7.4.1	EDS Spectra	284
		7.4.2	X-Ray Maps	284
		7.4.3	Line Scans	285
	7.5	Quantitative EDS Analysis		287
		7.5.1	Introduction	287
		7.5.2	EDS with Standards	289
		7.5.3	Examples of ZAF Correction Method	295
	7.6	Standardless EDS Analysis		296
		7.6.1	First Principles Standardless Analysis	298
		7.6.2	Fitted Standards Standardless Analysis	298
	7.7	Low-Voltage EDS		299
	7.8	Minimum Detectability Limit (MDL)		300
	7.9	Wavelength Dispersive X-Ray Spectroscopy (WDS)		300
		7.9.1	Instrumentation	300
		7.9.2	Working Principle	301
		7.9.3	Analytical Crystals	304
		7.9.4	Detection of X-Rays	304
		7.9.5	Advantages/Drawbacks of WDS Technique	305

		7.9.6	Qualitative WDS Analysis	306
	References			306
8	**Sample Preparation**			309
	8.1	Metals, Alloys, and Ceramics		309
		8.1.1	Sampling	309
		8.1.2	Sectioning	310
		8.1.3	Cleaning	310
		8.1.4	Embedding and Mounting	312
		8.1.5	Grinding, Lapping, and Polishing	312
		8.1.6	Impregnation	315
		8.1.7	Etching	315
		8.1.8	Fixing	316
		8.1.9	Fracturing	316
		8.1.10	Coating Process	318
		8.1.11	Marking Specimens	323
		8.1.12	Specimen Handling and Storage	323
	8.2	Geological Materials		324
		8.2.1	Preliminary Preparation	324
		8.2.2	Cleaning	324
		8.2.3	Drying	325
		8.2.4	Impregnation	325
		8.2.5	Replicas and Casts	326
		8.2.6	Rock Sample Cutting	326
		8.2.7	Mounting the Sample into the SEM Holder	326
		8.2.8	Polishing	327
		8.2.9	Etching	328
		8.2.10	Coating	328
	8.3	Building Materials		329
		8.3.1	Preparation of Cement Paste, Mortar, and Concrete Samples	329
		8.3.2	Cutting and Grinding	330
		8.3.3	Polishing	330
		8.3.4	Impregnation Techniques	331
		8.3.5	Drying the Specimen	332
		8.3.6	Coating the Specimen	332
		8.3.7	Cleaning the Surface of the Specimen	332
	8.4	Polymers		332
		8.4.1	Types of Polymers	333
		8.4.2	Morphology of Polymers	334
		8.4.3	Problems Associated with the SEM of Polymers	335
		8.4.4	General Aspects in Polymers Preparation for SEM	337
		8.4.5	Sample Preparation Techniques for Polymers	338

	8.4.6	Devices Used in Microtomy	339

        8.4.6   Devices Used in Microtomy 339
        8.4.7   Sample Preparation Procedure for Polymers 340
    8.5 Biological Materials 348
        8.5.1   Fixation 349
        8.5.2   Examples of Biological Sample Preparation 351
    References 358

Questions/Answers 361

Index 397

Abbreviations

BSE	Backscattered electron(s)
CAD	Computer-aided design
CCD	Charge-coupled device
CL	Cathodoluminescence
DBS detector	Distributed backscattered detector
E-beam lithography	Electron beam lithography
EBID	Electron beam-induced deposition
EBSD	Electron backscatter diffraction
EBSP	Electron backscatter pattern
EDS	Energy dispersive x-ray spectroscopy
EPMA	Electron probe microanalyzer
EsB detector	Energy selective Backscatter detector
ESEM	Environmental SEM
E-T detector	Everhart-Thornley detector
FEG	Field emission gun
FET	Field-effect transistor
FIB	Focused ion beam
FWHM	Full width at half maximum
FWTM	Full width at tenth maximum
GFIS	Gas field ion source
GPL	Gas path length
GSED	Gaseous secondary electron detector
GUI	Graphical user interface
HAADF	High-angle annular dark field
HSQ	Hydrogen silsequioxane
IBID	Ion beam-induced deposition
ICC	Incomplete charge collection
IPF map	Inverse pole figure map
LaB_6	Lanthanum hexaboride
LABe detector	Low-angle backscattered electron detector
LG	Light guide
LMIS	Liquid metal ion source
LV	Low vacuum

MCA	Multichannel x-ray analyzer
MCP	Microchannel plate
MDL	Minimum detectability limit
MFP	Mean free path
OM	Orientation map
PHA	Pulse-height analyzer
PLA	Pressure-limiting aperture
PMMA	Poly methyl methacrylate
PMT	Photomultiplier tube
Pre-Amp	Preamplifier
SACP	Selected area channeling pattern
SDD	Silicon drift detector
SE	Secondary electron(s)
SEM	Scanning electron microscope
Si(Li)	Lithium-drifted silicon
SNR	Signal-to-noise ratio
STEM	Scanning transmission electron microscope
t-EBSD	Transmission EBSD
TEM	Transmission electron microscope
TES	Transition edge x-ray sensor
TKD	Transmission Kikuchi diffraction
TTL	Through-the-lens
UED	Upper electron detector
UTW	Ultra-thin window
VPS	Volume plasma sources
WD	Working distance
WDS	Wavelength dispersive x-ray spectroscopy
XRD	X-ray diffractometer

Symbols List

σ_A	Standard deviation
$\left(\dfrac{\mu}{\rho}\right)$	Mass absorption coefficient
C_i	Mass fraction of element i
n_B	Number of incident beam electrons
n_{SE}	Number of secondary electrons
p_g	Pressure of the gas
σ_g	Total scattering cross section of gas molecule for electrons
ΔE	Energy spread
A	Area, Atomic weight
a, b, and c	Constants
A, C	Constants
A_A	Maximum intensity
B	Magnetic field
C	Contrast
C_c	Chromatic aberration coefficient
$C_{i\ (\text{standard})}$	Weight percent concentration of element i in the standard
$C_{i\ (\text{unknown})}$	Weight percent concentration of element i in unknown bulk specimen
C_s	Spherical aberration coefficient
C_Z	Z contrast
d	Escape depth of SE
d	Lattice plane spacing
d	Resolution
d	Beam diameter
d_0	Crossover diameter
d_A	Diameter of astigmatism disc
d_c	Diameter of the chromatic aberration disc
d_d	Diameter of diffraction disc
d_{opt}	Optimum probe diameter
d_p	Probe diameter
$d_{p,\text{min}}$	Minimum probe size
d_s	Diameter of the spherical aberration disc

E	Energy of the x-ray line
e	Electric charge in Coulomb
E	Kinetic energy
E_0	Incident beam energy
E_A	Average energy for the x-ray peak
E_{BSE}	Energy of the BSE
E_c	Critical energy of ionization
E_{exc}	Mean energy per excitation
E_K	Binding energy of K shell
E_L	Binding energy of L shell
E_m	Mean number of electron-hole pairs
E_v	Continuum x-ray photon energy at some point in the spectrum
F	Force
h	Planck's constant
I	Intensity of x-ray photons when leaving the specimen surface
I	Width of the intrinsic line of the detector
I_0	Original intensity of x-ray photons
i_B	Electron beam current entering the specimen
i_b	Beam current
I_b	Background intensity
i_{BSE}	Backscattered electron current moving out of the specimen
I_{cm}	Intensity of continuum x-ray
i_e	Emission current
i_f	Filament heating current
$I_{i\ (standard)}$	Intensity of characteristic x-ray peak emanating from element i in the standard
$I_{i\ (unknown)}$	Intensity of characteristic x-ray peak emanating from element i in unknown specimen
i_p	Probe current
$I_{p,max}$	Maximum probe current
IR	Infrared
J	Average loss in energy per event
J_b	Current density of electron beam
J_c	Current density of electron source
k	Boltzmann constant
LED	Light-emitting diode
$L_{monitor}$	Scan length on monitor
$L_{picture\ element}$	Length of picture element
L_{pixel}	Length of pixel
$L_{specimen}$	Scan length on specimen
m	Mass
n	Order of diffraction
n	Refractive index
N	Total number of atoms present in the irradiated volume

n		Total number of ionization events
nm		Nanometer
N_v		Avogadro's number
p		Probability
P		Quality indicator of the electronics used
Q		Cross section of ionization
R		Detector's energy resolution
R		Electron range
R		Extent of backscattering
r_s		Skirt radius
R_x		X-ray range
s		Distance travelled by an electron in the specimen
S		Stopping power
S_A		Signal emitted by the feature A
S_B		Signal emitted by the feature B
t		Thickness of specimen travelled
T		Absolute temperature (K)
t		Acquisition time
U		Overvoltage
v		Electromagnetic radiation frequency
v		Velocity of the particle
V_0		Accelerating voltage
W		Work function
X		The equivalent FWHM related to incomplete charge collection and leakage current of the detector
X_k		Number of emitted x-ray photons
Y		X-ray peak intensity
Z		Atomic number
α		Convergence angle of the beam
η		Backscatter coefficient
η_B		Number of incident beam electrons
η_{BSE}		Number of backscattered electrons
λ_{SWL}		Short wavelength limits
μ		Absorption coefficient
ρ		Density
Φ		Depth distribution function
φ		Work function
ψ		Take-off angle
ω		Fluorescence yield
Ω		Solid angle
α_{opt}		Optimum convergence angle
β		Brightness of gun
β_{max}		Maximum brightness of gun
δ		Secondary yield

λ	Wavelength, mean free path (escape depth of SE)
μm	Micron
θ	Bragg angle of diffraction
F	Frame scan time
q	Detector efficiency

Introduction

1.1 What Is the SEM

The word microscope is derived from Greek *micros* (small) and *skopeo* (look at). Just like any microscope, the primary function of the scanning electron microscope (SEM) is to enlarge small features or objects otherwise invisible to human sight. It does that by way of using electron beam rather than light which is used to form images in optical light microscopes. The images are obtained by scanning an electron beam of high energy on the sample surface, hence the name scanning electron microscope. By virtue of its smaller wavelength, electrons are able to resolve finer features/details of materials to a much greater extent compared with optical light. A modern day SEM can magnify objects up to one million times their original size and can resolve features smaller than 1 nm in dimension. Similarly, electron beam interaction with the specimen emits x-rays with unique energy that can be detected to determine the composition of material under examination. The SEM is, therefore, a tool used for materials characterization that provides information about the surface or near surface structure, composition, and defects in bulk materials. It allows scientists to observe surfaces at submicron and nano-level to elaborate material properties. It has emerged as one of the most powerful and versatile instruments equally valuable to materials and life scientists working in wide-ranging industries.

1.2 Image Resolution in the SEM

A human eye cannot distinguish objects smaller than 200 µm (0.2 mm). In other words, the resolution of a human eye is 200 µm, while a light microscope can typically magnify images up to 1000× to resolve details down to 0.2 µm. Resolution limit is defined as the smallest distinguishable distance separating two objects, i.e., minimum resolvable distance. For instance, two objects separated by a distance of less than 200 μm will appear as one object to the human eye since the latter is not able to resolve details that have dimensions smaller than 200 μm. Hence, 200 μm can be considered to be the resolution limit of the human eye. The same objects viewed under a light microscope will appear as two distinct entities since the light microscope can easily differentiate distances less than 200 μm. In fact, the objects can be brought closer together further to a distance of 0.2 μm and still maintain their separate identities under a light microscope. However, if the distance between the objects is decreased further to less than 0.2 μm, the light microscope will no longer be able to discern them as two separate objects, which will then appear as a single entity. Thus 0.2 μm can be defined as the resolution limit of the light microscope. It follows that the smaller is the value of minimum resolvable distance, the higher is the resolution of a microscope.

Both light microscope and humans use visible light as a means to probe into or interact with an object. The increased ability to observe details in a light microscope compared to the unaided eye is attributed to the lens/aperture system used to magnify the image of an object. It is theoretically possible to keep enlarging the image by increasing the magnification indefinitely. However, it is not possible to keep revealing newer details in an object by simply increasing the magnification. Fine details in an image cannot be resolved beyond a certain magnification. This is due to limitations imposed by the resolving power of the imaging technique as well as that of the human eye. The maximum useful magnification beyond which no further details are revealed is determined by the resolving power of a microscope. The following equation can be used to determine the typical useful magnification of a microscope:

$$\text{Useful Magnification} = \frac{\text{Resolution of the Human Eye}}{\text{Resolution of Microscope}} \quad (1.1)$$

For a light microscope, useful magnification $\left(\frac{200 \ \mu m}{0.2 \ \mu m}\right)$ is around 1000×. For a scanning electron microscope, useful magnification $\left(\frac{200 \ \mu m}{1 \ nm}\right)$ is typically 200,000×. Increase in the resolution of the instrument results in the increase of its useful magnification.

The ability of visible light to resolve image details is limited by its relatively large wavelength ($\lambda = 380$–760 nm) (see Fig. 1.1). Use of light with a shorter wavelength (such as ultraviolet) and a lens immersed in oil (high refractive index) improves resolution to around 0.1 μm. If the image is formed by using a radiation with a smaller wavelength, such as an electron beam, higher resolution limit can be achieved since the smaller the wavelength, the greater the resolving power and the greater the detail revealed in an image. Due to this fact, techniques like the SEM and TEM employ an electron beam to probe the material resulting in an image far superior in resolution compared to that of the light microscope. For example, an electron beam (λ of 0.000004 μm) with an accelerating voltage of 100 kV can achieve a resolution of 0.24 nm. The practical limit to resolution is determined by lens aberrations and defects. Modern-day field emission SEM typically operated at

1.2 Image Resolution in the SEM

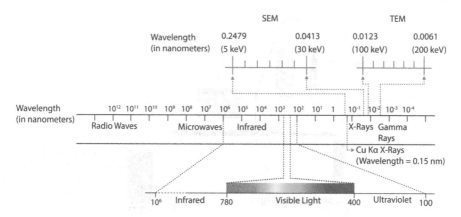

Fig. 1.1 Electromagnetic spectrum showing the size of the wavelength used in the light, scanning (SEM), and transmission electron microscope (TEM)

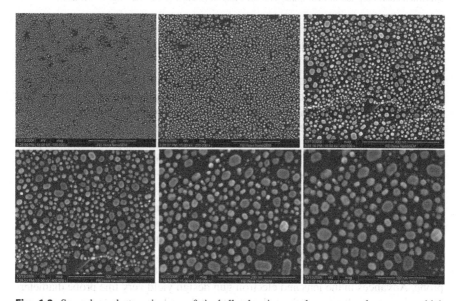

Fig. 1.2 Secondary electron images of tin balls showing good contrast at low to very high magnifications (100,000× to 1,000,000×)

20–30 kV accelerating voltages can achieve image resolution in the order of 1 nm or better. It is worth noting here that resolving power or resolution (a more commonly used term) of an instrument is demonstrated by manufactures using a specimen *ideally* suited for that instrument. For instance, tin balls/powder is routinely employed for the SEM since the former is conductive and has strong contrast (see Fig. 1.2). Details in real samples, however, are not usually revealed to that level of resolution.

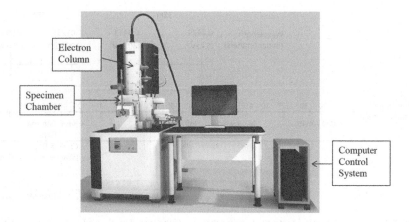

Fig. 1.3 A photograph showing three major sections of the SEM: the electron column, the specimen chamber, and the computer control system. (Courtesy of T. Siong, JEOL Ltd.)

1.3 Image Formation in the SEM

The SEM instrument can be considered to comprise of three major sections: the electron column, the specimen chamber, and the computer/electronic controls, as shown in Fig. 1.3. The topmost section of the electron column consists of an electron gun which generates an electron beam. Electromagnetic lenses located within the column focus the beam into a small diameter (few nanometers) probe. The scan coils in the column raster the probe over the surface of the sample present in the chamber that is located at the end of the column. The gun, the column, and the specimen chamber are kept under vacuum to allow electron beam generation and advancement. The electrons in the beam penetrate a few microns into the surface of a bulk sample, interact with its atoms and generate a variety of signals such as secondary and backscattered electrons and characteristic x-rays that are collected and processed to obtain images and chemistry of the specimen surface. The ultimate lateral resolution of the image obtained in the SEM corresponds to the diameter of the electron probe. Advances in the lens and electron gun design yield very fine probe diameters giving image resolutions of the order of <1 nm. In order to provide a perspective to the way the image is realized in the SEM, a comparison of how light and the transmission electron microscopes work compared to the SEM is shown in Table 1.1 and Fig. 1.4.

1.4 Information Obtained Using the SEM

The scanning electron microscope is used to observe and image the micro- and nanostructural surface details of a wide range of materials such as metals, alloys, ceramics, polymers, rock minerals, corrosion deposits, filters, membranes, foils,

Table 1.1 Comparison of various characteristics of the SEM with light and transmission electron microscope

Characteristics	Light microscope	Scanning electron microscope	Transmission electron microscope
Radiation used to form an image	Visible light	Electrons	Electrons
Wavelength of radiation	380–760 nm depending on the color of light	0.008 nm at 20 kV accelerating voltage[a]	0.0028 nm at 200 kV accelerating voltage[a]
Types of lenses used to focus the radiation	Glass	Electromagnetic	Electromagnetic
Useful magnification	1,000×	200,000×	2,000,000×
Possible magnification	Up to 2,000×	1,000,000×	10,000,000 or more
Magnification mode	With lenses	Without lenses	With lenses
Resolution	200 nm	1 nm	0.1 nm
Source of radiation	Tungsten-halogen lamp	Electron gun/emitter	Electron gun/emitter
Image formation	Light from the source is scattered by the sample surface and redirected by the objective lens to form an image onto the retina of the human eye. The image can also be displayed on an electronic display	Electrons originating from the source travel in vacuum within a column lined with electromagnetic lenses which focus these electrons into a small probe on the surface of the specimen. Electron-specimen interaction results in information emanating from the specimen which is passed through detectors and reconstituted as an image on an electronic display	Electrons originating from the source travel in vacuum within a column. Electron beam passes through a thin foil of sample and then focused and magnified by electromagnetic lenses to form an image on a fluorescent screen or transferred to an electronic display
Type of image	Real image. Color images. Images formed using visible light can be observed directly by the human eye	Processed/reconstituted image. Grayscale images (black and white). Images formed with electrons cannot be observed directly by humans	The real image is projected onto the screen which can be observed by the human eye. Grayscale images
Specimen preparation	Required	Can be omitted (based on specimen type)	Required, tedious
Specimen thickness	Thin, bulk	Bulk	Thin (Electron transparent, ≈100 nm)
Sample area examined	Large areas of a sample can be examined	Small areas of a sample are examined	Extremely small areas of a sample are examined

(continued)

Table 1.1 (continued)

Characteristics	Light microscope	Scanning electron microscope	Transmission electron microscope
Applications	Materials/life sciences	Materials/life sciences	Materials/life sciences
Examination of live specimens	Living organisms can be examined	Vacuum usage and high energy electron radiation precludes examination of live samples	Vacuum usage and high energy electron radiation precludes examination of live samples
Depth of field	Small[b] 15 μm (at 4×) 0.2 μm (1000×)	Large (3-D like images)[b] 4 mm (at 10×) 0.5 μm (500,000×)	Small
Lab size requirements	2 × 2 m	5 × 5 m	6 × 6 m
Capital and maintenance cost	Low	Medium	High
Features studied	Surface	Surface or subsurface	Microstructure
Interpretation of images	Easy	Moderate	Difficult

[a]Electron wavelength is calculated from De Broglie equation: $\frac{1.23\ nm}{\sqrt{\text{accelerating voltage in thousands}}}$

[b]The exact depth of field depends on the working distance (i.e., the distance between the lens and the imaged surface).

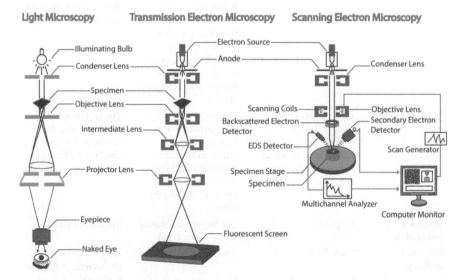

Fig. 1.4 Schematic comparing the modes of image formation in the light, transmission, and scanning electron microscopes

fractured/rough surfaces, biological samples, etc. The materials can be conductive or non-conductive either in solid or powder form and can be examined in an as-received or prepared (sectioned, ground, polished, etched, coated, etc.) condition. The SEM has the ability to examine materials in the dry or wet state as well as obtain microchemical information from fine structural details. The SEM equipped with a field emission gun can distinguish surface features that are only 1 nm apart (i.e., lateral spatial resolution = 1 nm). Extraordinary ability to depict large depths of field (10–100% of horizontal field width) allows large areas of a sample to remain in focus at one time and thus yield 3-D characteristics in SEM images (see Fig. 1.5). Imaging can be performed using both secondary electrons (for topographic contrast) and backscattered electrons (for topographic and/or compositional contrast). Microchemical information is generally obtained using energy dispersive x-ray spectrometer (EDS) detector attached to the SEM. Both qualitative and quantitative elemental information from microstructural features can be obtained from beryllium to uranium with limits of detection of approximately 0.2–0.5 wt%. The electron beam in the SEM can penetrate as much as a few microns into the sample depending

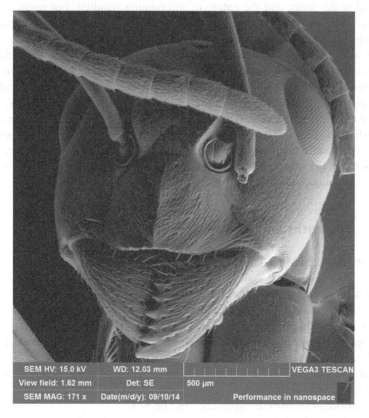

Fig. 1.5 Secondary electron SEM image of an ant showing detailed three-dimensional details. (Image courtesy of TESCAN)

on sample density, beam accelerating voltage, etc. Typical applications include observation of metallographically prepared samples (such as steel) to study surface morphology, grain size/shape, inclusions, precipitates, dendrites, grain boundaries, etc. It is also employed to observe materials in an unprepared (as-received) condition, e.g., fracture surfaces for metallurgical failure analysis, electronic devices for electronic failure analysis, corrosion deposits, catalyst shape, size and surface structure, polymer additives, rock mineral samples, etc. Apart from bulk samples, it is also used to examine coatings and thin films deposited on substrates. Table 1.2 highlights various applications and the range of information obtained using SEM and related techniques.

A combination of factors such as good resolution, large depth of field, compositional information, time-efficient analysis, as well as relative ease of use and image interpretation in both materials and life sciences has made the SEM one of the most heavily used instruments in academia, research, and industry. Although the SEM can generate rich morphological and compositional data for a wide variety of materials, it is often necessary to employ various other analytical tools to undertake complete materials characterization. Selection of these tools will depend on the type of material under study and the nature of information required. For instance, if the objective is to reveal true microstructure (not surface topography) of a material, transmission electron microscopy is employed. This technique can also extract chemical information from features at the nanometer scale. Likewise, if the bulk composition is desired, x-ray fluorescence is a better choice as it analyzes large volumes of material. Phase identification at a macro level is often carried out using x-ray powder diffraction analysis. Analysis of structure and composition of a thin surface constituting a few atomic monolayers is undertaken with Auger and x-ray photoelectron spectroscopy. All these techniques are tools available to a scientist to carry out essential materials characterization and information obtained from one complements the others. A brief comparison of the imaging and analytical abilities of these techniques is shown in Table 1.3.

1.5 Strengths and Limitations of the SEM

Strengths of the SEM include:

1. A wide variety of specimens can be examined.
2. Relatively easy and quick sample preparation.
3. Ease of use due to user-friendly and automated equipment.
4. Rapid imaging, quick results, time-efficient analysis, and fast turnaround time.
5. Relatively straightforward image interpretation.
6. Large depth of field (ability to focus large depths of samples at one time and produce 3-D like images)
7. Microchemical analysis capability from Be to U.
8. Samples can be dry or wet.
9. Nondestructive (some beam damage may result).

1.5 Strengths and Limitations of the SEM

Table 1.2 Summary of capabilities of the SEM and related techniques

Information obtained using SEM and related techniques	Morphology (grain shape, precipitate size, dimension, texture and volume fraction, and phase distribution of physical features) [using secondary electron (SE) imaging]
	Surface topography (distribution/arrangement of features at the surface of a sample, defects, cracks, voids, structure) [using secondary electron imaging]
	Atomic number or compositional contrast, grain boundaries, domains in magnetic materials [using backscattered electron (BSE) imaging]
	Microanalysis, elemental identification and quantification, line scans, x-ray maps [using energy dispersive x-ray spectroscopy (EDS)]
	Kikuchi maps of bulk samples—crystallographic orientation of grains, phase identification [using electron backscattered diffraction (EBSD)]
	Pore size/shape/distribution in rock minerals, organic samples [using cryogenic techniques]
	Image non-conductive charging samples such as polymers, rocks, etc., without coating [using low vacuum (LV) mode]
	Imaging fine details of surface structures [using low accelerating voltage]
	Preparation of precise plan and cross-sectional view TEM samples, fabrication of nano devices [using SEM combined with focused ion beam]
	In situ experiments under hydration, thermal cycling, or gaseous environment [using environmental SEM (ESEM)]
	Printing of circuit boards in the electronics industry [using E-beam lithography]
Applications	Materials identification, materials science, forensic science, metallurgical and electronic materials failure analysis, corrosion science, rock mineralogy, geo sciences, nano devices, polymer science, catalysis, semiconductor design, desalination, life sciences, oil and gas, mining
Types of samples examined	Metals, alloys, semiconductors, polymers, coatings, ceramics, rocks, sand, corrosion products, catalysts, membranes, carbon nanotubes, nanopowders, tissues, cells, insects, leaves
Industry	Academic and research, oil and gas, power generation, metals and alloys, industrial manufacturing, automobile, aero, aerospace, petrochemical, geosciences, nanotechnology, semiconductor, computer, chemical process industry, mining
Sample form/state/size	Form: Solid, bulk, thin, nanostructured, powder, pellet, organic, inorganic
	State: Dry, wet
	Sample size: 1 mm to 150 mm (typically 10–20 mm)
Spatial resolution	1 nm
Sampling depth	1–5 µm (depends on sample density and electron beam accelerating voltage)
Sample area analyzed	Submicron to >100 µm (depends on electron probe size, magnification)
Detection limit of elements	≈ 0.2 wt% (depends on relative elemental content and atomic weight)

Table 1.3 Comparison of imaging and analytical techniques generally used to study materials structure, chemical composition, and defects

Technique	Information/ applications	Spatial resolution	Sampling depth	Typical sample area analyzed	Elemental detection limits
Light microscopy	Imaging/materials and life sciences	200 nm	Surface	Millimeters	–
Scanning electron microscopy	Imaging, microchemical analysis/materials, and life sciences	1 nm	Submicron to several microns	Micron to millimeters	0.2–0.5 wt%
Transmission electron microscopy	Imaging, microchemical analysis, crystal structure/materials and life sciences	0.1 nm	<100 nm (thin foils)	<1 to few microns	0.01–0.1 wt%
Atomic force microscopy	Imaging, topography/ materials, and life sciences	Sub nanometer	Sub-nanometer	Tens to hundreds of microns	–
Electron probe microanalysis with wavelength dispersive x-ray spectroscopy	Imaging, microchemical analysis/materials science	1 nm	Submicron to several microns	Micron to millimeters	0.01 wt%
X-ray diffraction	Structural analysis, lattice parameters, phase constitution, crystal structure, crystallite size, pore size, stress, texture, lattice distortion, thin film analysis/materials science	–	Tens to hundreds of microns	Sq. cm	–
X-ray fluorescence	Bulk chemical analysis/materials science	–	Tens to hundreds of microns	Sq. cm	ppm
X-ray photoelectron spectroscopy	Surface chemistry (binding energy, chemical state)/ materials science	–	0.5–5 nm	Tens of sq. μm	0.001 wt% to ppm
Auger electron spectroscopy	Surface chemistry, imaging (electronic structure, depth profiling)/materials science	8 nm	<5 nm	Nanometers to microns	0.01–0.1 at%

10. High spatial resolution (1 nm) achieved by modern equipment.
11. A versatile platform that lends support to other sophisticated tools, devices, and techniques.
12. Capable of several modes of imaging, spectroscopy, and diffraction analysis.
13. Affordability and availability compared to more expensive equipment.

Limitations can be summarized as follows:

1. The sample size is limited.
2. Samples are solid.
3. EDS detector cannot detect H, He, or Li.
4. Poor detectability limit of elements (with EDS) compared to wet analytical methods.
5. Samples need to be examined under vacuum.
6. The instrument typically requires an installation space of 5 × 5 m.
7. Non-conductive samples need to be coated.

1.6 Brief History of the SEM Development

Limitation of light microscope in resolving fine details of organics cells provided the impetus to develop an electron microscope in the early twentieth century. The first electron microscope was a transmission electron microscope developed by German scientist Max Knoll and electrical engineer Ernst Ruska in Germany in 1931. It employed a working model similar to light microscope except that a beam of electrons, instead of visible light, was made to pass through the body of a sample to form an image on a fluorescent screen. Use of electrons as an imaging medium afforded a resolution of 10 nm compared to a resolution of 200 nm achievable by a light microscope. The resolution achieved at the time might seem modest today, but the real breakthrough was the fact that for the first time in history, electrons had been successfully employed to create images of matter. Subsequent improvements in the accelerating voltage, lens technology, vacuum systems, electron guns, power supplies, and overall design of the microscopes in the next few decades led to the imaging of atoms (i.e., atomic resolution was achieved). Due to his "fundamental work in electron optics and for the design of the first electron microscope," Ernst Ruska received a Nobel Prize in physics in 1986.

German physicist Max Knoll introduced the concept of a scanning electron microscope in 1935 [1]. He proposed that an image can be produced by scanning the surface of a sample with a finely focused electron beam. Another German physicist Manfred von Ardenne explained the principles of the technique and elaborated upon beam-specimen interactions. He went on to produce the earliest scanning electron microscope in 1937 [2–4] as shown in Fig. 1.6.

Later, the SEM with a resolution of 50 nm was built by American scientists Zworykin, Hillier, and Snijder in 1942 [5] and later by Professor Sir Charles W. Oatley and his postgraduate student D. McMullan at the University of Cambridge

Fig. 1.6 First scanning electron microscope built by Manfred von Ardenne in 1937

in 1952 [6]. Scintillator-based secondary electron detector was developed by Everhart and Thornley in 1960 [7]. Further improvements in the technology led to the development of first commercial SEM known as "Stereoscan" by Cambridge Scientific Instruments in 1965 [8, 9]. The SEMs built in the 1960s had a resolution of about 15–20 nm. In the 1970s and 1980s, the resolution was improved to 7 nm and 5 nm (at 1 kV), respectively. Next couple of decades saw resolution improvements down to 3 nm and then to 1 nm. Currently, manufacturers claim resolutions of 0.5 nm in the SEM. Although the scanning electron microscope was developed subsequent to the transmission electron microscope, the former quickly became popular due to its ease of use, simple sample preparation, and ability to generate 3-D like images of the sample topography. A detailed history of the SEM development has been documented by some authors [10–13]. Main events in the development of SEM techniques and instrumentation are listed in a chronological order in Table 1.4 [10–14]:

1.6 Brief History of the SEM Development

Table 1.4 Development of SEM instrumentation and techniques in a chronological order [10–14]

Year	Key developments	Contributors
1935	The concept of the SEM	M. Knoll (1935)
1938	Concept and development of the scanning transmission electron microscope (STEM)	M. von Ardenne
1942	Development of the SEM with 50 nm resolution	Zworykin et al.
1952	Development of the SEM with 50 nm resolution	D. McMullan and Prof. C. Oatley
1956	Signal processing, double deflection scanning, stigmator (improved image quality)	K. C. A. Smith [15]
1960	Scintillating secondary electron detector, photomultiplier tube (improved signal-to-noise ratio)	Everhart and Thornley [7]
1957	Observation of voltage contrast	Oatley and Everhart [16]
1960	Stereographic 3-D SEM images	O. C. Wells [17]
1963	Development of "SEM V" microscope with three magnetic lenses and Everhart-Thornley (ET) detector	R. F. W. Pease [8]
1963–1965	Development of first commercial SEM "Stereoscan"	R. F. W Pease and Nixon [8, 9]
1970s	Microchemical analysis using energy dispersive x-ray spectrometer (EDS) coupled with the SEM	Contributions from various individuals and commercial manufacturers
1970s	Lanthanum hexaboride (LaB_6) cathode gun	
1970s	Field emission gun	
1970s	Electron backscattered diffraction (crystal structure and grain orientation)	
1970s	Cathodoluminescence in the SEM	
1973	Hot stage to examine samples at elevated temperatures	
1970s, 1980s	Large specimen stages to hold sample sizes up to 23 cm	
1980s	Auto focus and auto stigmator functions	
1980s	Yttrium doped silicate scintillator as backscattered electron detector	
1980s	Low temperature (Cryo) stage	
1990s	Variable pressure SEM (for charging samples)	
Since 1990s	Improved automation and analysis due to computers	
2000 to date	High resolution microscopes to study nanomaterials	

References

1. Knoll M (1935) Aufladepotentiel und Sekundäremission elektronenbestrahlter Körper. Z Tech Phys 16:467–475
2. von Ardenne M (1937) Improvements in electron microscopes. GB Patent No. 511204
3. von Ardenne M (1938) Das Elektronen-Rastermikroskop, Practische Ausführung. Z Tech Phys 19:407–416
4. von Ardenne M (1939) Das Elektronen-Rastermikroskop, Theoretische Grundlagen. Z Phys 108(9–10):553–572
5. Zworykin VK, Hiller J, Snyder RL (1942) A scanning electron microscope. ASTM Bull 117:15–23
6. McMullan D (1952) Ph.D. Dissertation, University of Cambridge, Cambridge, UK
7. Everhart TE, Thornley RFM (1960) Wide-band detector for micro-microampere low-energy electron currents. J Sci Instr 37(7):246–248
8. Pease RFW (1963) Ph.D. Dissertation, University of Cambridge, Cambridge, UK
9. Pease RFW, Nixon WC (1965) High resolution scanning electron microscopy. J Sci Instr 42:31–35
10. Oatley CW (1972) The scanning Electron microscope. Cambridge University Press, Cambridge
11. Thomas G (1999) The impact of electron microscopy on materials research. In: Rickerby D, Valdre G, Valdre U (eds) Proceedings of the NATO advanced study institute on impact of electron and scanning probe microscopy on materials research (1999). Springer Science+Business Media, Dordrecht. Originally published by Kluwer Academic Publishers in 1999, pp 1–24. https://doi.org/10.1007/978-94-011-4451-3
12. McMullan D (2006) Scanning electron microscopy 1928–1965. Scanning 17(3):175
13. McMullan D (1988) Von Ardenne and the scanning electron microscope. Proc Roy Microsc Soc 2:283
14. Goldstein J et al (2003) Scanning Electron microscopy and X-ray microanalysis, 3rd edn. Kluwer Academic/Plenum Publishers, New York
15. Smith KCA (1956) Ph.D. Dissertation, University of Cambridge, Cambridge, UK
16. Oatley CW, Everhart TE (1957) The examination of p-n junctions with the scanning electron microscope. J Electron 2:568–570
17. Wells OC (1960) Correction of errors in stereomicroscopy. Br J Appl Phys 11:199–201

Components of the SEM

2

The primary components of the SEM are electron column, specimen chamber, and computer control system as shown in the photograph of Fig. 2.1. These components are used to carry out various functions of microscopy and microchemical analysis. The SEM instrumentation may include secondary and backscattered electron detectors, energy-dispersive x-ray spectrometer (EDS), low vacuum detector, electron backscattered diffraction (EBSD) detector, etc. Some of this instrumentation may not be necessary for basic imaging but play an increasingly important role in more demanding microscopy applications. A user has a continual interaction with the primary components of the SEM, which has a direct bearing on the quality of images and analyses obtained. In addition to these components, secondary/miscellaneous equipment such as vacuum pumps, water chiller, and electronics form an essential part of the overall system without which the SEM cannot function. However, this equipment runs seamlessly in the background and hardly needs any input from the user. Modern day SEMs are controlled with computers. However, the quality of images obtained largely depends on the input parameters as determined by the operator. This necessitates the study of SEM and its various components and the way it can be used to produce high-quality images and reliable analytical data.

2.1 Electron Column

An electron column of the SEM is the long cylindrical body located above the specimen chamber as can be seen in Fig. 2.1. An electron column holds an electron gun, two or more electron lenses, scan coils, and condenser and objective apertures within its body as illustrated by a schematic diagram in Fig. 2.2. An electron column is kept under vacuum at all times.

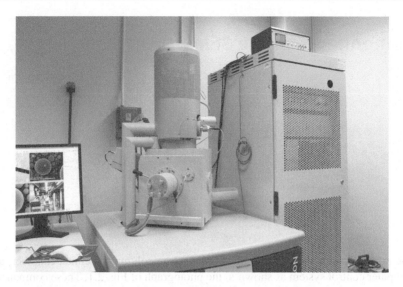

Fig. 2.1 Photograph of a typical SEM with electron column, specimen chamber, and electronics/computer control system

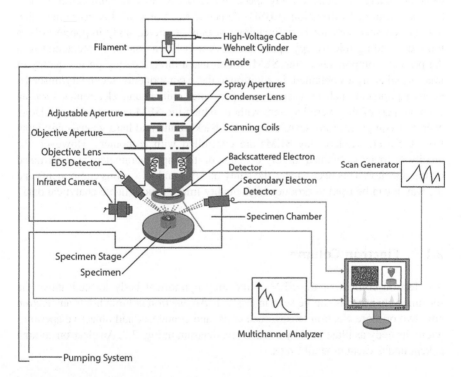

Fig. 2.2 Schematic showing construction and working of various components located within the electron column and specimen chamber of the SEM

2.1.1 Electron Gun

The upper section of the column is occupied by the assembly of cathode and associated electrodes termed as electron gun, which is connected from the outside to a high-voltage (30–40 kV) power cable. A schematic diagram of the inner section of an electron column in Fig. 2.2 shows the location of the electron gun at the topmost section of the electron column. The electron gun serves as the source of electrons that can be varied in magnitude and acceleration. The primary function of the electron gun is to generate electrons which are then accelerated downward through the column by virtue of a potential difference that exists within the gun assembly. The force by which the electrons travel through the column (and ultimately strike the sample) depends on the accelerating voltage used, which is typically varied from 2 to 30 kV depending on the type of sample examined, nature of analysis, and the required information. An electron gun operates in vacuum to avoid scattering of emitted electrons. Four important types of electron guns are tungsten filament, LaB_6 emitter, Schottky field emission, and cold field emission gun. Ideally, an electron gun should produce a stable electron beam with high brightness, fine source size, good beam stability, and small energy spread. Based on these criteria, the abovementioned guns show considerable variation in performance, as seen below.

2.1.1.1 Current Density and Brightness of the Electron Source

Current density and brightness are two important parameters used to express the performance of an electron source. Current density J_b of a beam is given as:

$$J_b = \frac{\text{Beam Current}}{\text{Area}} = \frac{i_b}{\pi \left(\frac{d}{2}\right)^2} \qquad (2.1)$$

where i_b and d are the current and diameter of the beam, respectively.

The current density of the beam changes as it moves through the column since the electrons are blocked by various components such as anode, apertures, etc., resulting in the reduction of the beam current. To take into account the beam divergence, brightness is used as another criterion to indicate the performance of an electron gun. Brightness (β) of the electron source is defined in terms of the total number of electrons (current, I) emitted from a unit area (A) of the source and the solid angle (Ω) of emission subtended by those electrons. This relationship can be expressed as follows:

$$\beta = \frac{\text{Current}}{\text{area} \times \text{solid angle}} = \frac{I}{A\,\Omega} = \frac{j_c}{\pi \alpha^2} \qquad (2.2)$$

where j_c is the current density expressed in A/cm^2 and α is the convergence angle of the beam in radians. As seen above, brightness is the beam current per unit area per solid angle and is measured in $A.cm^{-2}.sr^{-1}$ units. The solid angle is expressed in a dimensionless unit called steradian and equals $\pi \alpha^2$. This relationship takes into account the angular spread of the beam during its movement through the column. Due to this reason, brightness can be considered to remain constant throughout the SEM column, if any reduction due to lens defects is ignored. Therefore, brightness can also be thought of as the measure of total current that can be focused on the surface of a sample as represented by the following equation:

$$\beta = \frac{4i_p}{\pi^2 d_p^2 \alpha_p^2} \quad (2.3)$$

where i_p is the probe current and d_p and α_p are the diameter and convergence angle of the probe, respectively. These parameters cannot be modified independently of each other. Any adjustment in the value of one of the parameters i_p, d_p or α_p will be compensated by a change in the other two variables, thus keeping the brightness conserved. For example, if the probe is demagnified (i.e., d_p is decreased), then convergence angle α_p will increase to keep brightness constant. If an aperture is inserted to keep α_p unchanged, then current i_p will decrease in proportion to keep the brightness constant.

The maximum theoretical brightness of thermionic source is given by the following equation:

$$\beta_{max} = \frac{J_c e V_0}{\pi k T} \quad (2.4)$$

where J_c is the current density at the source, V_0 is accelerating voltage, e is the electric charge in coulomb, k is Boltzmann's constant, and T is the absolute temperature (K). It can be seen that brightness depends on the accelerating voltage at which the scanning electron microscope is used. For any gun type, brightness increases linearly with the accelerating voltage, i.e., brightness at 30 kV will be 3× than that at 10 kV. The above equation also indicates that brightness decreases with increasing filament temperature.

Brightness determines the amount of current in a probe of given size. It is a measure of the number of electrons that can be focused at a given point of a specimen in 1 s. It is one of the most important criteria of an electron gun as it directly affects the quality of the images obtained. Level of brightness obtained varies considerably with various types of guns employed. It can be seen from Table 2.1 that the brightness value of W filament is the lowest at 1×10^4 A.cm^{-2}.sr^{-1}. LaB$_6$ emitter shows 10× brightness, while Schottky field emitter is 1000× and the cold emitter is 2000× brighter than the W filament.

2.1.1.2 Size of the Electron Source

Electrons emitted from an electron gun come together to form an electron beam. The diameter of this beam (at the first electron crossover) is taken as the electron source size which varies depending on the type of electron gun used. Electron guns that emit electrons from tightly controlled small areas (i.e., tip) of the filament/emitter will

Table 2.1 Overall comparison of important properties of various electron sources [1]

Emitter type	Thermionic	Thermionic	Schottky FE	Cold FE
Cathode material	W	LaB$_6$	ZrO/W (100)	W(310)
Cathode shape				
Operating temperature [°C]	2,500	1,600	1,500	25
Cathode radius [nm]	60,000	10,000	<1,000	<100
Effective source radius [μm]	15	5	0.015(^{a})	0.0025(^{a})
Emission current density [A/cm^2]	3	30	5,300	17,000
Total emission current [μA]	200	80	200	5
Normalized brightness [A/cm^2.sr.kV]	1.10^4	1.10^5	1.10^7	2.10^7
Maximum probe current [nA]	1,000	1,000	10	0.2
Energy spread at gun exit [eV]	1.5–2.5	1.3–2.5	0.35–0.7	0.3–0.7
Beam noise [%]	1	1	1	5–10
Emission current drift [%/h]	0.1	0.2	<0.5	5
Operating vacuum, Pa	1×10^{-3}	1×10^{-5}	1×10^{-7}	1×10^{-9}
Typical Cathode life [h]	100	>1,000	>3,000	>3,000
Cathode regeneration	Not required	Not required	Not required	Every 6–8 h
Sensitivity to external influence	Minimal	Minimal	Low	High

^{a}Virtual source that would appear to form looking toward the gun from a point beyond the anode within the SEM column

result in smaller source size. Electrons coming out of a large area of a gun (not just from the tip) will be difficult to converge at a fine point and result in a large source size and reduced brightness. The source diameter is reduced to a much smaller probe size (typically a few nanometers or less in scale) within the SEM column by the time it reaches the surface of the specimen. The smaller is the probe size, the higher is the spatial resolution achieved by a microscope. Such demagnification within the SEM column is easier to achieve if the electron source is small to begin with. Large electron source requires considerable demagnification. The larger the demagnification required, the smaller the current present in the probe ultimately used for imaging. This is undesirable since adequate current is required in a probe for the generation of a strong image signal. Advanced guns achieve higher brightness by virtue of focusing electrons in the beam into a comparatively smaller probe. Table 2.1 shows that the electron source of LaB$_6$ is 3× while that of field emission guns is up to 1000× smaller than that of W filament, respectively.

2.1.1.3 Stability of the Electron Source

Steady electron emission current over a period of time is a measure of electron beam or source stability. Table 2.1 shows that thermionic emission sources are most stable, while cold field emitters show considerable change/drift in current over a period of time. The most stable source is the W filament. It shows a small drift of 0.1% per hour, which is considerably less than 5% per hour change depicted by cold field emitter. Drift can occur over a short (seconds) or long (hours) interval. Short-term instability can prompt image flicker during an image scan, while long-term instability can cause deterioration in the quality of x-ray scans. Stability becomes worse with mechanical drift, degrading vacuum and filament contamination. Use of new filament or switching on a filament can cause beam instabilities in the initial few minutes of operation.

2.1.1.4 Energy Spread of Electrons

An electron source operating at an accelerating voltage of 20 kV will emit most electrons at an energy of 20 keV. However, the energy of a finite number of electrons will vary slightly from this value. This variation in electron energies is called energy spread ΔE. Electrons with different energy (i.e., velocity) will be focused at different planes by the electron lenses as they pass through the SEM column resulting in an increased probe size and a blurry image. Electron guns with low energy spread will emit electrons with energy that varies only slightly within a small range, resulting in a comparatively clear image. Energy spread is inversely proportional to accelerating voltage. Energy spread in the electron beam determines the extent of probe enlargement due to lens defects. The lower energy spread results in smaller probe enlargement. Large energy spread also limits the theoretical current density in the final probe. Tungsten filament has the highest spread, while field emitters have the lowest energy spread (see Table 2.1), the latter ensuring formation of fine probe size. Beam instability can also cause energy spread.

Based on the mode of electron generation, electron guns are classified into (i) thermionic and (ii) field emission guns.

2.2 Thermionic Emission Electron Guns

Due to attractive forces that exist between negatively charged electrons and positive nuclei of the metal atoms, the electrons require energy to leave a metal surface. If a metal is heated to a high enough temperature, some of the electrons acquire sufficient energy to overcome the natural potential energy barrier that prevents them from leaving this metal. This natural barrier is termed the *work function*. The work function is therefore defined as the minimum amount of energy that is required to remove an electron to infinity from the surface of a given solid, usually a metal. The work function is a characteristic of a material and has a value in the order of a few electron volts (see Fig. 2.3a). Such electron emission that results from heating of a material was termed thermionic emission by British physicist Owen Richardson who later received a Nobel Prize in Physics for this particular body of work in 1928. It

2.2 Thermionic Emission Electron Guns

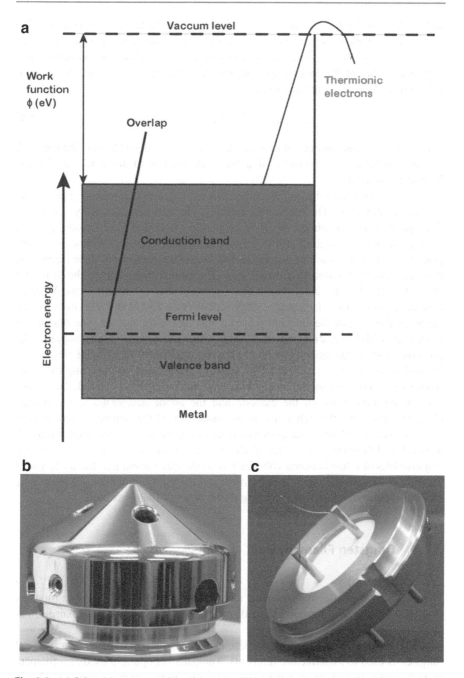

Fig. 2.3 (a) Schematic representation of the potential barrier (work function ϕ, eV) that electrons need to overcome to escape the metal surface in the absence of the extraction field. Photographs of (b) Wehnelt cylinder and (c) W filament. These two components are assembled together by inserting the filament into the Wehnelt cylinder to form a W filament gun

follows that the thermionic current can be increased by decreasing the work function of a material. This is often undertaken by coating the material surface with an oxide. The material is often shaped in the form of a wire. Richardson postulated that thermionic current increases exponentially with the temperature of the wire as depicted by the following equation [2]:

$$J = A_G T^2 e^{\left(\frac{-W}{kT}\right)} \left(\text{units}: \text{A/cm}^2\right) \tag{2.5}$$

where J is the emission current density, A_G is a constant dependent on the material, T is the temperature of the metal, W is the work function of the metal, and k is the Boltzmann constant.

The phenomenon of thermionic emission is exploited in a thermionic gun in order to generate electrons. The thermionic emission electron gun essentially consists of three parts: emitter (cathode, negative electrode), grounded plate (anode, positive electrode), and surrounding grid cover with a circular aperture (Wehnelt cylinder, grid cover, control electrode) (see Fig. 2.3b). Due to the presence of three electrodes, it is occasionally referred to as triode gun. The emitter cathode is in the form of a filament/wire (see Fig. 2.3c) whose tip is positioned at the center of the Wehnelt cylinder aperture. The filament is heated by a current to enable electrons to overcome the work function of the filament material and escape its surface. The filament is placed at a high negative potential using the high-voltage power supply. The potential at the filament corresponds to the required beam energy. For instance, if the desired beam energy is 20 keV, the negative potential of 20 kV is applied to the filament cathode. The anode, meanwhile, is grounded at zero potential. The difference in potential between the cathode and the anode accelerates the generated electrons downward through a hole in the anode toward the sample. Kinetic energy (E) of 1 electron volt (eV) is acquired by electrons when they are accelerated through a potential difference of 1 V. The Wehnelt cylinder or the grid cover is kept at a negative bias of a few hundred volts relative to the cathode emitter. Such a negative bias serves to focus the generated electrons into a beam. Commonly used thermionic emissions guns are classified according to the types of filaments used.

2.2.1 Tungsten Filament Gun

2.2.1.1 Material

In this gun, the emitter is a filament made of fully annealed strain-free tungsten (W) wire (≥ 100 μm diameter) shaped into the form of a V-shaped "hairpin," as shown in Fig. 2.4a, b.

The tip is approximately 100 μm diameter in size. High-grade refractory metal tungsten is used to manufacture filaments. Tungsten is popularly used as a filament material due to its unique properties, i.e., it has the highest melting point (3422 °C), the lowest vapor pressure at high operating temperatures, the highest tensile strength, and the lowest coefficient of thermal expansion compared to all metals in pure form. Tungsten has a reasonable value of work function and high current density that

2.2 Thermionic Emission Electron Guns

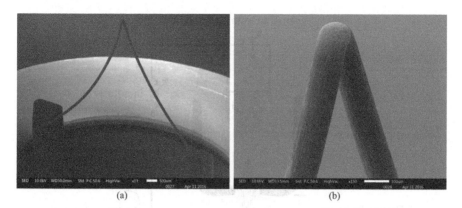

Fig. 2.4 (a) Low- and (b) high-magnification SEM images of V-shaped tungsten (W) filament cathode. (Courtesy of JEOL Ltd.)

allows it to operate well below its melting temperature. Tungsten can also be alloyed with 3% rhenium to improve wire ductility and its resistance to sagging and vibration. The filament is attached to legs mounted on an insulator (i.e., glass or ceramic base). In this gun, the filament itself is the cathode and is heated to emit the electrons. This arrangement is referred to as the *directly heated* cathode.

2.2.1.2 Working Principle

Working principle of W filament electron gun is depicted in the schematic diagram shown in Fig. 2.5. Electric current is used to resistively heat the V-shaped filament to the temperature of up to 2500 °C. The tip offers the highest resistance to the flow of current and thus becomes the hottest part of the filament. Electrons at the tip receive sufficient thermal energy to overcome the relatively large work function (4.5 eV) of the W metal and leave the filament surface. The area of electron emission is around 100×150 μm. This comparatively large emission area results in large source size for this gun. At an accelerating voltage of 20 kV, the filament is set at a negative potential of -20 kV with respect to the anode that is at ground (zero) potential. Most of the microscope itself is at ground potential. The filament is surrounded by Wehnelt cylinder (or grid cap) which connects to the former by a self-biasing variable resistor that is also connected in series with the high-voltage power supply to the filament. Emission of electrons sets the Wehnelt cylinder at slightly more negative potential than the filament creating a negative bias between the two. This creates a voltage drop across the resistor that sets the bias to constantly replenish electrons lost from the filament due to emission. Emitted electrons from the source tend to spread in all directions. Negative bias between the Wehnelt cylinder and filament cathode ensures that electrons are emitted from an area of the filament close to the tip only. It focuses the beam and controls electron emission and divergence angle α_0. The combined effect of electric field formed due to the cathode, Wehnelt cylinder and anode acts as an electrostatic lens and focuses the emitted electrons to form a crossover of diameter d_0 (≈ 50 μm) just below the Wehnelt cylinder. This

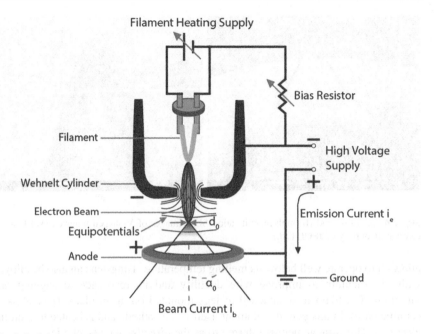

Fig. 2.5 Schematic of a typical tungsten thermionic emission electron gun [3]

could be thought of as an image of the area of the filament from where electrons are emitted. Electrons at crossover are characterized by Gaussian intensity distribution. The image of the crossover is used to set the saturation level of the filament.

The current used to heat the filament is called *filament heating current* i_f. The total emitted current (typically 50–100 μA) from the filament is called *emission current* i_e. Amount of emission current leaving through the anode is called *beam current* i_b. Anode allows only fraction of electrons to pass through it. Current decreases at every lens and aperture. Final current measured at specimen surface is called *probe current* i_p (i_p could be as low as ≈1 pA). This indicates a large decrease in the current from the filament to the specimen surface. All electrons that leave the tip must be replenished by "traveling" through the circuit.

The size of crossover is too large to undertake imaging as the required final probe diameter is 10 nm or less. Demagnification of the probe diameter takes places within the ensuing electron column. The potential difference between the filament and the anode results in acceleration of electrons that enables them to reach the sample surface sitting in the specimen chamber at the end of the electron column. Electrons are not accelerated any further beyond the anode.

2.2.1.3 Role of Self-Regulating Bias Resistor

The nature and magnitude of electron emission are controlled by the bias of an electron gun as depicted in the schematic diagram shown in Fig. 2.6. It should be noted that the Wehnelt cylinder would always be at some negative potential (bias) relative to the filament. The magnitude of this bias can be adjusted. Figures 2.5 and

2.2 Thermionic Emission Electron Guns

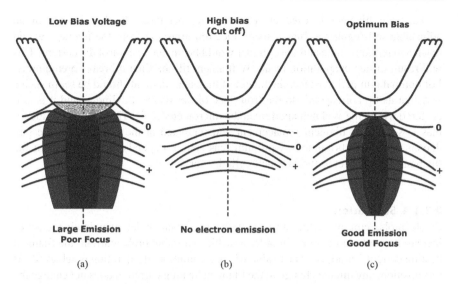

Fig. 2.6 Schematic showing the effect of voltage bias on electron emission and probe size (e.g., focus). (**a**) Low bias will result in large emission and poor focus. (**b**) High bias will give no electron emission. (**c**) Good emission and focus is obtained at an optimum level of bias

2.6 show the presence of *electrostatic field lines* (also called *electrostatic potentials* or *equipotentials*) between the Wehnelt cylinder and the anode. Electric field lines are formed due to an electrostatic field formed by the voltages maintained between filament, Wehnelt cylinder, and anode. The voltage of equipotentials is represented relative to the filament and can vary from zero at the filament to negative at the Wehnelt cylinder and positive at the anode. The configuration of these electrostatic potentials is controlled through a bias resistor (with 1–10 MΩ resistance) attached to the high-voltage power supply and the filament. The function of this self-regulating bias resistor is to control the emission of the electron gun and keep it at optimum levels at all times during its operation.

At low bias (see Fig. 2.6a), negative voltage applied across the Wehnelt cylinder will be small making the negative field lines weak. The presence of positive field gradient toward the cathode will result in high emission current since electrons will be emitted not only from the tip of the filament but also from the sides of the tip. Electrons emitted from a large area of the filament are difficult to converge and hence ultimately result in poor focus and low brightness. At high bias (see Fig. 2.6b), the negative voltage applied across the cylinder will be high resulting in strong negative equipotentials around the filament that will discourage electron emission. Any emitted electrons will be attracted back into the filament due to high negative electric field surrounding the filament. Brightness, in this case, will become zero. Optimum bias (optimum negative potential between the filament and the cylinder) results in good emission only from the tip of the filament with optimal focus and high brightness (see Fig. 2.6c). Electrons emitted from the sides of the tip tend to be pushed back into the filament.

Tungsten filament is used at white hot temperatures. In the absence of an adjustable self-regulating bias resistor, electron emission from the filament would reach a maximum value and the cylinder would be unable to control the current. This will result in fast evaporation of the W filament till the time it breaks prematurely. For optimal gun performance, the height of tungsten filament should be set in such a way that its tip corresponds to the front face of the Wehnelt cylinder, and it is also centered within the Wehnelt aperture at the micrometer scale. The optimum value for bias is set during filament replacement and regulated at the point where the gun depicts maximum brightness. Optimum bias setting depends on the height of the filament with respect to the Wehnelt cylinder surface.

2.2.1.4 Saturation

Stable probe current at the surface of the specimen under examination ensures increased quality in imaging. In order to achieve a stable probe current, the filament heating current i_f is adjusted at a value where a condition of saturation is achieved. At this position, any minute change in the filament heating current i_f does not change the electron beam current i_b since it reaches a plateau (see Fig. 2.7). Saturation is thought to occur due to the negative feedback effect of the series resistor used to generate the bias. Tungsten filaments are operated at near saturation point above which level any further increase in current only serves to greatly decrease the service lifetime of the filament without any concurrent increase in brightness. Use at saturation helps to achieve a small probe, while in an unsaturated condition, electrons are emitted from the tip as well as from the sides of the filament contributing to an increase in source size. Operation at saturation also makes a filament produce stable beam current despite small variations in temperature. Saturation point can be determined by either

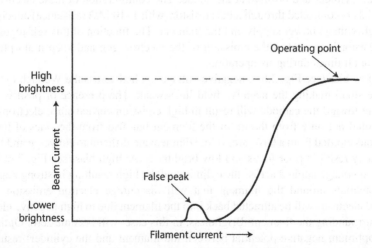

Fig. 2.7 Beam current increases with filament current to a saturation point beyond which any further heating will not result in increased beam current, only in decreased filament life. This is why a W electron gun should always be used at the saturation point

looking at the image display or reading the current values displayed on the meter. Saturation value for a filament at each accelerating voltage will be different. Most SEMs are capable of setting these values in the background unobtrusively. Saturation is self-regulated by a process where if the filament current i_f increases resulting in an increase in electron emission, negative bias on the Wehnelt cylinder is also raised by the auto-bias resistor. This slows down emission ensuring stable beam current.

Figure 2.7 shows the presence of a "false peak," which is a momentary increase of brightness as the beam current is increased. This false peak should not be mistaken for saturation point. The filament cannot be set at this value as the beam will be highly unstable and not suitable for imaging. The exact nature of this false peak varies with the filament, gun, and the microscope. Its origin is unclear and is thought to depend on electrostatic field lines created during filament heating.

2.2.1.5 Advantages/Drawbacks

Due to its low cost and reliability, W filament has been the most commonly used filament since the SEM was invented. It is widely used for low-magnification imaging and microchemical analysis. It is quite sturdy and its service lifetime depends on the emission current used. It can last up to 100 h at medium currents. Its replacement is straightforward and quick and is usually undertaken by the SEM operator without the help of service personnel. It requires low vacuum (i.e., 10^{-3} Pa is enough to prevent oxidation of heated cathode) to work. Its disadvantages are relatively high work function of 4.5 eV, insufficient emission current density (low brightness, 1×10^4 A/cm^2.sr), large energy spread (ΔE is 1.5–2.5 eV), and a comparatively short service lifetime. Its large energy spread, and low brightness make it unsuitable for microscopy at low beam energies (e.g., 1–5 keV).

2.2.1.6 Service Lifetime

Filament service lifetime depends on the rate of W evaporation and oxidation and it decreases as the filament temperature increases. The temperature at which filament is saturated can be decreased by increasing the distance between the filament tip and the Wehnelt cylinder during filament installation. This tends to increase the service lifetime of the filament but affects its ability to form a fine probe, thus making the use of this option undesirable. Maintenance of good vacuum within the SEM and preventing contamination buildup in the column also ensures longer service lifetime for the filament. Similar to any parts used in vacuum, filament cathodes should be handled with clean-room gloves to keep surfaces free of fingerprints and other contaminants. Gun chamber should not be vented while the filament is still hot to avoid oxidation of the filament material. Yellow or brownish color on the base (where filament is mounted) usually suggests a vacuum leak. Filament that has failed due to evaporation retains its original wire thickness and depicts a rounded blob at the region of the break. Any wire thinning toward the broken region suggests vacuum leak. These precautions along with avoiding over-saturation of the hairpin can serve to prolong service lifetime of the filament cathodes.

2.2.2 Lanthanum Hexaboride (LaB₆) Emitter

2.2.2.1 Material

The emitter is made of refractory ceramic single crystal LaB$_6$ that is intense purple-violet in color and has a melting point of 2210 °C. Lanthanum hexaboride is selected due to its low work function, high melting point, low vapor pressure (that inhibits evaporation), and chemical and high temperature stability. The crystal is 100 µm in cross section and 0.5 mm in length, while its tips are shaped as flat (microflat, truncated), round (standard), or sharp (see Fig. 2.8a, b). LaB$_6$ ceramic has high electrical resistance and cannot be heated directly. Therefore, crystal emitter is mounted on a single-piece stress-free nonreactive graphite carbon or rhenium which is heated resistively to raise the temperature of the emitter. This arrangement, where the LaB$_6$ emitter/cathode is heated by a separate heater, is called *indirectly heated* cathode. A carbon ferrule is used to provide support and hold this assembly in place. Nonreactive carbon is employed since LaB$_6$ is susceptible to carbon contamination.

The crystal is usually produced with an orientation of <100> due to its superior brightness and symmetry. The plane symmetric shape of the crystal ensures even

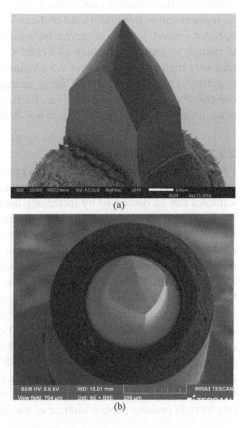

Fig. 2.8 (a) SEM images of the LaB$_6$ emitter with (a) sharp tip and (b) flat tip

evaporation of the cathode relative to its axis at high temperature. This enables the emitter surface to remain flat and centered and retain consistent emission pattern throughout its service lifetime. Inert gas arc float zone refining process is usually used to produce single crystals of LaB_6 cathode. Impurities present within the starting cathode material during production are strictly controlled down to tens of ppm as their presence tends to reduce the brightness and lifetime of the emitter.

Proper alignment is crucial for optimal performance and longer lifetimes of LaB_6 emitters. Cathode temperature should be optimized for desired emission current. LaB_6 is a stable material, thereby providing a comparatively steady emission. Continuous use at operating temperature increases the thermal stability of the gun resulting in improved electron beam stability. The cathode is used at the minimum power necessary for adequate emission at saturation. Operation at an excessive temperature (overheating) should be avoided as it reduces the lifetime of the emitter. LaB_6 gradually evaporates from the cathode at excessive temperature and tends to form a deposit on the Wehnelt cylinder and aperture. Low operating temperature and a poor vacuum will result in the formation of a rough white powder-like deposit on the cathode itself. In order to reduce contamination buildup, LaB_6 emitters are generally kept in operation mode at slightly elevated temperatures at all times. This also prevents frequent exposure to thermal shock due to intermittent cooling and heating and serves to increase lifetimes. Poor gun vacuum can lead to short cathode lifetime.

2.2.2.2 Tip Design

Beam current, brightness, and spot size depend on the design of the cathode tip, which also affects the performance and service lifetime of the emitter. The most popular design of the cathode is a conical tip with a flat emitting surface at the end. Small cone angle (60°) produces higher brightness, while a larger angle (90°) imparts longer lifetime and convenient alignment. Flats with small diameters offer smaller source size with higher brightness, but longer lifetimes and larger beam currents are afforded by larger flats. For SEM applications, a cathode with 90° cone angle and a 15 μm radius flat tip provides a good combination of small source size, high brightness, and long service lifetime. Tip radii for flat tips can range from 15 to 100 μm, while sharp tips are polished to 5 μm finish.

2.2.2.3 Advantages/Drawbacks

Lanthanum hexaboride has a lower work function (2.66 eV) than W enabling higher emission at lower temperatures of around 1400–1600 °C (see Table 2.1). A current of 1.7–2.1 A is employed to resistively heat the cathode. The LaB_6 emitter is 5–10 times brighter than W filament and has 10x longer service lifetime (typically 1000 h). Source size (i.e., crossover diameter) is 5 μm resulting in a relatively small probe size with more beam current at the specimen surface and an improved image quality.

Lanthanum hexaboride emitters are about 10 times more expensive than W filaments and also require high vacuum ($\approx 10^{-5}$ Pa) environment to avoid the formation of volatile oxides. The higher vacuum within the gun is realized by getter-ion pump installed toward the upper side of the SEM column. Tungsten filament can generally be replaced by a LaB_6 emitter in conventional SEMs provided differential pumping of the gun chamber is available to meet more stringent vacuum requirements and to avoid emitter contamination.

Cerium hexaboride (CeB_6) crystal is also used as an emitter for microscopy applications. CeB_6 is known to resist degradation due to C contamination better than LaB_6 emitter does. However, the LaB_6 crystal is by far the most commonly used hexaboride crystal in microscopy.

Thermionic emission guns are not suited for imaging at low accelerating voltages ($\leq 5\,kV$) due to low gun brightness at low electron energies. Thermionic guns are also characterized by large energy spread, which accentuates lens defects at low accelerating voltage. These drawbacks along with large probe size and limited service lifetimes associated with thermionic emission guns instigated the development of field emission guns, which tend to overcome the abovementioned shortcomings.

2.3 Field Emission Electron Guns

2.3.1 Working Principle

In field emission electron guns, cathode emitter is in the shape of a sharp tip (≤ 100 nm to 1 μm in radius) made of single crystal tungsten wire and spot welded to a tungsten hairpin, as shown in Fig. 2.9.

Fine cathode tips are produced using electrochemical anodic etch method. In this method, the etched tungsten wire (anode, set at specific current and voltage) is dipped with high precision into liquid electrolyte held within a grounded cylinder (cathode). The bottom of the wire "drops off" due to anodic dissolution leaving behind a sharpened tip.

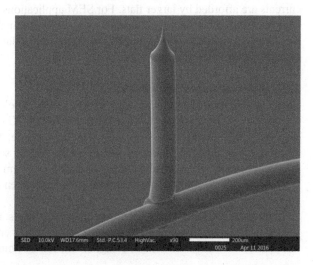

Fig. 2.9 Sharpened cathode tip used for field emission guns

2.3 Field Emission Electron Guns

In field emission guns, electrons are emitted by applying a negative potential to the cathode tip. Due to the fine size of the tip, the electric field becomes highly concentrated at this small region. The strong electric field in the tip region reduces the height of the potential barrier and makes it narrow enough for the electrons to overcome the barrier without ever requiring an increase in the kinetic energy of the electrons. This phenomenon is termed *quantum tunneling* whereby a particle can pass through a potential barrier higher than the energy of the particle. A negative potential of 3–5 kV applied to the cathode is adequate to produce an electric field of the order of 10^3 V/μm at the tip that makes the potential barrier thin enough for the electrons to escape the tip surface via tunneling effect. This is in contrast to thermionic emission where the kinetic energy of electrons is increased through heating to overcome the potential barrier. In field emission, fine cathode tip allows electrons to tunnel through directly and generate emission required for SEM imaging. Generation of large electric field at the tip generates high magnitude of stresses at the tip. Tungsten is the material of choice as it can withstand these stresses during operation.

2.3.2 Advantages/Drawbacks

Emission from a fine tip results in a small virtual electron source that does not require large demagnification to produce a small probe size (see Table 2.1). Fewer lenses are thus required for demagnification simplifying microscope design. Crossover point does not exist within a field emission gun, and the source is thought to be the apparent size of the virtual disc within the tip from where the electrons are generated. The spot produced on the specimen is an image of the source itself (and not that of the crossover as in the case of thermionic sources). The omission of electron interactions that typically occur within the crossover region results in decreased chromatic aberrations and probe size. This makes the SEMs equipped with field emission guns most suitable for high-resolution imaging. In fact, machines equipped with field emission guns are usually classified as high-resolution microscopes. Since the virtual source is small and its angle of emission is also narrow, a larger proportion of the emitted current reaches the specimen. Use of a smaller final aperture in microscopes with field emission guns also ensures greater depth of field. The ability to focus electrons into small probes also results in high brightness (1000× higher than W filament). Another important advantage of field emission guns is small energy spread compared to thermionic emission guns (4× less than W filament) resulting in smaller probe size. Lastly, service lifetime of field emission guns (with operational times of more than 3000 h) is much longer than thermionic emission guns.

Field emission guns are invariably more expensive and require more time consuming and tedious service maintenance compared to thermionic emitters. They also need more stringent vacuum conditions (10^{-7} to 10^{-9} Pa) to be sustained for their operation. High vacuum is required to keep the cathode clean and to prevent damage to the tip due to possible ion bombardment from the residual gas. Such degradation effects can diminish the sharpness and surface finish of the tip, which then eventually needs to be replaced. Field emission guns are less suited for an application that requires a large probe size (of the order of ≥ 30 nm) since adequate emission current will not be available for this kind of work. Thermionic emission guns with large total emission currents are devices of choice for such applications.

Two commonly used emitter types are cold field emitter and Schottky field emitter.

2.3.3 Cold Field Emitter (Cold FEG)

2.3.3.1 Working Principle

In this type of emitter, electrons are drawn out of tungsten cathode tip through quantum-mechanical tunneling without any contribution from thermionic emission. The emitter relies solely on electric field for electron emission and is not heated during its operation, hence the term cold field emitter. A schematic representation of cold field emission is shown in Fig. 2.10a. The tip used for the cold emitter is very

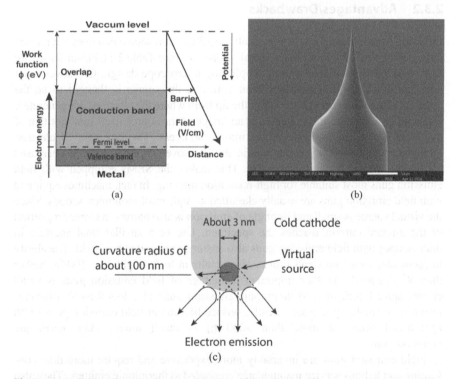

Fig. 2.10 (a) Schematic representation of the potential barrier that electrons overcome by application of an electric field during cold field emission. (b) SEM image of sharply pointed conical cold field emitter (courtesy JEOL Ltd). (c) Schematic representation of the virtual electron source formed within the cold field emitter tip produced from single crystal tungsten in <310> orientation

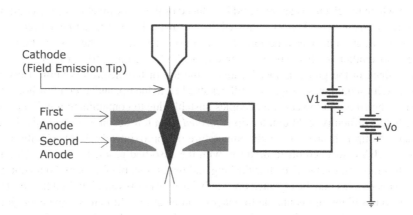

Fig. 2.11 Schematic of the Butler triode FE source. V_1 is the extraction voltage (few kV) and V_0 is the accelerating voltage [4]

fine in size (≈100 nm radius, <310> orientation; see Fig. 2.10b), and the area of the tip from where the emission occurs is still smaller (i.e., of the order of a few nanometers). A schematic representation of the shape of the cold emitter tip is shown in Fig. 2.10c.

Schematic diagram of the cold field emitter setup is shown in Fig. 2.11. It is essentially a triode gun configuration where the first anode is set at 3–5 kV of potential (V_1) relative to the cathode, which generates an electric field strong enough to extract electrons from the tip without heating it up. The second anode is used to produce the required acceleration of electrons between the cathode and the anode. The potential (V_0) is set at the required accelerating voltage. Total emission current produced is in the order of 5 µA which is quite small compared to that produced in W filament (i.e., 200 µA). However, due to its small source size and its ability to focus the beam into a small area, the brightness of cold emitter is much higher (2000×) than that of W filament (see Table 2.1). The higher the temperature of the emitting cathode, the larger is the resulting energy spread (i.e., $\Delta E = 2.5\ kT_c$) where k is Boltzmann constant and T_c is the temperature of the cathode tip. Energy spread of the cold emitter is the lowest of all guns, as it is not heated during operation.

2.3.3.2 Service Lifetime

Gas adsorption at the cathode tip can be prevented by keeping it at elevated temperature (800–1200 °C) even when it is not in operation. The cold emitter, however, is not heated during emission for which reason it is highly susceptible to contamination that can take the form of adsorption of column gases and ion sputtering at the tip surface. Large electric fields present in the gun can accelerate the positively charged gas molecules toward the cathode resulting in sputtering of emitter surface. This damages the surface of the emitter and changes its geometry. Interaction of electron beam with the first anode can also produce ions that can damage the tip. Because of adsorbed gases and surface damage, the effective area from where the electrons are released is reduced or the work function of the emitting surface is increased. This is followed by an unstable or fluctuating current and decreased emission over a period of time. In order to reduce the susceptibility of the cold emitter to this type of degradation, high vacuum (10^{-8} to 10^{-9} Pa) is maintained in the gun chamber located at the top of the electron column. Despite such high vacuum, gas molecules still get attached to the tip, and after a few hours of usage, the emission lowers and becomes unstable due to contamination buildup. To tackle this difficulty, cold emitter tip is *flashed* for a few seconds on a regular basis where the tip is heated to a high temperature of approximately 2200 °C with high voltage (HV) switched off to do away with the adsorbed gases. It takes 10–15 min for the emission to stabilize after flashing and remains stable for many hours during use. This process may be repeated on a daily basis. After several hundred flashes, the tip becomes blunt and is unable to generate an electric field that is large enough to cause emission through the tunneling effect. At this stage, the cathode tip is replaced but not before it has served for several years.

Recent advances in technology allow the creation of an ultra-high vacuum environment (10^{-10} Pa) within the gun chamber that slows down the adsorption of ions to an extent where the cold emitter can be used for many hours in a highly "clean" condition. This results in an increased current emission and stability. Flashing technology has also seen improvements where automatic cleaning of the tip is carried out in the background, while high voltage remains switched on, thereby increasing the duration for which the emitter can be used at a stretch before regular flashing. This way high probe currents can be maintained for a longer duration.

2.3.3.3 Advantages/Drawbacks

Long service lifetime, high brightness, small energy spread due to low temperature operation, and small probe size due to a restricted area of emission are the primary strengths of cold field emitters. Microscopes fitted with cold field emitters are the equipment of choice for high-resolution microscopy. High brightness and low energy spread make it a desirable electron source for low beam energy microscopy. Both short and long term instability and limited availability of current can be regarded as its main shortcomings. It is not suitable for applications that require the use of high beam currents. Additionally, beam current feedback stabilization mechanism needs to be incorporated if quantitative data is collected. Stringent vacuum conditions must be maintained at all times. Low degree of demagnification of electron source required in cold emitters makes them very sensitive to vibration and outside electromagnetic fields complicating microscope design.

2.3.4 Schottky Field Emitter

This is a field-assisted thermionic emission gun in which use is made of Schottky effect where a considerable amount of extraction voltage is applied to the emitter and the latter is heated at the same time. The applied field serves to lower the work function considerably assisting in the generation of electrons through thermionic

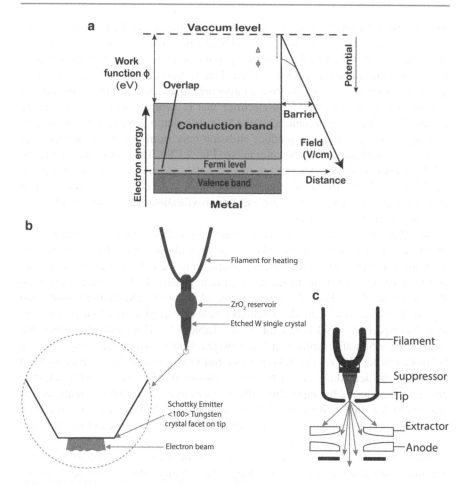

Fig. 2.12 (a) Schematic representation of potential barrier that electrons overcome by application of heat along with an electric field during Schottky field emission. (b) Schematic diagram of emission patterns from a Schottky emitter. (Adapted from Tuggle et al. [5]). (c) Schematic representation of Schottky field emission gun setup

emission as shown in the schematic diagram of Fig. 2.12a. Schottky field emitter is made of tungsten wire welded to tungsten filament. The wire is 1 mm in length and 125 μm in diameter and has a tip radius of ≈ 0.5–1 μm. The tip is produced using electrolytic etching with {100} planes set perpendicular to the wire axis as shown in the schematic diagram in Fig. 2.12b. This crystallographic orientation gives lower work function than other configurations.

The cathode is heated to a temperature of around 1500 °C to produce electron emission. A current of approximately 2 A is used to heat the tungsten wire. The fine size of the tip serves to reduce the work function barrier when an electric field is applied to it. The tungsten tip is also coated with ZrO_x, which further reduces the work function from 4.5 eV to 2.7 eV. The Zr reservoir is provided at the shank of the tungsten wire. During heating, Zr metal thermally diffuses to the tip of the cathode where it reacts with oxygen to form a layer of ZrO_x. The presence of Zr reservoir allows re-formation of ZrO_x layer on the tip as it is sputtered away during use. Such a tip is generally known as ZrO/W emitter. Due to filament heating, the cathode is susceptible to emitting currents from a larger area than desirable. The emission area is contained by a suppressor grid electrode that is a metal cup with a hole in the center. The cathode tip is centered in the hole such that it protrudes out of the cup. The suppressor grid helps to produce emission only from the tip region and stops stray emission to pass down the column (see Fig. 2.12c). The emission is controlled by the potential on the extractor electrode. Total emission current of 200 µA is available which is 40x higher than that available with cold field emitters. High emission current decreases the susceptibility to vibration effects and affords more flexibility in microscope design.

Schottky field emitter is kept on at all times to avoid contamination buildup at the tip surface. This results in a stable beam current without the need to flash the gun. Vacuum conditions are not as stringent compared to the cold emitter, but a vacuum of 10^{-7} Pa is desirable to maintain stable current, produce high brightness, and keep ZrO_x reservoir clean. Typical service lifetime of Schottky field emitter is more than 2 years. Use of thermal cathode in Schottky field emitter gives rise to higher energy spread compared to cold field emitter (see Table 2.1). This limits the achievable resolution for Schottky emitters at low voltages. Comparatively larger tip radius in Schottky emitter also results in bigger electron virtual source than that of the cold emitter, again degrading attainable spatial resolution. Microscopes with Schottky emitters are preferred for applications that require large stable probe currents such as electron backscattered diffraction (EBSD) and wavelength dispersive x-ray spectroscopy (WDS). These microscopes are suitable where long-term carefree operation without adjustments is required.

In summary, field emitters are preferred over thermionic guns due to their high brightness, small source size, greater long-term stability, and long service lifetime. Field emission microscopes have become the equipment of choice for demanding applications due to their long-term high-end overall performance. On the other hand, microscopes with thermionic emission guns are selected for their cost-effectiveness, applications that require high currents and simpler service maintenance routine.

2.3.5 Recent Advances

Field emission gun technology continues to improve. In one of the advances, the gun is placed within the magnetic field of a low aberration condenser lens. The anode of the electron source is integrated into the condenser lens. This results in a tight constricted beam of electrons with the minimal spread as they accelerate toward the specimen. An improved objective lens in conjunction with such an electron source ensures probe of small size and high current. The probe current thus obtained is 200–400 nA, which is 10x more than that obtained for probes of similar size produced by conventional FEG. With this technology, it is possible to obtain 5 nm spot size with 5 nA current at an accelerating voltage of 3 kV [6]. In another development, carbon nanotubes (CNTs) have been tested as a promising candidate material for the development of nano-sized field emission cathode tips. However, it will be some time before these tips are produced on a commercial scale.

2.4 Electromagnetic Lenses

The electron optical system in the scanning electron microscope consists of electromagnetic lenses (condenser and objective lens), apertures, stigmator, and scan coils. The main function of these components is to form a fine electron probe that scans the area of interest on the specimen in the form of a raster. The electromagnetic lenses used in the SEM do not form images in the conventional sense as customary in light optical or transmission electron microscopes. Lenses in the SEM are used to demagnify or focus the electron beam generated by the electron gun. This demagnification or re-convergence is initially accomplished by typically two condenser lenses followed by formation of a fine probe on the specimen surface by the final lens called objective lens. The electron beam emanating from a typical tungsten filament gun measures 50 μm in diameter. This diameter needs to be reduced by 100 to 5000 times to a probe size of approximately 0.5 μm to 10 nm in order for it to be useful for imaging and microchemical analysis. Large demagnification requires the presence of up to three condenser lenses, while two condenser lenses suffice to demagnify small beam diameters produced by field emission guns.

Electromagnetic lenses are analogous to thin convex lenses used to focus visible light in optical microscopy. Electromagnetic lenses are made of Cu coil enclosed in an iron casing having cylindrical pole pieces, as shown in the schematic illustration of Fig. 2.13a.

Direct current running through the coil produces concentrated magnetic field restricted within the iron shroud except for a gap between lens pole pieces that allows the field to exert a force upon the electrons traveling down the electron column. The magnitude of force F experienced by an electron of charge e and velocity v while passing through an electromagnetic lens is given as:

$$F = Bev \sin \theta \qquad (2.6)$$

where B is the magnetic field and θ is the angle between B and v.

The magnetic field has a radial B_R (perpendicular to the optic axis) and axial/longitudinal B_L (parallel to optic axis) component. The radial field causes the off-axis electrons to spiral down the optic axis in a helical path as seen in Fig. 2.13b. The interaction between the electrons emanating from the gun and the magnetic field deflect the electrons toward the axis. The electrons away from the optic axis are deflected with greater force compared to the ones that are closer, thus creating a focusing effect. Unlike optical lens where the focal length is fixed, varying current in

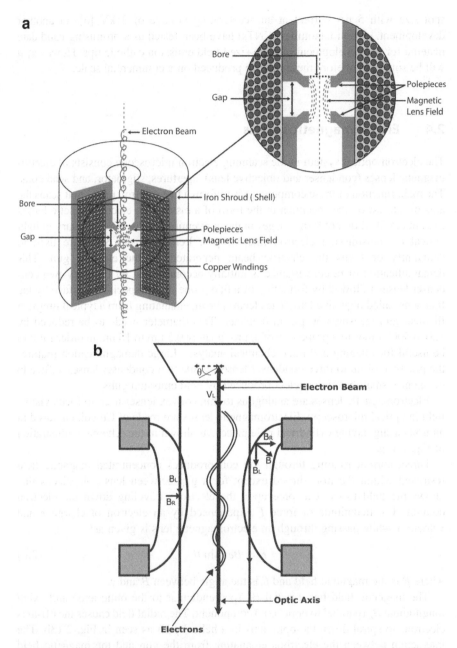

Fig. 2.13 (a) Schematic representation of an electromagnetic lens. Copper wires are held within iron shroud. Electromagnetic field interacts with the electron beam through the gap between the pole pieces. (b) Working of an electromagnetic lens: The radial component of the magnetic field makes the off-axis electrons go down the optic axis in a spiral

the coils of the electromagnetic lens can vary focal length. The degree of focus can be controlled by changing the current in the coil. The probe diameter upon exit from an electromagnetic lens can be calculated using the following equation:

$$d_0 = \frac{1}{2B}\sqrt{\frac{2mV_0}{e}} \quad (2.7)$$

where d_0 is the probe diameter in mm, B is the magnetic field strength in Tesla and m is the mass (kg), e is the charge (Coulomb), and V_0 is the accelerating voltage (kV) of the electrons.

Helical path of electrons can result in image rotation, θ. In modern SEMs, image rotation is minimized by controlling longitudinal component of the magnetic field by placing successive lenses in a way that nullifies θ. Present-day SEMs also keep the probe focused through changing accelerating voltages by automatically adjusting for the current in the electromagnetic lenses.

Different types of lenses housed in the SEM column and their functions are summarized below:

2.4.1 Condenser Lens

Two to three condenser lenses can be present directly below the electron gun in an electron column of the SEM depending upon the extent of demagnification required. These are used in combination and are controlled through one control knob (labeled as spot size, beam current, or probe diameter) to demagnify the electron beam by regulating the current in the lens coils. Increase in the current of the lens coil results in a strong lens with strong focusing action (e.g., electrons focused closer to the lens/ small focal length). A strong condenser spreads out the electrons such that a large number of them are blocked by the aperture and cannot reach the next lens. This results in small probe size and current on the specimen (Fig. 2.14a). Decreasing the current in the lens coil makes the lens weak resulting in weak focusing action (e.g., electrons focused away from the lens/long focal length). The trajectory of electrons through a weak condenser lens allows a large number to reach the second lens resulting in large current and spot size (see Fig. 2.14b).

Analogous to optical lenses, the demagnification is given as the distance between the object plane and the center of the lens (p in Fig. 2.14) divided by the distance between the center of the lens and the image plane (q in Fig. 2.14).

$$\text{Demagnification} = \frac{p}{q} \quad (2.8)$$

The crossover diameter d_0 is the original object, while d_1 formed after the condenser lens is considered its image. It can be seen in Fig. 2.14 that d_1 is reduced in size compared to d_0 because of demagnification. It is also rotated. For the final

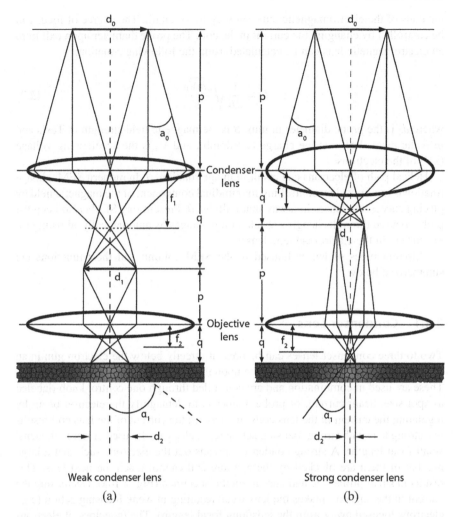

Fig. 2.14 Schematic diagram depicting the use of lenses to demagnify the electron beam. The first lens is the condenser lens, while the second lens is the objective lens. (**a**) Strong condenser lens results in small spot size, i.e., demagnification $=\frac{p}{q}$ is large. (**b**) Weak condenser lens results in large probe size, i.e., demagnification $=\frac{p}{q}$ is small (Note that all electrons emanating from a single point in the object plane come together in a single point in the image plane. Parallel electron beams emanating from the object plane focus on a point called the focal point of the lens. Electrons passing through the center of the lens do not change trajectory) Adapted from [7]

lens, d_1 becomes the object and d_2 is its image. There is a further reduction in the size of the probe d_1 after passing through the final lens resulting in d_2. Selection of a small spot size reduces image brightness but facilitates the formation of fine probe size that improves image resolution.

2.4.2 Apertures

An aperture strip is a thin rectangular piece of molybdenum or 95% Pt–5% Ir alloy with precisely drilled central holes to allow the beam to pass through. These holes are termed as apertures. A strip is located within the SEM column and houses several apertures that can be interchanged by the SEM operator. The radii of these apertures range from 10 μm to 500 μm.

The function of an aperture is to control the number and the convergence angle of electrons passing through the column. Apertures prevent the off-axis electrons from reaching the specimen surface thus decreasing the effect of lens defects (i.e., spherical aberration) and improving image resolution. Small aperture will result in a small probe size with less current. This will increase the image resolution but decrease the signal strength. For high image resolution, the smallest aperture with adequate/necessary signal-to-noise ratio is normally used. For chemical analysis that requires large currents, large aperture can be employed. Spray apertures block off-axis electrons, while beam-limiting apertures reduce the beam convergence angle and the probe current.

The apertures need to be kept clean of dust and carbon contamination. If the aperture surface is dirty, the electron beam is distorted due to its interaction with the nonconductive contamination on the aperture disc. This results in a defect called astigmatism. A high level of astigmatism cannot be corrected by the stigmator coil and needs cleaning or replacement of the aperture.

2.4.3 Objective Lens

The final lens in the column is called the objective lens. It is designated as the probe forming lens in the SEM. It is controlled by adjusting the "*focus*" knob during the SEM operation. The current in the objective lens coil is adjusted to demagnify and focus the electron beam on the surface of the specimen. This lens needs to be water-cooled to prevent over-heating due to the high magnitude of current passing through it. Three types of objective lenses are as follows:

2.4.3.1 Pinhole Lens

This is the most commonly used objective lens with a small bore diameter. It is also referred to as the conventional or out-lens objective lens. The specimen sits outside and at a considerable distance away from the lens. It is used for conventional imaging at relatively low magnifications. The lens is asymmetric with upper pole piece depicting large bore, while lower pole piece has a smaller bore size to reduce the effect of magnetic field on the specimen. This provides flexibility in the use of specimens with large dimensions (up to several centimeters) and also imaging at large working distances of up to 40 mm resulting in high depths of field. The lower part of the lens is conical to allow tilting of specimens and placement of detectors close to the specimen surface to increase solid angle of signal collection. The end of the lens facing the specimen is flat to allow mounting of a backscattered detector directly

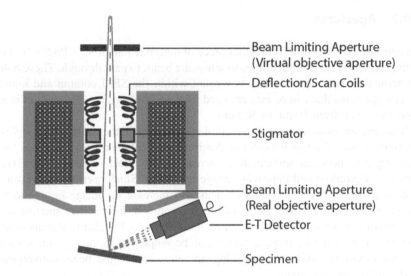

Fig. 2.15 Schematic diagram of pinhole lens (Adapted from [7])

below the pole piece. Due to a large distance between the lens and the specimen, the latter can be manipulated freely, but the focal length of the lens is large. As a result, large aberrations/defects occur which prevent the ability to undertake high-resolution microscopy. Figure 2.15 shows an illustration of a pinhole lens.

2.4.3.2 Immersion Lens

In this design, the specimen to be examined is placed within the bore of the lens. The specimen is raised up to its position. This restricts the size of the sample that can be examined to a few millimeters. It also makes it more challenging to collect the signals generated by the specimen. In immersion lens mode, the magnetic field of the lens engulfs the specimen giving a small focal length of 2–5 mm. Small working distance and focal length yield lowest lens aberration, finer probe size, and highest image resolution. This type of lens design is currently used in modern SEMs to image nanomaterials. Figure 2.16 shows an illustration of an immersion lens.

2.4.3.3 Snorkel Lens

In this design, the specimen is placed immediately below the objective lens. The relative placement of the sample is such that the magnetic field of the lens extends to the surface of the sample. This design combines the benefits of both types of lenses mentioned above. It enables high-resolution imaging while allowing for the examination of relatively large specimens. Figure 2.17 shows an illustration of a snorkel lens.

As stated earlier, an objective aperture is inserted into the beam path to limit the beam diameter that reaches the specimen thus reducing the effect of spherical aberration. By blocking most of the electrons, this beam limiting aperture also effectively controls the current that passes through down to the specimen surface.

2.4 Electromagnetic Lenses

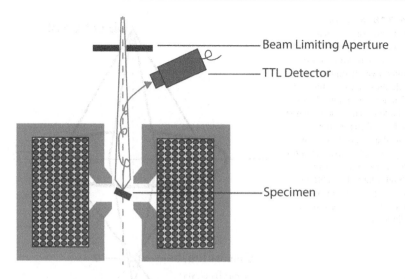

Fig. 2.16 Schematic diagram of immersion lens (Adapted from [7])

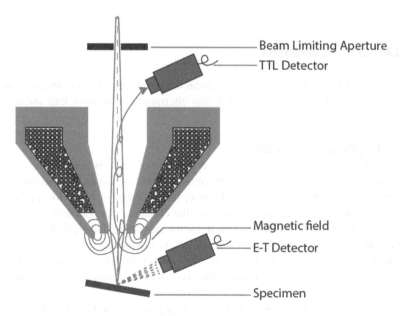

Fig. 2.17 Schematic diagram of a snorkel lens (Adapted from [7])

An objective aperture also has a direct influence on the depth of field. The smaller the radius of an aperture, the greater the depth of field observed in an SEM image. If the aperture is located within the gap of an objective lens, it is called *real aperture*. If its location is between the condenser and objective lens, it is termed as *virtual aperture*, which is usually employed for immersion and snorkel lenses.

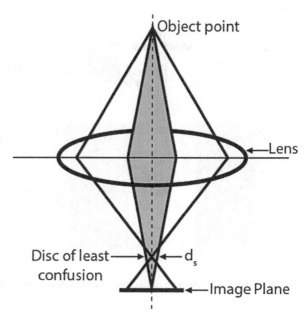

Fig. 2.18 Schematic illustration of spherical aberration. Electrons close to the optic axis are focused farther away from the lens, while those close to the edge of the lens are focused closer to the lens. This phenomenon results in different points of focus thus blurring the image. Weak pinhole lenses used for large working distances exhibit a large C_s value of 20–40 mm, while in other lenses, C_s is reduced to a few millimeters

2.4.4 Lens Aberrations

The ability of electromagnetic lenses to focus the beam into a fine symmetrical probe is primarily limited by defects called lens aberrations. Common lens aberrations include:

2.4.4.1 Spherical Aberration

The electromagnetic field is stronger near the edge of the lens and weaker near the optic axis. As a result, electrons near the edge are bent by the lens more strongly than the ones close to the axis. This results in different points of focus for an electron beam depending on which part of the lens the electrons pass through as shown in Fig. 2.18. Consequently, rays converge to form a disc rather than a point at the image plane causing the image to blur. This defect is known as *spherical aberration*. Electrons passing closer to the optic axis will form the best image without aberrations. Electrons farthest from the axis will be characterized by highest aberrations.

The smallest disc formed just above the image plane is referred to as the spherical aberration disc of least confusion. The diameter of this disc d_s is given as:

$$d_s = \frac{1}{2} C_s \alpha^3 \qquad (2.9)$$

where d_s is the diameter of the spherical aberration disc, C_s is the spherical aberration coefficient, and α is the angle of outer beam convergence. C_s is nearly proportional to lens focal length; large focal length will result in large disc as rays will be focused

2.4 Electromagnetic Lenses

Fig. 2.19 Spherical aberration is reduced by inserting an aperture after the objective lens, which serves to block the off-axis electrons and stops them from forming several focal points

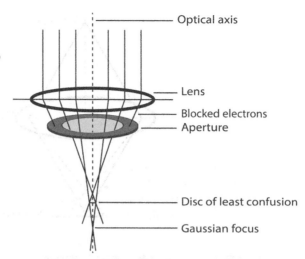

farther away from the lens. The beam diameter at the Gaussian image plane is measured as $2\ C_s\alpha^3$.

Spherical aberration is controlled by introducing a small aperture after the objective lens, which blocks the off-axis electrons preventing them from contributing to the image formation and reducing the beam convergence angle, as seen in the illustration of Fig. 2.19. This decreases the beam diameter and limits the number of electrons that reach the specimen surface, also reducing brightness. Spherical aberration decreases at smaller focal lengths. Immersion lens design uses small focal lengths of 2–5 mm by reducing the working distance between the sample and objective lens. This reduces spherical aberrations making the design suitable for high-resolution imaging.

2.4.4.2 Chromatic Aberration

At any given accelerating voltage, not all electrons in the beam possess the same energy. Electrons with smaller energy will be focused strongly by the lens and will cross the optic axis closer to the lens. Electrons with higher energy will be focused at a greater distance from the lens. Different points of focus will make the electrons converge in the form of a disc instead of a point at the image plane resulting in a defect known as the *chromatic aberration*, as seen in Fig. 2.20a. Chromatic aberration increases the probe size and degrades the image contrast. Causes of chromatic aberration include energy spread of electron source and use of large beam convergence angle. Fluctuations in accelerating voltage and lens current can also lead to this defect.

The smallest disc formed immediately above the image plane is called the chromatic aberration disc of least confusion. The diameter of this disc d_c is given as:

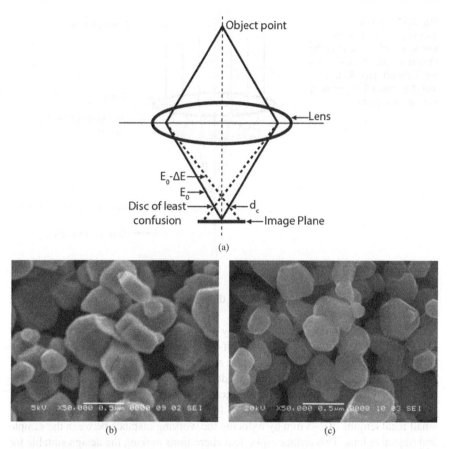

Fig. 2.20 (a) Schematic illustrating chromatic aberration produced due to electrons with different energies being focused at different points of the optic axis. E_0 represents electrons with higher energy and E_0-ΔE are electrons with lesser energy. The beam converged at the image plane is in the form of a disc rather than a point resulting in increased probe size. SEM images showing chromatic aberration at (**b**) 5 kV and its removal at (**c**) 20 kV

$$d_c = C_c \alpha \left(\frac{\Delta E}{E_0} \right) \qquad (2.10)$$

where d_c is the diameter of the chromatic aberration disc, C_c is the chromatic aberration coefficient, α is the angle of outer beam convergence, ΔE is the energy spread, and E_0 is the electron beam energy. C_c is almost proportional to the focal length of the lens. The fractional variation $\frac{\Delta E}{E_0}$ in the electron beam energy indicates that the effect of chromatic aberration will be higher at low beam energies. For example, if ΔE is taken as 2.5 eV for a tungsten filament, at a beam energy of 25 keV, $\frac{\Delta E}{E_0}$ is 1×10^{-4}, and at 2.5 keV, $\frac{\Delta E}{E_0}$ will be ten times larger, i.e., 1×10^{-3}. Therefore, at low accelerating voltages, chromatic aberration becomes a limiting

2.4 Electromagnetic Lenses

factor in acquiring fine probe size. This is especially true with thermionic emission guns that exhibit typically large energy spread. Therefore, the effects of chromatic aberration in thermionic emission guns are more pronounced. For a cold field emitter, the energy spread E_0 is 0.3 eV. The fractional variation $\frac{\Delta E}{E_0}$ at 2.5 keV will be 1.2×10^{-4} which is almost as low as that of a tungsten filament used at 25 keV. Since field emitters have low energy spread, their use reduces the contribution of this defect to the broadening of probe size. Chromatic aberration becomes seriously noticeable at accelerating voltages below 10 kV. SEM images shown in Fig. 2.20b, c demonstrate the removal of chromatic aberration as accelerating voltage is increased from 5 to 20 kV.

Present-day technology has allowed commercialization of microscopes that employ a stack of nonsymmetrical multipole lenses to correct spherical and chromatic aberrations. Currently, these specialized microscopes are too expensive to afford in a typical laboratory.

2.4.4.3 Diffraction at Aperture

Due to their wave nature, electrons can diffract at the edge of the small aperture used to reduce spherical and chromatic aberrations. The *Fraunhofer* diffraction pattern is formed at the Gaussian image plane and takes the form of an "airy disc" with the highest intensity at the optical axis surrounded by additional maxima as shown in Fig. 2.21. The contribution of this error to the probe broadening is given as half-width (d_d) of the disc:

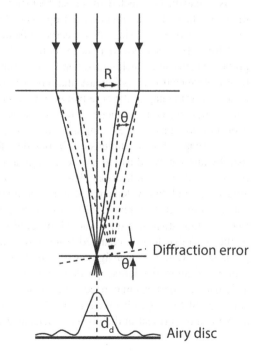

Fig. 2.21 Diffraction at the edge of the final aperture resulting in the formation of an "airy disc" at the image plane

$$d_{\rm d} = 0.61 \frac{\lambda}{\alpha} \qquad (2.11)$$

where λ is the wavelength of the electrons and α is the beam convergence angle. The above equation shows that the effect of aperture diffraction can be reduced by using a large convergence angle. On the other hand, spherical and chromatic aberrations increase with increasing convergence angle. An optimum convergence angle needs to be determined to strike a balance between these set of defects.

2.4.4.4 Astigmatism

Small imperfections in lens construction, asymmetry in copper windings, or contaminated apertures can cause the lens to generate inconsistent electromagnetic field. This defect is known as *astigmatism* from the Latin word *stigma* meaning "mark" or "spot." In astigmatism, two line foci perpendicular to each other are formed at two different focal lengths such that image is stretched into one direction on one side of focus and stretched perpendicular to it on the other side as illustrated in Fig. 2.22. Astigmatism introduces a distortion in the shape of the probe in a way that it changes from round to elliptical with a change in focus. The diameter of the disc of least confusion is given as:

$$d_{\rm A} = \Delta f_{\rm A}\, \alpha \qquad (2.12)$$

where $\Delta f_{\rm A}$ is the distance between the two focal points.

Astigmatism is detected by under-focusing and over-focusing the beam. If the image stretches to one direction during under-focusing and stretches perpendicular to it during over-focusing, this clearly indicates the presence of this defect. The stretching effect is removed at the exact focus; however, the image is still a blur since probe diameter in an astigmated condition is bigger than the optimal size. Astigmatism is corrected using stigmator that consists of an electromagnetic octupole lens located near the gap of the pole pieces within the objective lens (see Fig. 2.15). This coil applies an electromagnetic field of appropriate strength at 90° relative to the field distortion to cancel out existing astigmatism. The aim is to bring the probe into a circular shape of small dimension. In practice, the SEM operator corrects astigmatism by alternately adjusting x- and y-stigmator control knobs and focus knob at a relatively high magnification. The focus is adjusted after each x- or y-control adjustment and this cycle is repeated until a satisfactory image is obtained. Astigmatism is corrected at a high magnification such as 10,000x. An image can be termed as free of astigmatism if it does not exhibit unidirectional elongation when the power of the objective lens is changed from under- to over-focus at a magnification of $\geq$10,000x. Removal of astigmatism usually results in sharply focused images. If a properly aligned microscope does not yield sharp images after the removal of astigmatism, apertures may need to be cleaned. An example showing astigmated and corrected SEM images is given in Fig. 2.23. The effect of lens aberrations described above is more critical in the objective lens as it determines the final probe size.

2.4 Electromagnetic Lenses

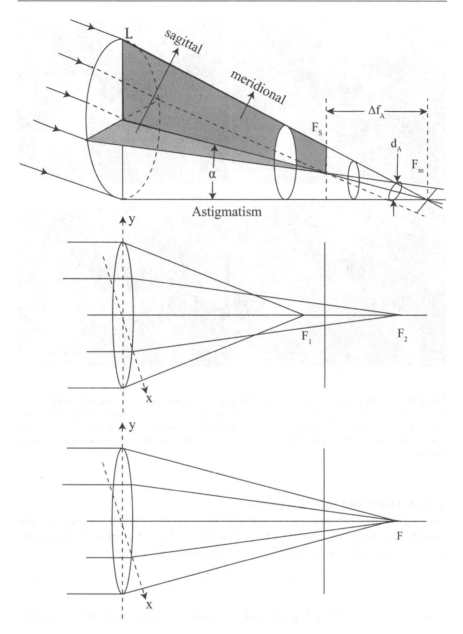

Fig. 2.22 Schematic representation of astigmatism where electrons passing from two planes of the lens perpendicular to each other focus at different distances from the lens

Fig. 2.23 The effect of astigmatism in the objective lens: (**a**) focused condition, astigmated image; (**b**) under-focus condition, image stretched to one direction; (**c**) over-focus condition, image stretched to the perpendicular direction; and (**d**) focused condition, astigmatism removed. The insets at the top right-hand side of the images show the approximate probe shape for each image condition

2.4.4.5 Effective Probe Diameter

The relationship between probe diameter (d_p) and the current, brightness, and convergence angle was shown by Eq. 2.2 in Sect. 2.1.1.1. Effective electron probe diameter (d_{pe}) will be influenced by lens aberrations and can be given as:

$$d_{pe}^2 = d_p^2 + d_s^2 + d_c^2 + d_d^2 \qquad (2.13)$$

where $d_p^2 = \dfrac{4i_p}{\pi^2 \beta \alpha_p^2}$ (from Eq. 2.2) and d_s, d_c, and d_d are the diameters of the discs formed due to spherical, chromatic, and diffraction aberrations, respectively. For conventional microscopy undertaken between an accelerating voltage of 10 and 30 kV, the effects of chromatic and diffraction aberrations can be ignored. The most appropriate aperture (to control half angle α) for generating the smallest probe size is given as:

2.4 Electromagnetic Lenses

$$\alpha_{opt} = \left(\frac{d_p}{C_s}\right)^{1/3} \tag{2.14}$$

When the above aperture size is used, the minimum effective probe size is given as:

$$d_{p,min} = \left(\frac{4}{3}\right)^{3/8} \left[\left(4\frac{i_p}{\pi^2 \beta}\right)^{3/2} C_s\right]^{1/4} \tag{2.15}$$

It can be seen from the above equation that minimum probe size ($d_{p,min}$) decreases as probe current (i_p) and spherical aberration (C_s) decrease and brightness (β) increases. Small probe current and large brightness can be achieved by field emission guns underlying their capability of high-resolution microscopy. A typical probe diameter achieved during conventional microscopy is approximately 10 nm in size.

Minimum probe size can also be represented by the following equation:

$$d_{p,min} = KC_s^{1/4} \lambda^{3/4} \left(\frac{i_p}{\beta \lambda^2} + 1\right)^{3/8} \tag{2.16}$$

where K is a constant whose value is close to 1. The above equation shows that $d_{p,min}$ decreases as electron wavelength λ decreases (or accelerating voltage increases). In theory, the smallest probe size is achieved when i_p reaches the limit zero. At this point, the size of the probe will equal $C_s^{1/4} \lambda^{3/4}$ which can be regarded as the theoretical resolution of a microscope. Another way to express the theoretical resolution of the electron microscope is by modifying Abbe's equation $\left(\frac{0.612\lambda}{n \sin \alpha}\right)$ for a light microscope, as follows:

$$d_1 = \frac{0.612\lambda}{\alpha} \tag{2.17}$$

where d_1 is resolution in nm, λ is the electron wavelength at the given beam energy, and α is the angle of convergence in radians. From Eq. 2.12, the minimum spot size using the optimum aperture is given by $d_2 = \alpha^3 C_s$. By adding d_1 and d_2, we can assume to get the net resolution:

$$d_1 + d_2 = \frac{0.61\lambda}{\alpha} + \alpha^3 C_s \tag{2.18}$$

As described above, the use of a small aperture (small α) decreases spherical aberration, while it increases aperture diffraction. In order to find the best compromise between the spherical and diffraction related aberration, the above expression is differentiated with respect to α to find the optimum convergence angle α_{opt}:

$$\alpha_{opt} = (0.203)^{\frac{1}{4}} \left(\frac{\lambda}{C_s}\right)^{\frac{1}{4}} \tag{2.19}$$

The above equation shows that α_{opt} is proportional to $\left(\frac{\lambda}{C_s}\right)^{\frac{1}{4}}$. By substituting this value into Eq. 2.16, we get the optimum resolution d_{opt}:

$$d_{opt} \alpha \lambda^{\frac{3}{4}} C_s^{\frac{1}{4}} \tag{2.20}$$

The above conclusion is similar to that obtained in Eq. 2.14.

Maximum probe current can be given by the following equation:

$$I_{p,max} = \frac{3}{16} \pi^2 \frac{\beta}{C_s^{2/3}} d_{p,min}^{8/3} \tag{2.21}$$

It is clear from the above equation that $I_{p,max}$ becomes proportional to $d_{p,min}^{8/3}$ when spherical aberration is accounted for. Large probe current is important and is required for some applications such as x-ray microanalysis because only a small proportion of electron beam interactions within the specimen result in the emission of x-rays. From the above equation, we can see that for imaging at a low magnification, an SEM user can increase the probe current by increasing the spot size using condenser lens. Another way is to increase the brightness by increasing the accelerating voltage. Reducing the spherical aberration by employing shorter working distance is yet another option available to maximize the probe current.

An increase in probe current will result in an increase in probe diameter. For a high-resolution image, d_p must be small but with sufficient beam current to attain adequate signal-to-noise ratio that makes a feature of interest visible. A spot size could be small enough to resolve fine detail, but without enough contrast, that detail will not be visible.

2.4.5 Scan Coils

The electron beam is scanned from left to right point by point across the surface of the specimen. The signal generated from a discrete location of a specimen is processed in a detector and synchronously displayed on a corresponding pixel of the viewing monitor. This scan across the specimen surface is accomplished through two sets of deflection electromagnetic coils located within the bore of the objective lens assembly in the electron column (see Fig. 2.15). The coils are connected to a scan generator that creates the raster on the specimen (see Figs. 2.2 and 2.24). The beam is swept across the area of interest on a specimen from left to right (x-direction) and returns to the starting point. It then moves down a line (y-direction) to scan again from left to right, thus carrying out a complete two-dimensional rectangular raster of that specific area. Once the scan of the selected area is finished, the beam goes back to the starting position of the scan to initiate scanning for another image frame. The time required to complete one image frame can be calculated by multiplying the total number of points in the frame with the dwell time at each point. Current in the coils is changed as a function of time to create the raster. The beam is deflected off the

2.4 Electromagnetic Lenses

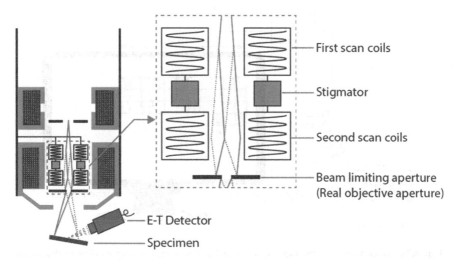

Fig. 2.24 Schematic diagram showing the working of scan coils that create a raster on the specimen surface

course of optic axis by the first pair of coils and brought back to the axis by the second set of coils at the pivot point of the raster.

The beam scans are digitally controlled through a digital-to-analog converter. The beam scans discrete locations one after the other. The beam stays at a location for a predetermined dwell time and then quickly moves to the next location. The speed of the scan can be varied using controls in the software. Slow scans take longer but give clear images with less electronic noise. On the other hand, image drift can occur during a slow scan that can result in blurred images. An appropriate scan speed, therefore, needs to be selected for each image. The number of lines in the image can vary from 500 to 2,000 depending on the ratio of frame and line scan frequencies. The beam can also be scanned on the specimen in a single line by using only one set of deflection coils, called *line scan*. The coils can also be used to generate scans confined to localized regions called *spot analysis* or carry out scans of irregular shaped features of a specimen during chemical composition analysis. By placing the scan coils within the objective lens, the specimens can be viewed at short working distances reducing the deleterious effects of lens aberrations.

2.4.6 Magnification

The electron beam which is focused into a small probe strikes the specimen surface at a single location. In order to obtain an image, the probe is rastered across the area of interest in a specimen. Image of the scanned area is displayed on the viewing screen as seen in Fig. 2.25. Magnification of the image is the ratio of the length of scan on the viewing screen to that of the specimen:

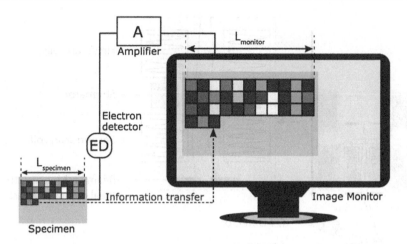

Fig. 2.25 Magnification is the ratio of the scan lengths of the viewing monitor to that of the specimen

$$\text{Magnification} = \frac{\text{Length of scan on the monitor}}{\text{Length of scan on the specimen}} = \frac{L_{\text{monitor}}}{L_{\text{specimen}}} \quad (2.22)$$

From above it is clear that magnification is controlled by the scan coils, and it depends on the size of the raster and that of the display. The smaller the raster dimensions and/or the larger the size of the display, the greater is the magnification obtained. Magnification of an image displayed on 20 cm screen will be twice of the same image displayed on 10 cm screen length. A screen of typically 10 cm in length and a specimen scan length of 10 μm will produce a magnification of 10,000×.

Since the maximum length of the viewing monitor is more or less fixed, magnification is increased by decreasing the length of the specimen scan. The smaller the specimen area scanned, the higher is the magnification achieved. Magnification is changed by varying the current in the scan deflection coils through the knob labeled *Magnification*. Counterclockwise movement of this knob changes the current flowing through the scan coils, in turn increasing the length of the scan on the specimen, thus decreasing the magnification. Magnifications of up to one million times can be produced in microscopes in use today. Small magnifications are generated through large beam deflections by the scan coils and hence are susceptible to lens aberrations that can distort the raster on the specimen. Small beam deflections used during large magnifications diminish the possibility of distortion.

Magnification also depends on the distance between the objective pole piece and the specimen surface known as the *working distance*. With the current in the scan deflection coils kept constant, if the working distance is reduced (i.e., specimen brought closer to the objective lens), the magnification will increase, as the length of the scan on the specimen will become comparatively smaller, as shown in Fig. 2.26a.

2.4 Electromagnetic Lenses

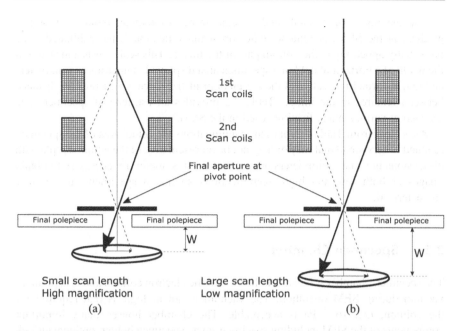

Fig. 2.26 Working distance (WD) is the distance between the objective lens and the specimen surface. With constant current in the scan coils, (**a**) small working distance results in a small specimen scan length (higher magnification), and (**b**) large working distance leads to large specimen scan length (lower magnification)

Table 2.2 Table showing the scanned area on the specimen surface at various magnifications used in the SEM

Magnification	Area scanned on specimen surface
10×	100 mm^2
100×	1 mm^2
1,000×	100 µm^2
2,000×	50 µm^2
5,000×	20 µm^2
10,000×	10 µm^2
50,000×	2 µm^2
100,000×	1 µm^2
500,000×	200 nm^2
1,000,000×	100 nm^2

On the other hand, if the working distance is increased (Fig. 2.26b), the scan length will correspondingly increase, resulting in decreased magnification. Correct magnification is automatically displayed by the SEM as the working distance is changed. It follows that the minimum attainable magnification decreases as the working distance is increased. Therefore, in order to view a specimen area at the smallest possible magnification, the working distance should be increased.

The magnification is calibrated by examining commercial replicas of optical gratings in the SEM. A reliable indicator of magnification is the calibrated scale bar usually appearing at the bottom part of the image. This scale bar is a measure of the width of a horizontal field of a specimen area displayed in microns or nanometers and remains true irrespective of the screen or print size since it measures the distance between features in the image. Table 2.2 provides an estimate of specimen area scanned at various magnifications used in the SEM.

As is clear from Table 2.2, at high magnifications the actual area of the specimen scanned is very small that may or may not be representative of the whole sample with dimensions in square millimeters or centimeters. It is, therefore, necessary to obtain images at both low and high magnifications so that a specimen can be fully characterized.

2.5 Specimen Chamber

The specimen chamber is located at the base of the electron column and is kept under vacuum during SEM operation. The vacuum is not as high as that required for the column, i.e., 10^{-3} Pa is acceptable. The chamber houses many important components of the SEM including specimen stage, specimen holder, optional airlock chamber, CCD camera, and several detectors used for imaging as well as microchemical analysis. A brief description of these components is given below.

2.5.1 Specimen Stage

The sample to be examined is loaded onto a holder/stub that is placed in a specimen stage. Photographs of the specimen stage and inside of the SEM specimen chamber are shown in Fig. 2.27a, b, respectively. Specimen holders of various designs are

Fig. 2.27 Photographs of (**a**) open specimen chamber showing specimen stage and holder pulled out of the chamber. (**b**) Inside of another specimen chamber showing sample holder (on a stage) positioned below the lower objective pole piece

2.5 Specimen Chamber

Fig. 2.28 Photographs of various forms of holders used for scanning electron microscopy. (**a, b**) Multiple sample holders. (**c**) Cup-like holder to hold a single sample with a screw

shown in Fig. 2.28a–c. The wiring provided within the stage can allow measurement of specimen current, whereas the specimen itself is at ground potential. The stage has the ability to move in x, y (perpendicular to column optic axis), and z (parallel to column optic axis) directions as well as to tilt (15–75°) and rotate (360°). The x and y translations are used to move the specimen under the electron probe to select the areas of interest for imaging. Additionally, z translation is used to lower the stage and move it away from the objective lens pole piece toward the chamber door to make sample exchange safe and convenient. Movement in z direction is also used to change the working distance between the sample and the objective lens pole piece. The stage may be tilted if the sample is required to face a detector in order to improve signal collection. Stage rotation can be employed to rotate the sample in order to realign features in the SEM image. The stage can be tilted and rotated in a eucentric (centered about the column optic axis) fashion such that the area of interest remains under the electron probe during movement. The working distance and magnification do not change during eucentric tilt. Stage movements can be controlled manually or are motorized in modern microscopes. Some SEM designs employ an airlock chamber to keep the vacuum in the specimen chamber intact during sample exchange. The maximum size of a sample that can be examined in a microscope is determined by the specific dimensions of the specimen chamber, door, and stage. Some chambers and stages are capable of handling large samples, while others can accommodate only small samples. Chamber size of up to roughly 35 cm in horizontal diameter and height is available. Size, shape, and type of holders/stubs to secure samples vary and depend on the type and size of the samples to be examined. Some holders can support several samples simultaneously, which forgoes the need to exchange samples frequently keeping the vacuum in the chamber secure for longer periods.

Various types of specialized specimen stages have been developed to support research and development activities [8]. For example, advanced SEMs are equipped with highly stable specimen stages to allow for high-resolution microscopy. In addition, hot stages to perform in situ experiments at elevated temperature are available. Specimens can be heated to around 1100 °C in such a stage. Cold stages are available to examine specimens at low temperatures. This type of stage is suitable for the study of organic, biological, cryo-electronic, and superconducting materials. Liquid nitrogen or helium is used for cooling. In another type of stage, specimens can be loaded under stress to study the static and dynamic response of materials to deformation. Several types of loading including tension, compression, bending, etc. can be incorporated in the experiments. In situ microhardness tests of specimens can be undertaken. Deformation of the specimen at elevated temperatures can be carried out. High precision stages equipped with laser interferometer technology are available and are used to undertake electron beam lithography, metrology, and failure analysis of semiconductors. To perform scanning at very low magnification, motor-driven stages are available that move back and forth to get the specimen surface scanned by the beam. This enables imaging of large surface areas of the specimen.

2.5.2 Infrared Camera

Modern SEMs are fitted with an infrared (IR) camera inside the specimen chamber. The live image from this camera is displayed on the computer monitor. This is a very convenient tool as the movement of the specimen stage and holder can be monitored in real time. This allows the user to keep a watchful eye on the distance between the specimen and the objective lens pole piece to avoid their collision during microscopy at small working distances. Interior of the chamber is lit using IR light-emitting diodes (LED). The backscattered detector is sensitive to IR rays, while EDS detector can also show a false peak and/or an increase in the background in the EDS spectrum when IR LED are turned on. Therefore, IR camera and LED are switched off when these detectors are in use.

2.6 Detectors

The detector is a device that collects and converts the signal that leaves the specimen into an electrical pulse for processing and eventual display by the SEM electronics in the form of an SEM image or energy dispersive x-ray (EDS) spectrum. There are several detectors that are essential to the imaging and microanalytical capability of the SEM and are installed inside the specimen chamber. Some of these are fitted permanently within the chamber, while others can be mounted and dismounted on the pole piece as per requirement, while others are retractable. The "through-the-lens" (TTL) detector is located above the objective lens within the electron column and not within the specimen chamber of the SEM. The function of all these detectors is to collect and analyze the signal emanating from within the specimen. This signal primarily consists of secondary and backscattered electrons that are used to generate secondary and backscattered images, respectively and x-rays that are used to analyze the composition of the specimen. Various commonly used detectors in the SEM are described in the following sections.

2.6.1 Everhart-Thornley Detector

Everhart-Thornley (E-T) detector is capable of detecting both secondary and backscattered electrons emanating from the specimen in order to undertake imaging. However, its widespread use is to generate secondary electron images of the specimen surface, not backscattered images for which a separate detector is usually employed. This scintillator-photomultiplier.

type detector is named after their designers Thomas E. Everhart and Richard F. M. Thornley [9] who improved upon the previously existing detectors. The efficiency of secondary electron collection and its role in the generation of SEM images has made the E-T detector the most frequently used component of the microscope. In fact, if only one reason was to be suggested for the widespread use of scanning electron microscopy by the scientific community, it could be identified as the capability and performance of E-T detector.

2.6.1.1 Working Principle
The Everhart-Thornley detector is placed inside the specimen chamber close to the specimen that is located directly below the objective pole piece, as shown by the photograph in Fig. 2.29.

The E-T detector is built in the form of a tube with its front end facing the specimen surface at an angle of around 30°. Secondary electrons emanating from the specimen possess low energy (0–50 eV) and need to be pulled toward the detector by applying a positive bias (+200–300 V) to a wire mesh/collector grid (called Faraday cage) that covers the face of the detector (see schematic in Fig. 2.30). Positive bias at the Faraday cage serves to attract the majority of secondary electrons including those having initial trajectories pointed away from the detector. The sample can also be tilted toward the detector to increase the secondary electron signal collection.

Fig. 2.29 Photograph of the inside of the specimen chamber showing objective pole piece at the top and specimen stage and holder at the bottom. Immediately to the right-hand side of the pole piece is the collimator attached to the front of EDS detector. The upper rightmost section shows E-T detector with the front end covered by Faraday cage to attract low-energy secondary electrons emanating from the specimen surface

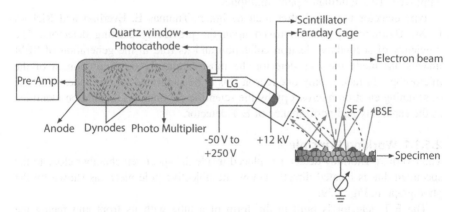

Fig. 2.30 Schematic showing the working of an E-T detector. SE, secondary electrons (0–10 eV); BSE backscattered electrons (10–30 keV); LG, light guide. Faraday cage, pre-amp, preamplifier component: At +250 V E-T detector will collect all secondary and line-of-sight and low-energy backscattered electrons. At −50 V, E-T detector will reject all secondary and collect only line-of-sight backscattered electrons

The electrons collected by the Faraday cage strike the phosphorescent (cathodoluminescent) scintillator surface of the detector. The energy of secondary electrons reaching the detector is only a few electron volts (eV) which is inadequate to produce a viable output from the scintillator. The scintillator material can be a doped light-emitting plastic, lithium-activated glass, crystalline CaF_2 doped with europium, P-47 powder, or YAG/YAP single crystals or garnet (oxide) material. It is 8–20 mm in diameter, and its surface is covered by a thin metal coating (e.g., Al: nanometers in thickness) with a +10–12 kV positive bias that serves to accelerate the incoming electrons onto the scintillator. The scintillator material has a high refractive index that serves to focus the electrons toward the light guide. Faraday cage is insulated from the scintillator bias. Due to this reason, Faraday cage protects the electron beam from unwanted deflection and distortion from the large potential present on scintillator surface. The bias on the Faraday cage itself is too small to affect the electron beam. The charged electrons are converted into a burst of visible light (photons, scintilla) upon striking the surface of the scintillator. The number of photons created by the incoming secondary electrons depends on the kinetic energy of the electrons. An electron with an energy of 10 keV can create roughly 100 photons. The light then passes through a total reflection plastic (Perspex) or glass pipe called light guide. The electrons are converted into light so that it can pass through a quartz window which acts a barrier between the light guide placed inside the evacuated specimen chamber and the photomultiplier tube (PMT) positioned outside the SEM chamber. The quartz window serves to keep the vacuum inside the specimen chamber intact while concurrently allowing the light to pass through to the PMT. Use of electron-to-light conversion (i.e., scintillation) and the quartz barrier allows the signal to be transported to the outside of the SEM chamber. The design of light guide allows the scintillator to be placed at a position suitable for secondary electron collection and for the PMT to be placed outside the SEM specimen chamber. The PMT itself is a sealed glass container that is evacuated.

The light strikes the thin layer of photocathode placed at the front end of the PMT. The photocathode material has low work function and, once struck by light, releases electrons from its conduction or valence atomic shells. These low-energy electrons are released into the PMT as photoelectrons. Thus, the PMT converts light back into electrons which then travel along the former lined with a series of dynode electrode, typically 8 in number. The first dynode has a positive bias (100–200 V) with respect to the photocathode, which tends to accelerate the photoelectrons toward this electrode. Secondary electrons are emitted when photoelectrons strike the surface of the first dynode. The dynode is coated with a material that has a high secondary yield (δ) to promote the creation of SE whose number could range from 2 to 10 for every photoelectron that collides with the electrode surface. The second dynode is biased positive by at least 100 V with respect to the first dynode. The SE ejected from the first dynode accelerate toward the second electrode and create further SE upon striking its surface. Each electron may produce at least two secondary electrons. The number of electrons increases, and the electric signal is proportionally amplified when it strikes the successive electrodes resulting in a cascading effect. Up to $\times 10^6$–10^8 gain in signal is achieved without an appreciable increase in noise. Any gain in the photomultiplier is translated into an increase in the contrast of the image displayed on the screen. The gain in PMT is controlled by the contrast knob available to the SEM operator. Increasing gain through contrast knob brightens light areas more than the dark areas in the image. The fact that the PMT is sealed and placed outside of the chamber protects it from external contaminants. The PMT is a low cost robust component and processes signals of varying intensity in an efficient manner.

The electric signal at the output of the photomultiplier is further amplified by preamplifier (Pre-Amp) device that forms part of the E-T detector. The output from the preamplifier is controlled through the brightness knob present in the SEM. An increase in preamplifier gain increases the brightness of the image (i.e., dark and light areas brighten evenly). After passing through the E-T detector, the signal is fed into an amplifier where the signal strength is further increased to a level suitable for image formation on a computer monitor. Before reaching the monitor though, the signal passes through the video control unit and an image processing board for analog-to-digital image conversion. Signal amplification with PMT, preamplifier, and the amplifier is necessary since the original current (i.e., number) of low-energy secondary electrons is very small (i.e., in the order of 10^{-12} A).

2.6.1.2 Efficiency of Signal Collection

The efficiency with which the signal emanating from the specimen is collected by the detector depends on the position and size of the detector. The position is defined in terms of take-off angle (ψ—psi) which is the angle from the surface of the specimen to the center of the detector face (see Fig. 2.31). High take-off angles can collect a larger signal from an un-tilted specimen, but due to space restrictions within the chamber, ψ is usually set by the manufacturer at around 30°. Another important factor that influences the collection efficiency is the size of the detector described by the solid angle (Ω—omega) which is the area (A) of the surface of the detector divided by the square of the radial distance (r^2) to the point where the beam strikes the sample surface, $\Omega = A/r^2$. The solid angle is measured in steradian (sr) units. It is clear from Fig. 2.31 that a large diameter of the detector surface and also its close proximity to the specimen surface will favor collection of the larger signal emanating from the specimen. The typical distance between the E-T detector and the specimen can be 4 cm.

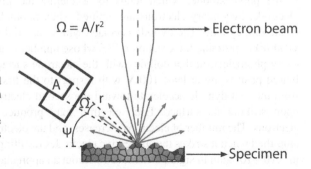

Fig. 2.31 Schematic showing the take-off angle (ψ, set at around 30°) and solid angle of collection $\Omega = A/r^2$, where A is the area of the face of the detector and r is its distance from the beam impact point. Detectors with large areas placed close to the specimen collect more signals

2.6.1.3 Types of Signals Collected

The amount and type of signal collected by the E-T detector can be controlled by varying the potential on the Faraday cage between typical values of -50 V and $+250$ V. At -50 V, all secondary electrons (SE) will be rejected, and only line-of-sight backscattered electrons (BSE) will be collected. At negative bias, E-T detector can be used as a backscattered detector. However, a separate dedicated backscattered detector is usually employed for this purpose. At $+250$ V, all secondary electrons and line-of-sight and low-energy backscattered electrons will be collected with 100% efficiency. It can be seen that the SE images generated under a positive bias will have a significant contribution from the BSE.

A large proportion of SE is also generated by the BSE in an indirect manner as can be seen in Fig. 2.32. SE_1 is used to denote the direct secondary electrons that are emitted close to the probe due to beam-specimen interaction. SE_2 are the indirect secondary electrons that are generated by the BSE when the latter is exiting the specimen away from the probe. SE_3 and SE_4 are indirectly generated when the BSEs emanating from the specimen surface interact with the SEM chamber walls/objective pole piece and final aperture, respectively.

It is important to note that these indirect SE will represent characteristics of BSE that were used to generate them. Indirect SE will add to the BSE component in the secondary electron images and serve to degrade the resolution of an SE image. This apparent contradiction can be understood by considering a situation where SE images are taken from Al particle and then from Au particle in a specimen. Materials with a high average atomic number (in this case Au) generate the higher magnitude of BSE which in turn will produce the larger magnitude of SE_3 signal compared to

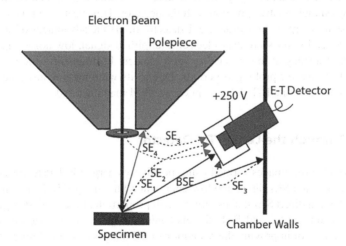

Fig. 2.32 Schematic showing the generation of direct secondary electrons SE_1 and indirect secondary electrons SE_2 from within the specimen. Indirect SE_3 and SE_4 signals are produced when high-energy BSEs emanated from the specimen strike various microscope components inside the specimen chamber. SE_3 are generated from the walls of the specimen chamber. SE_4 signal originates from the final aperture

Table 2.3 The magnitude of SE signal collected by E-T detector from Au specimen

Type of signal	Percentage (%) of the total signal
Direct SE (SE_1)	9
SE produced by BSE within the specimen away from the probe (SE_2)	28
SE produced by BSE upon striking the chamber walls/pole piece (SE_3)	61
SE produced by BSE upon striking the final aperture (SE_4)	2

that in Al. The increased SE_3 signal in Au will be due to increased BSE component and not due to an increase in SE_1 generation at the Au surface close to the probe. The SE image produced under these conditions will thus include information collected by BSE that is reflected from the SEM chamber walls. The E-T detector, therefore, at a positive bias will collect all direct SE_1; indirect SE_2, SE_3, and SE_4; and low-energy and directional BSE. The SE image produced as a result will have significant BSE component. For Au specimen, the proportion of various types of secondary electrons collected by E-T detector is estimated in Table 2.3 [10]:

2.6.1.4 Advantages of E-T Detector

The electron beam strikes the specimen surface from the top and the E-T detector is placed asymmetrically at one side of the specimen to collect the secondary electrons. This arrangement is analogous to the operator looking down the column onto the specimen surface while the latter being illuminated by a light source from the side. This gives rise to a topography where features are either highlighted or shadowed depending on their relative position with the detector. This creates an image contrast that is easy to interpret. In addition, E-T detector offers the advantages of large solid angle of collection for secondary electrons, amplified signal, low noise, high signal collection efficiency at low (1 keV or less) incident beam energies, low cost, and continued service for prolonged periods. Due to its wide-ranging advantages, it is customary to have one E-T detector in every SEM manufactured.

2.6.2 Through-the-Lens (TTL) Detector

Secondary electron images formed using positively charged E-T detector comprise SE_1, SE_2, SE_3, and SE_4 signal and low-energy and line-of-sight BSE. It can be seen in Table 2.3 that direct SE_1 makes only a small proportion of the total SE signal. The presence of indirect SE and BSE signals tends to degrade the resolution of an SE image. One way of improving the resolution of SE images is to use through-the-lens (TTL) detector in combination with field emission scanning electron microscopes generally employed for high-resolution imaging. These microscopes are equipped with immersion or snorkel objective lens (see Sect. 2.4.3). The magnetic field produced by these types of lens is strong and reaches the specimen plane. This effect is augmented due to the close proximity of specimens to the objective lens

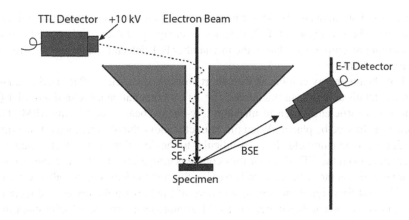

Fig. 2.33 Through-the-lens (TTL) detector is placed within the column above the objective lens and is used with a scintillator bias of 10 kV that attracts SE_1 and SE_2 secondary electrons emanating from the specimen. Due to the exclusion of SE_3 and SE_4 signal, TTL detector is able to form a secondary electron image with a strong SE contrast

usually employed to shorten the working distance in order to reduce the deleterious effects of objective lens aberrations. Because of this configuration (see Fig. 2.33), the SE_1 and SE_2 secondary electrons are trapped by this magnetic field in an efficient manner. Due to their axial velocity component, electrons are made to spiral up the column through the bore of the objective lens. This signal is collected by an E-T detector placed above the objective lens within the SEM column. This detector is termed through-the-lens (TTL) detector and is used with a scintillator bias of 10 kV that attracts SE_1 and SE_2 secondary electrons. The indirect SE_2 signal is dependent on and is characteristic of the magnitude of BSE leaving the specimen surface, and it serves to add BSE contrast to secondary electron image, thus degrading its resolution. The maximum number of secondary electrons is generated normal to the specimen surface.

Low-resolution signal comprising of SE_3 is not collected by the TTL detector since it is generated far from the optic axis as a result of BSE collision with chamber walls and final aperture/pole piece (see Fig. 2.32). Off-axis BSEs are not influenced by the magnetic field of the objective lens and are not collected by the TTL detector. On-axis BSEs that travel upward through the lens bore may have too high an energy to be influenced by the 10 kV potential on the TTL scintillator. In this way, the SE signal is effectively separated from the BSE signal. Due to the complete elimination of SE_3 signal and a large reduction in BSE component, the image formed using TTL detector has high signal-to-noise ratio and is rich in SE contrast compared to that produced by an E-T detector. The high collection efficiency of high-resolution SE signal irrespective of its emission direction, elimination of low-resolution SE signal, and reduction of BSE are the main advantages of TTL detector. Such type of detector is especially useful for imaging of pits, cavities, or voids. A standard E-T detector within the specimen chamber is nevertheless incorporated in the SEM design to undertake imaging at large working distance. In fact, both E-T and TTL detectors can be used simultaneously. This can prove useful for specimens that charge during imaging. The sensitivity of TTL detector to charging effects can be countered by combining its output with that of the in-chamber E-T detector that is less sensitive to charge buildup.

Low beam energy combined with short working distance is often used to image surface details at high magnifications. The 10 kV potential on the scintillator of TTL detector is sufficiently high to misalign low-energy beams used in the SEM. This issue is addressed by placing Wien filter directly above the objective lens. Wien filter produces a magnetic field (B) perpendicular to the direction of the electric field (E) emanating from the TTL detector thus compensating each other to keep the electron beam exactly on the optic axis while passing through the lens. On the other hand, for the SE_1 and SE_2 signal moving up the bore of the lens (in the opposite direction), these fields are added thus deflecting the SE signal toward the biased scintillator by a magnitude whose value depends on SE energy.

2.6.3 Backscattered Electron Detector

The energy of backscattered electrons (BSE) emanating from the specimen can be several keV (depending on the accelerating voltage used) which is much higher than that of secondary electrons (≈ 10 eV). This large difference in energy can be easily exploited to separate the two types of electrons and collect only backscattered electrons using the Everhart-Thornley detector. Small negative bias (≈ -50 eV) on the E-T detector will reject all secondary electrons and exclusively collect line-of-sight backscattered electrons. However, most of the backscattered electrons are ejected upward along the beam direction and scattering probability decreases toward the sides. The in-chamber E-T detector is placed at a relatively small angle toward one side. In this geometric configuration, the solid angle of collection subtended by the detector is small and does not lend itself to the collection of a large number of backscattered electrons. It is therefore customary to have a separate dedicated backscattered electron detector installed in a scanning electron microscope. One possible arrangement is to use a retractable E-T detector that can be extended closer to and above the specimen surface, thus improving the collection efficiency of backscattered electrons. However, it is more common to position a fixed BSE detector directly below the pole piece and above the specimen to increase the solid angle of collection. A photograph of the BSE detector is shown in Fig. 2.34.

2.6.3.1 Working Principle

The most commonly employed backscattered detector is a solid-state diode (SSD) detector consisting of two electrodes making up p-n junction. It is made of p-type (Si/Ge doped with boron/gallium) and n-type (Si/Ge doped with arsenic/phosphorus/antimony) semiconducting electrodes fused to each other. Pure (intrinsic) Si has a diamond cubic structure where each Si atom is surrounded by four neighboring atoms. Silicon is placed in the group 4 of the periodic table. Silicon atom has 14 electrons with 4 being in the outer electron orbital or valence shell. These four

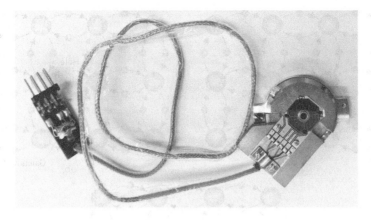

Fig. 2.34 Photograph of backscattered electron (BSE) detector along with associated electronics. The BSE detector is the dark wafer chip with a hole inside to allow passage of electrons. The chip is fixed on the brass mount. Four quadrants are visible on the chip

valence shell electrons are shared by four nearest neighboring atoms forming four covalent bonds. Every Si atom shares one electron with one of the four surrounding atoms. Therefore, eight electrons are shared by each atom and its surrounding four atoms as shown in Fig. 2.35a. In this manner, the valence band of Si is fully occupied with electrons while the conduction band is empty. Silicon atom is electrically neutral, i.e., the number of protons equals the number of electrons. There are no excess electrons or holes; their numbers being the same. Pure Si, therefore, is almost an insulator as no appreciable current flows through it under an electric field. This can be changed through doping or addition of an impurity into the Si lattice.

If one of the atoms in the Si lattice is replaced (i.e., doped) by a group 3 element such as B or Ga that has three valence electrons, it will serve to create an excess hole in the lattice (see Fig. 2.35b). Dopants that create excess holes are termed *acceptors* since holes are deemed to "accept" electrons and are considered positively charged. Semiconductor doped with an acceptor is called a *p*-type semiconductor (where *p* stands for positive). The material after doping still remains electrically neutral. The current in *p*-type semiconductors is primarily carried by excess holes by readily accepting electrons from its neighbor resulting in movement of the holes in the presence of an electric field.

If the atoms of Si are replaced by group 5 elements (As or P) with five valence electrons, it will result in excess/free electrons in the lattice (see Fig. 2.35c). The dopant, in this case, is called *donor* as it "donates" excess electrons. This type of semiconductor is known as *n*-type semiconductor (where *n* stands for negative) since the lattice has excess negatively charged electrons. These electrons move into the conduction band and are primarily responsible for carrying current in the *n*-type semiconductors in an applied field. The *p*- and *n*-type semiconductors are put together to form a *p-n* junction. A band gap (or an electric field) exists across this junction that allows the current to flow in one direction only which is the primary function of a diode.

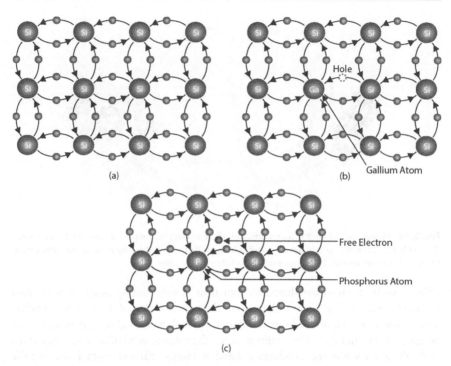

Fig. 2.35 (a) Schematic representation of the Si crystal lattice in an intrinsic semiconductor showing covalent bonds. Each Si atom has four valence shell electrons which it shares with its four neighboring atoms. Arrowheads on the lines connecting the atoms represent the origin of an electron that is shared between the two neighboring atoms. (b) A *p*-type semiconductor doped with an element that has three valence electrons results in the creation of excess holes that carry a charge. (c) An *n*-type semiconductor is created when the dopant with five valence electrons is used. In this case, excess (free) electrons in the Si lattice serve to carry the current

High-energy backscattered electrons emanating from the specimen enter the backscattered detector attached to the end of the objective pole piece directly above the specimen (see Fig. 2.36) and are scattered inelastically within the semiconductor material. The front end of the detector is Au electrical contact ($\approx$ 10–20 nm in thickness) coated on a thin (few tenths of a micron) layer of Si (called "dead layer") followed by Si semiconductor. The gold contact layer is used on both surfaces of the detector to apply the bias voltage. The BSE from the specimen interact with the electrons of the Si lattice and are scattered inelastically resulting in the movement of electrons from valence into the conduction band of Si leaving behind a hole. Thus, an electron-hole pair is created. The energy required to create such a pair in Si is ≈ 3.6 eV. The mean number of electron-hole pairs (E_m) generated is given by:

$$E_m = \frac{E_{BSE}}{E_{exc}} \quad (2.23)$$

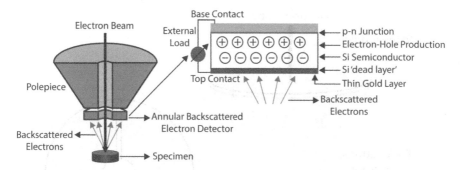

Fig. 2.36 Backscattered detector is attached to the lower end of the pole-piece directly above the specimen. Design of BSE detector is illustrated in a schematic seen at the top right-hand side. Backscattered electrons create electron-hole pairs upon striking the solid-state semiconductor detector surface. The number of pairs created is proportional to the number of BSE interacting with the detector. A large number of BSE will give rise to strong contrast and a small number of BSE will produce weak BSE contrast in the image

where E_{BSE} is the energy of the BSE striking the semiconductor surface and E_{exc} is the mean energy per excitation that equals 3.6 eV. The current produced in the detector is proportional to the number and energy of backscattered electrons that interact with it. The higher the number of the BSE that strike the surface of the detector and the greater the BSE energy, the larger the number of electron-hole pairs produced. A single backscattered electron with an energy of 5 keV can produce 1,400 electron-hole pairs.

Before the electrons can jump back into their vacated spaces that will result in the recombination of electrons and holes, they are separated by the application of an electric field. In a semiconductor device, this purpose is served by the *p-n* junction that acts as an internal field to keep the electrons and holes apart and allow the current to move in one direction only. The current is fed into an amplifier to generate a signal with adequate strength to form an image. This sensitivity of the detector to the number and energy of electrons emanating from the specimen gives rise to the contrast observed in the backscattered image. Areas of the specimen with a high average atomic number will produce more backscattered electrons (i.e., large current) which in turn will generate a higher number of electron-hole pairs within the detector making these areas appear bright in the BSE image displayed on the viewing monitor. Due to their low energy, secondary electrons exiting from the specimens are absorbed in the outer thin Au electrical contact and are not detected by BSE detector. Similarly, the magnitude of x-rays produced due to the interaction of the electron beam with the specimen is too low to affect the performance of the BSE detector.

2.6.3.2 Advantages/Drawbacks

Solid-state BSE detector is popular due to its low cost and maintenance free service. It occupies very little space within the specimen chamber as it takes the form of a thin (i.e., few mm in thickness) circular semiconductor wafer and is mounted directly

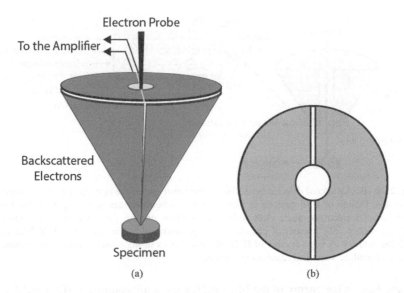

Fig. 2.37 (a) Placement of BSE detector directly above the specimen increases the efficiency of electron collection due to the large solid angle subtended by the detector on the specimen. (b) The detector is made of two segments which can receive signals independently. These signals can be viewed separately or in combined form

under the objective pole piece. It has a hole in the middle to allow the electron beam to pass through. The diameter of the detector surface is relatively large (>25 mm), and its placement directly above the specimen results in high BSE collection efficiency due to the large solid angle subtended by the BSE detector over the specimen (see Fig. 2.37a).

Low-energy BSEs (e.g., <5 keV) exiting from the specimen do not reach the active detector surface due to the presence of Au contact and dead layer. This lowers the detection sensitivity of the SSD and precludes its use in low voltage microscopy. On the other hand, elimination of low-energy BSE from the image improves contrast and resolution, as these electrons do not generally carry specimen-specific information since they undergo a high degree of scattering within the specimen. Detector response time is low and imaging at fast scan rates is not possible with solid state detectors.

The simple design of solid-state detectors allows them to be produced in a configuration where an annular SSD has two segments where each segment is able to receive and process discrete signals. This arrangement is equivalent to having two detectors or semiconductor chips in one annular panel as shown in Fig. 2.37b. The signals received by one segment can be added to or subtracted from the other to view compositional or topographic images, respectively.

2.6.3.3 Scintillator BSE Detector

Another type of BSE detector is scintillator type. This detector has a scintillator, light guide, and photomultiplier and works on the same principle as described for the E-T detector in Sect. 2.6.1. The detector can be mounted below the objective pole piece. It has a hole that allows passage of electron beam toward the specimen. Scintillator detector can also be designed as a retractable detector. Scintillator has a higher detector response time allowing faster BSE signal processing than a solid-state BSE detector. This enables the use of rapid specimen scan rates with scintillator detector. Sensitivity (to detect low-energy BSE) of scintillator type detector is also comparatively higher. The size of the scintillator detector is large compared to the thin semiconductor chip used as the solid-state detector. The aim of such design is to maximize the BSE collection efficiency. However, this makes scintillator detector bulky and spacious, which may restrict imaging of specimens at a short working distance. Such a detector may need to be retracted to allow detection of x-rays by the energy dispersive x-ray (EDS) detector.

2.6.3.4 Channel Plate Detector

This type of detector is composed of a large number of parallel and small electron multiplier tubes a few millimeters in length. It is a thin detector placed between the specimen and the objective pole piece. Such type of detector is known as Channel Plate Detector. BSE to SE conversion can be undertaken within the detector itself. Backscattered electrons emanating from the specimen enter one end of the capillary that is grounded. The high-energy electrons strike the walls of the tube which then emits secondary electrons that are accelerated through the capillary as its farther end is under an applied potential. The set of capillaries acts as a multiplier resulting in a cascade of secondary electrons. It is capable of detecting both SE and BSE. The SE yield at low beam energy increases making this detector suitable for use in low voltage and low current imaging.

2.7 Miscellaneous Components

Various components in the SEM perform their tasks in the background and their utility may not be obvious to the operator. However, their presence and function are crucial to maintaining an optimum level of SEM operation. These components can be considered its computer control system, vacuum system, chiller, heater, electronics assembly, and anti-vibration table.

2.7.1 Computer Control System

Modern day SEMs are primarily controlled with computers equipped with appropriate software systems. Most functions of the SEM ranging from venting/evacuating the specimen chamber, specimen stage movements, and switching on electron beam to imaging, microchemical analysis, storing data and networking, etc. are all

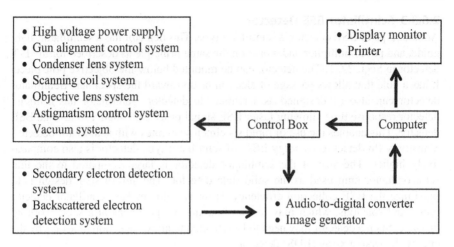

Fig. 2.38 Schematic of computer control system used in the SEM. Numerous parameters of operation are controlled by the user through the computer

controlled through the computer. A schematic in Fig. 2.38 shows an example of various functions controlled by the computer in the SEM.

The operator is provided with a user-friendly graphical user interface that allows easy access and control of most imaging and analytical functions through both computers. Same operations can be performed using conventional operations control panel. Different levels of user accounts are available for SEM users having a range of expertise. Access to different functions such as alignment can be restricted based on user levels. Automation of the SEM functions using computers has improved safety, increased ease of use, and has rendered analysis less time-consuming.

2.7.2 Vacuum System

The SEM needs to be kept under vacuum to increase the mean free path of electrons and allow them to pass through the electron column and specimen chamber without getting scattered by air molecules. Any water vapor and organic contaminant present in the specimen chamber needs to be pumped out since it deposits on the specimen surface and interacts with the electron beam to impair visualization of fine surface details. Vacuum is also required to inhibit oxidation damage to the electron source. It prevents high-voltage discharges between the anode and the filament that can result in filament failure. The electron column is maintained under a high vacuum of the order of 10^{-4} Pa, while vacuum in the specimen chamber is around 10^{-3} Pa. An isolation valve between the electron column and the specimen chamber enables evacuation of both sections independent of each other. The very high vacuum of 10^{-6}–10^{-7} Pa is maintained in the columns of field emissions SEMs.

The vacuum system is composed of pumps, valves, airlocks, vacuum gauges, pipes, valves, etc. Initially, the vacuum in the chamber is obtained through a rotary

Fig. 2.39 Photograph of a typical rotary pump

mechanical pump (also called rough pump or forepump; see Fig. 2.39) that drives air out through a pipe connected to an outlet in the chamber.

Other low vacuum pumps that may be used are the scroll pump and the diaphragm pump. After initial vacuum (approx. 10^{-1} Pa) is established, an oil diffusion pump is activated to obtain final high vacuum (10^{-3}–10^{-4} Pa) in the chamber. The diffusion pump is relatively simple in design and has no moving parts. Schematic of a simple vacuum system used in the SEM and photograph of a rotary pump is shown in Fig. 2.40, respectively.

The oil-free turbomolecular pump is used instead of a diffusion pump as the latter can contaminate the specimen chamber by introducing oil vapor into it due to possible back pressure. Both the turbomolecular and the diffusion pumps have similar vacuum ranges, but the turbomolecular pump generates "cleaner" vacuum compared to a diffusion pump. Due to this reason, it has increasingly replaced diffusion pumps in the SEM. The turbomolecular pump has a design that is more complex with blades moving at >20,000 rpm. The ultra-high vacuum within the column of a field emission SEM can be attained using up to two ion getter pumps (IGP) connected to the column. A gun-isolation valve is present to enable the gun to be vented and evacuated independently during filament replacement. The presence of specimen exchange airlock chamber (load lock) allows specimen exchange without having to vent the chamber. The specimen chamber can be vented with dry nitrogen gas instead of air to minimize the introduction of water vapor and other contamination in the chamber and to reduce the duration of the pump down. The vacuum system is automated and has inbuilt safeguards against accidental de-evacuation, power failure, stoppage of water supply, contamination of chamber due to back pressure, etc. Vacuum levels are measured using a cold cathode gauge or a Pirani gauge.

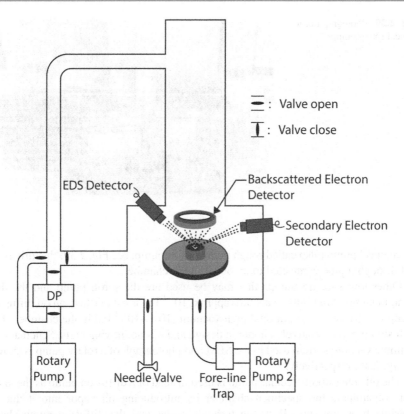

Fig. 2.40 Rough pumping down to 10^{-1} Pa is undertaken by a rotary pump(s). Once an initial vacuum is established, the diffusion pump takes over to bring the pressure down to 10^{-3} to 10^{-4} Pa. The diffusion pump is always backed up by a rotary pump

2.7.3 High-Voltage Power Supply (HT Tank)

High-voltage power supply or high-tension (HT) tank is the apparatus that is used to generate a high voltage of up to 30 kV that accelerates electrons between the filament and the anode. High-voltage tank is insulated with a pressurized gas (around 0.5 MPa) such as sulfur hexafluoride (SF_6) or Freon to avoid flashovers. The electron gun chamber is also insulated in a similar manner.

2.7.4 Water Chiller

Usually, an air-cooled water chiller is used to circulate cold water continuously through the walls of the electron column to dissipate heat from the electron gun, lenses, diffusion pump heater, and other components. A photograph of a typical SEM chiller is shown in Fig. 2.41. The water temperature is maintained at about

2.7 Miscellaneous Components

Fig. 2.41 Typical water chiller used to remove heat from various components of the SEM

18 °C. The electron beam cannot be switched on unless an adequate rate of water flow is maintained within the equipment. The chiller is usually kept outside the SEM room to avoid noise pollution.

2.7.5 Heater

High vacuum in the electron column of a field emission SEM is continuously maintained during day-to-day operation of the microscope. However, disruption in vacuum may occur in case of a power failure or during a maintenance shutdown. Subsequent evacuation of the column is undertaken only after the column is subjected to an overnight bakeout in order to remove contamination properly. This bakeout is carried out by switching on the heaters attached to the exterior of the column near the electron gun area. The temperature of bakeout is approximately 60–70 °C.

2.7.6 Anti-vibration Platform

Stability of the whole microscope, including electron column, specimen stage, etc., is of utmost importance to undertake high-resolution imaging. The site selected for installation of the microscope needs to meet strict specifications concerning floor vibration, electromagnetic interference, etc. The microscope is usually installed at basement or ground floor levels to minimize instability caused due to vibrations. The specimen chamber is mounted on a platform that is isolated from the ground by using an air table or shock absorbers. The microscope should also be free from the interference of electromagnetic radiation emanating from electrical equipment present near the lab. Depending on the distance and intensity of the source, interference starts to distort features in images, especially those taken at relatively high magnifications. Electromagnetic interference cancellation systems that are programmed to negate the intruding frequencies can be installed to overcome this problem.

References

1. https://www.tedpella.com/apertures-and-filaments_html/tungsten-filaments.htm
2. Richardson OW (1901) On the negative radiation from hot platinum, vol 11. Philosophical Magazine of the Cambridge Philosophical Society, Cambridge, United Kingdom, pp 286–295
3. Hall CE (1966) Introduction to electron microscopy. McGraw-Hill, New York
4. Crewe AV, Wall J (1970) A scanning microscope with 5 Å resolution. J Mol Biol 48:375–393
5. Tuggle DW, Li JZ, Swanson LW (1985) Point cathodes for use in virtual source electron optics. J Microsc 140:293
6. Erdman N, Bell DC (2013) SEM instrumentation developments for low kV imaging and microanalysis. In: Low voltage electron microscopy: principles and applications. John Wiley & Sons Ltd, West Sussex, United Kingdom
7. Goldstein JI, Newbury DE, Joy DC, Lyman C, Echlin P, Lifshin E, Sawyer L, Micheal JR (2003) Scanning electron microscopy and X-Ray microanalysis, 3rd edn. Springer, New York
8. Reichelt R (2007) Scanning electron microscopy. In: Hawkes PW, Spence JCH (eds) Science of microscopy, vol 1. Springer, p 155
9. Everhart TE, Thornley RFM (1960) Wide-band detector for micro-microampere low-energy electron currents. J Sci Instrum 37(7):246–248
10. Peters K-R (1984) In: Kyser DF, Niedrig H, Newbury DE, Shimizu R (eds) Electron beam interactions with solids for microscopy, microanalysis, and microlithography. SEM, Inc., AMF O'Hare, Illinois, p 363

Contrast Formation in the SEM   3

An electron beam is rastered across the specimen surface resulting in the generation of secondary and backscattered electrons which are used to form images while x-rays are used to obtain elemental constitution of the specimen material. Present-day technology allows imaging of features as small as 1 nm in the SEM. The predominant use of the SEM is to generate SE and BSE images showing topographic and compositional contrast, respectively. This chapter deals with the mechanism of contrast formation, factors that affect its development, and how its interplay contributes to the appearance of various features observed in the images.

3.1 Image Formation

The electron gun located at the topmost section of the column generates an electron beam with an energy range of few hundred eV to 30 keV, which is focused into a fine probe by electromagnetic lenses located within the column. The fine electron probe is then rastered over specimen surface in a rectangular area by scan coils also present within the column. The scanned sample sits in the chamber located at the end of the microscope column. The gun, the column, and the specimen chamber are kept under vacuum to allow the electron beam to travel through the column and interact with the specimen. The electron beam penetrates into the specimen in the form of a teardrop extending from 100 nm to 5 μm depending on beam energy and specimen density. Interaction of beam with the specimen produces a variety of signals including secondary and backscattered electrons and x-rays, which are collected and used to produce images as well as to determine the elemental composition of the specimen material. In present-day microscopes, images are digitally processed, displayed on computer screens, and saved on hard drives.

3.1.1 Digital Imaging

The electron probe with a specific diameter, current, and convergence angle is scanned from one point to the other across the specimen surface. In a digital scan, the probe dwells at a discrete location on the specimen for a set duration before it moves on. The electrons emanating from a discrete location are detected by the detector as a signal with a specific intensity. Each of these discrete locations, therefore, has stored values of x and y (position) and I (intensity). For each point where the electron beam interacts with the specimen and produces a signal, a corresponding point on the viewing screen (at x, y) is exhibited with intensity I. The detector acts as an interface between the signal emitted from the specimen and the corresponding image displayed on the screen. The level of brightness of the point on the display monitor is proportional to the strength of the signal emanating from the corresponding point in the specimen as measured and amplified by the detector. High signal strength from a location in the specimen will result in greater intensity at the corresponding location on the screen. The image on the screen, therefore, represents an intensity distribution map of the electron signals derived from the specimen. During imaging, the signal strength is proportional to the number of electrons ejected from a point on the specimen. During EDS analysis, the magnitude of the signal corresponds to the number of characteristic x-rays emitted from that discrete location and is indicative of the concentration of chemical element at that location. The length ($L_{monitor}$) of the scan on the viewing monitor is larger than the actual scan length ($L_{specimen}$) on the specimen resulting in magnification of the image that equals $\frac{L_{monitor}}{L_{specimen}}$ (Eq. 2.22). As shown in Fig. 3.1a one-to-one correspondence is established between the beam impact points on the specimen surface and the points on the display screen.

Discrete beam locations on the specimen from where the signal is generated are called *picture elements* with a roughly corresponding number of *pixels* in the image displayed on the viewing screen. The analog signal obtained from each picture element of the specimen is measured, amplified, and converted into a voltage signal by the detector electronics. It is then fed into an image processor board with an integrated analog-to-digital converter (ADC) and video control unit. The voltage signal is digitized into a video signal for each picture element, which is assigned a digital address with a specific numeric value (called the *index*) related to its x, y, I value. The video signal is then fed into the video capture card installed in the computer. The digital video signal is processed by this card to be displayed through a graphical user interface (GUI) on the flat panel screen typically used at present. Later it can be saved in various image formats in a variety of recording media. During the digitization process, each pixel is assigned a value for detected signal electronically which determines its intensity level. This is accomplished by assigning a gray level to each pixel. Typically, 256 gray levels are possible when working with 8 bit (i.e., $2^8 = 256$) information systems. The gray level of 0 is assigned for black and 256 for white, and the rest of the numbers represent various levels of gray. The human eye can discern up to 20 shades of gray only. Thus, 256 gray levels are deemed more

3.1 Image Formation

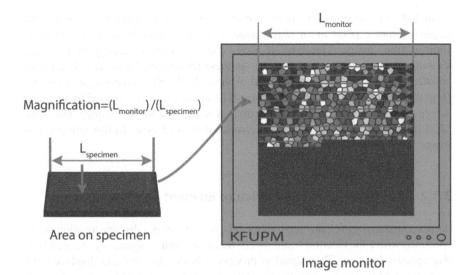

Fig. 3.1 Schematic illustrating the one-to-one correspondence between the beam locations on the specimen and the points on the monitor. Magnification is the scan length of the monitor to that of the specimen, i.e., $\frac{L_{monitor}}{L_{specimen}}$

than adequate for producing good-quality images. The number of pixels used in the image is specified as the digital image resolution such as 512 × 512, 1024 × 1024, etc. The image with 512 × 512 pixels will have 512 pixels in the horizontal direction and 512 pixels in the vertical direction. A higher number of pixels per unit area will produce images with higher resolution with correspondingly large file size.

A range of scan rates is available to the user to allow from slow to medium to fast scans. Initially, the specimen can be scanned quickly (several image frames per second) to search for the area of interest. Once the feature of interest has been located, the scan rate can be slowed down to several seconds per frame to capture a noise-free image. The dwell time at each location should be sufficient to record a signal that provides ample statistical accuracy. The time for the beam to move from one location to the next is short compared to the dwell time, making it appear as a continuous scan. The total time to capture an image frame is given by the product of the total number of points scanned in a field of view and the dwell time at each point. Due to fast scan rates, small discrete points blend into a continuous image composed of various levels of gray produced by differences in signal strength from one location to the other.

Digital imaging technology is efficient and cost-effective and easily lends itself to storage and further handling or processing. Also, several images or frames can be acquired and averaged to reduce noise or charging effects. Additionally, digital images are stored in memory and can be displayed on the screen without having to continuously scan the beam on the specimen surface, thus reducing the probability of beam-induced contamination or damage in sensitive specimens.

In light or transmission electron microscope, rays of light or electrons that originate from a point in an object plane meet at a point in the image plane to form a "true" image. This type of image can be observed directly by placing a viewing screen in the path of the rays. In these techniques, image data is acquired simultaneously from the selected field of view. In the SEM, the image is *enacted* from the signals emitted by the specimen. The image cannot be viewed by placing a screen anywhere in the path of the electrons. The data is collected sequentially in the SEM during the raster scan across the selected field of view. In this context, *true* image is not formed in the SEM.

3.1.2 Relationship Between Picture Element and Pixel

As described in the previous section, the picture element is a discrete location on the specimen where the electron beam dwells and as a result, a signal is generated from that specific location. That signal is processed in the detector and displayed at a corresponding location (called pixel) on the display monitor (see Fig. 3.1). The gray level of that pixel is proportional to the strength of the signal obtained from the corresponding picture element. Since one-to-one correspondence is established between the picture elements on the specimen and the pixels on the screen, it is clear that the number of picture elements will be equal to that of the pixels for a given specimen scan and the resulting image. Let us assume that the digital resolution used for a particular image is 2,000 × 2,000 pixels and the size of the display is 20 × 20 cm. The size (given in horizontal edge length) of one pixel in the image is given as $\frac{20 \text{ cm}}{2,000} = 0.01$ cm $= 100$ µm. The size of the picture element at the corresponding location of the specimen will be smaller by a magnitude that equals the magnification at which the image was taken. In other words, the length of the picture element can be obtained by dividing the length of the pixel with the value of magnification as shown below:

$$L_{\text{picture element}} = \frac{L_{\text{pixel}}}{M} \qquad (3.1)$$

It was stated in Sect. 2.4.6 that the magnification in the SEM is changed by varying the size of the scan on the specimen. The smaller scan produces higher magnification and vice versa. As the size of the scan gets smaller at higher magnification, the size of the picture elements is proportionally reduced to accommodate picture elements whose number is governed by the pixels selected for the image. Conversely, the size of the picture element can be increased by using a smaller magnification during imaging as shown in Table 3.1.

The scattering of the electron beam within the specimen results in an interaction volume much larger than the probe diameter. The signals are generated from a much wider area within the specimen and are not limited to those emanating closer to the probe at the surface. If the diameter of the area that encompasses the scattering within the specimen is projected to the surface and is found to be smaller

3.1 Image Formation

Table 3.1 Size of the picture elements at various magnifications for 2,000 × 2,000 pixel image resolution and 20 × 20 cm display size. The pixel size in the image is 100 μm and does not need to be any smaller as the human eye can discern details down to 200 μm only

Magnification	Size of picture element on the specimen, $L_{\text{picture element}}$
10×	10 μm
100×	1 μm
1000×	100 nm
10,000×	10 nm
50,000×	2 nm
100,000×	1 nm
1000,000×	0.1 nm

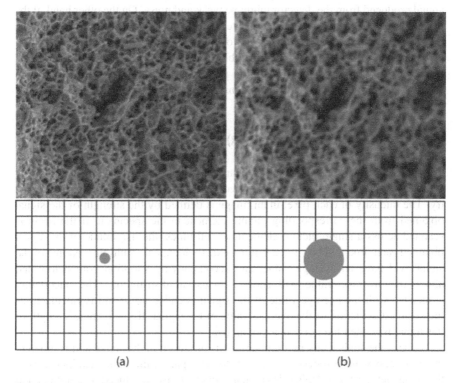

Fig. 3.2 (a) If the signal-generating area within the specimen is smaller than the picture element at a particular magnification, the image will appear in sharp focus. (b) If the size of the area becomes large, signals from neighboring picture elements will overlap and make the image blur

than the size of the picture element attained at that particular magnification, the image will be in sharp focus. If the probe diameter at the surface or the resulting sampling region within the specimen exceeds the size of the picture element, the image will appear blurred (see Fig. 3.2). The blurring occurs because signals from adjacent or neighboring picture elements overlap. At high magnifications, the size of picture elements becomes small which readily results in overlap of information and blurring. This directly affects the ability of a microscope to resolve fine details in specimens. The area that produces the signal will vary with the atomic number of material under examination. Light elements will generate large areas, while in heavy elements the signal-generating area within the specimen will be relatively restricted. This has a direct consequence on the ability to produce acceptable images at high magnifications with the same probe diameter. Imaging of light elements at high magnifications is challenging due to greater delocalization of the signal within the specimen.

In summary, the higher is the required magnification the smaller is the pixel size on the specimen surface. If the probe diameter is smaller than the pixel size, the signal produced from the detector might be weak and noisy. On the other hand, if the probe diameter is larger than the pixel size, information from adjacent pixels will be detected, and the sharpness or the resolution of the image will be reduced. Therefore, the optimum probe diameter is comparable to the size of the specimen pixel. This means that in order to achieve optimum performance from an SEM, the spot size should be adjusted according to the magnification of the SEM.

3.1.3 Signal-to-Noise Ratio (SNR)

The interaction between the electron beam and the sample generates a signal that provides useful information about the specimen's topography, structure, or chemistry. The intensity of each pixel in an image is defined by the strength of the signal obtained from the corresponding picture element in the specimen. If the beam is scanned across a single picture element several times, the magnitude of the signal obtained each time will be similar but not exactly identical. Multiple readings of the signal will have an average value, and accompanying variation both below and above that average value will show a Gaussian distribution. The uncertainty associated with the average value of the signal is called noise and is calculated as the standard deviation of the average value of the signal. If the average strength of the signal is n, then the noise is given as $\sqrt{n}$.

Noise is the statistically random variations produced in the signal and is inherent in the process which produces the signal. Noise is part of the signal and increases as the signal gets stronger. Noise also contains fluctuations introduced by amplification and signal processing. Quality of an image is determined by the relative occurrence of signal and noise which is expressed as signal-to-noise ratio (SRN):

$$\text{SNR} = \frac{S}{N} = \frac{n}{\sqrt{n}} = \sqrt{n} \qquad (3.2)$$

where $S = n = $ signal and $N = \sqrt{n} = $ noise

When noise increases to an unacceptable level, the image is said to have a low SNR. In order to detect a feature against a background of random noise, signal

3.1 Image Formation

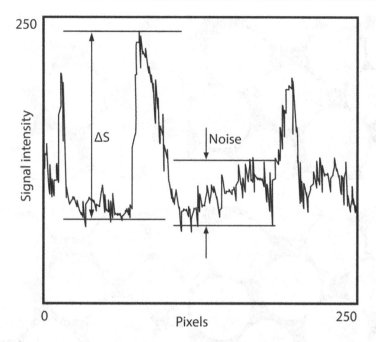

Fig. 3.3 Schematic illustration of statistical fluctuation in a signal giving rise to noise that forms the background of an image. Such a plot is obtained when the beam is scanned on the surface of the specimen once from left to right (called line scan). ΔS is the change in the signal which needs to be roughly 5× of noise to show ample contrast [1]

strength from that feature should be roughly 3× that of the noise level (see the plot in Fig. 3.3) [1]. Signal-to-noise ratio determines the quality of the image. If the SNR is acceptable, the image appears smooth. If the SNR is low, the image looks grainy.

High probe current and slow scan rate set over an optimum range results in high signal-to-noise ratio. As an example, image shown in Fig. 3.4a is taken using small probe current (low SNR) resulting in a grainy image, while that in Fig. 3.4b is taken with large probe current (high SNR) resulting in a smooth image. In another example, the SEM image in Fig. 3.4c was taken at a high scan rate (low SNR) which resulted in a grainy image. The same region appeared smooth (Fig. 3.4d, high SNR) when a small scan rate was used. During a fast scan, the number of probe electrons interacting with each picture element of a specimen is small resulting in a weaker signal emitted that gives rise to low SNR. In practice, the fastest scan rate that gives a smooth image is used in order to prevent charge-up of the specimen which can occur during slow scans. The increased solid angle of collection of the detector will improve SNR. Another way to increase SNR is to take several frames at a high scan rate and undertake frame averaging to reduce the noise. As a post-acquisition procedure, software programs can be used to reduce grainy appearance of an image by *smoothing* operation.

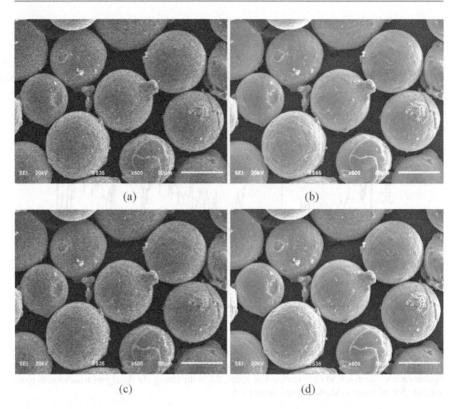

Fig. 3.4 Images of NiCrAlY powder taken under different acquisition conditions: (**a**) small probe current (low SNR), (**b**) large probe current (high SNR), (**c**) high scan rate (low SNR), and (**d**) low scan rate (high SNR). The images depicting high signal-to-noise ratio appear smooth, while those with low SNR show grainy appearance

The SNR of the image is a function of the specimen material, its topography, and the beam current. A filter is needed to remove the noise from the images [2]. For the design of the filter, SNR needs to be estimated. Earliest proposed method involved taking the cross-correlation of two images with the same specimen area [3]. This technique was used by many researchers [4]. However, it requires perfect alignment of the two images, and this method cannot be used for already captured images. The cross-correlation coefficient is given by [5]:

$$\rho_{12} = \frac{r_{12}(0,0) - \mu_1\mu_2}{\sigma_1\sigma_2} \quad (3.3)$$

where $r_{12}(0,0)$ is the peak of the cross-correlation function, μ is the image mean, and σ is the variance of the image. The SNR is then calculated as:

3.1 Image Formation

$$\text{SNR} = \frac{\rho_{12}}{1 - \rho_{12}} \quad (3.4)$$

Single image method based on autocorrelation was also proposed [6]. This approach involved having the noise be white and adaptive, and to get the noise-free autocorrelation peak function, an estimation technique is used. Two basic techniques were used, the nearest neighborhood (NN) technique and the first-order interpolation (FOI) technique. The autocorrelation coefficient is calculated as:

$$\rho_{12} = \frac{r_{12}(0,0) - \mu_1^2}{\sigma_1^2} \quad (3.5)$$

Another recent method involves the use of B-spline, cubic spline, statistical autoregressive (AR) model, and mixed Lagrange time delay autoregressive (MLTDEAR) model [5].

Several processes degrade the SNR of an image taken in an SEM, like the secondary emission noise, primary beam noise, and noise from the final detection system. Shot noise in the primary beam arises due to random fluctuation of the number of electrons emitted from the gun. This type of noise is most prominent in thermionic emission guns. The field emission guns are also susceptible to shot noise and flicker noise. The fluctuation in the generation of the secondary electrons emitted per incident beam electron gives rise to secondary electron noise. The detection system that is composed of a photomultiplier tube (PMT) and scintillator contributes to additional noise sources in the SEM image. The noise produced by the detection system is less prominent in comparison to the shot noise and the secondary emission noise [7].

3.1.4 Contrast Formation

Contrast makes a feature in an SEM image distinguishable from other features and the background. It is a measure of visibility of a particular feature in an image. The information contained in a signal emanating from a specimen is interpreted through the formation of contrast. The magnitude of contrast depends on the strength of signals derived from the feature relative to that obtained from its background. The greater the difference in signals, the higher the contrast. Contrast is determined by the difference in the magnitude of signals emitted at two points on the specimen during the beam scan. If the signal from one point is compared with the signal from another point, then the contrast C from the specimen can be written as:

$$C = \frac{S_A - S_B}{S_A} \quad (3.6)$$

where S_A is the signal emitted by the feature A and S_B is the signal emitted by the background. C is always positive (i.e., $S_A > S_B$). Maximum contrast is obtained when S_B equals zero. The contrast level ranges from 0 to 1.

Nature and type of contrast depend on the type of specimen and its interaction with the electron beam and the number of electrons emitted from the specimen based on the operating conditions employed during microscopy. The difference in the signal between the two points may arise due to many factors including change in specimen topography, the difference in composition, crystal orientation, magnetic or electric domains, surface potential, and electrical conductivity. Information about the specimen is contained in the contrast formed by the signals that are emitted from the specimen prior to detection and amplification by the detector. Subsequently, the contrast is modified by the detector without adding new information content. The position and type of detector and its efficiency in collecting and recording the arriving electrons and amplification of the signal play an important role. The strength of the signal collected by the detector depends on the number, energy, and trajectory of electrons, which directly affects the brightness of the pixels in the image.

3.2 Beam-Specimen Interaction

The electrons in the scanning electron microscope are generated by either a thermionic or field emission electron gun. The electrons are transformed into an electron beam by applying a bias voltage that is used to accelerate and focus electrons at a point known as "crossover point." The electron beam then travels in vacuum through electromagnetic lenses, apertures, and scanning coil present within the SEM column to reach the chamber where the specimen to be examined is placed. The electron beam is focused onto the surface of the specimen and scanned across its surface in a raster-like pattern. When an electron beam hits and penetrates a specimen, it is deflected by the specimen in an elastic scattering or inelastic scattering mode. This results in the generation of a variety of signals including secondary electrons, backscattered electrons, Auger electrons, characteristic x-rays, photons, etc. Some of these signals are used to generate SEM images and obtain microchemical information from the specimen. The study of electron beam-specimen interactions that produce such signals is therefore important in order to understand and better control the information obtained from the SEM and also to extract high-quality data using this technique.

3.2.1 Atom Model

The solid specimen to be examined within the SEM is composed of atoms. The nuclei of the atoms, composed of protons and neutrons, are characterized by a concentrated positive charge. The negatively charged electrons are placed around the nucleus in orbits that are grouped together into shells known as K, L, M, etc., as shown in the schematic in Fig. 3.5. The negative charge or energy of the electrons is

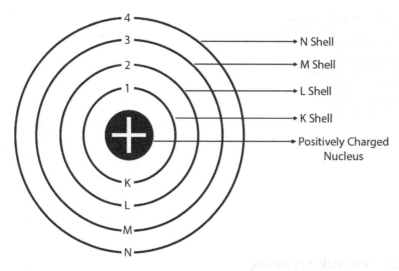

Fig. 3.5 Atom model showing the presence of K, L, M, and N shells at various energy levels around the nucleus

distributed in the way they are placed in orbits. Large atoms or heavy elements contain a larger number of electrons and electron orbits. Each shell contains a large number of energy levels.

3.2.2 Elastic Scattering

The electrons collide and penetrate the specimen surface at an accelerating voltage of 5–20 kV typically used in an SEM. If the electrons are deflected into a new direction by the atomic nuclei without losing any kinetic energy, the mode of beam-specimen interaction is termed as *elastic scattering* as shown in Fig. 3.6.

This deflection of an electron from its original path occurs due to the attractive force exerted on it by a comparatively large positively charged nucleus of the specimen. In this mode of electrostatic interaction with the atomic nuclei, the electrons are deflected at large angles with no significant transfer of energy from the electron to nuclei due to latter's comparatively large mass. The probability of high angle elastic scattering is greater in specimens with heavy/large atoms (roughly proportional to Z^2) due to their stronger positive charge. The tendency of the electron beam to scatter elastically decreases as its energy is increased (proportional to $1/E^2$) since the beam can now more effectively overcome the positive charge of atomic nuclei of the specimen.

Fig. 3.6 Schematic showing elastic scattering where beam electrons change direction without losing any appreciable energy upon interaction with the Coulombic field of the specimen material

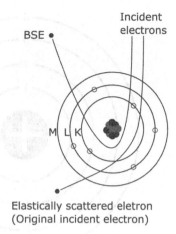

Elastically scattered eletron
(Original incident electron)

3.2.3 Inelastic Scattering

In addition to elastic scattering, electrons in the electron beam also interact with the specimen in such a way that they lose and transfer their kinetic energy to the atomic nuclei. This type of beam-specimen interaction is known as "inelastic scattering" as shown in Fig. 3.7. Electrons that undergo inelastic scattering are deflected at small angles. The beam electron strikes the electron of an atom in the specimen and knocks it out of its orbit. In the process, beam electron loses energy that is equivalent to that required to create the vacancy in the orbital. Energy dissipation in this type of event is random and gradual and mean energy loss per event is small. The energy of the electrons decreases as a function of distance traveled. Beam electrons keep moving through the specimen lattice till they lose all of their kinetic energy and are brought to rest within the specimen. In metals specimens, they could become conduction electrons. The rate at which these electrons lose energy depends on the density of the specimen and the distance traveled. The higher the density of the specimen material and the greater the distance already traversed by the beam within the material, the greater the energy loss. The rate of energy loss is defined by the "stopping power" S of the target material as shown in the following equation:

$$S = -\left(\frac{1}{\rho}\right)\frac{dE_0}{ds} \qquad (3.7)$$

where ρ is the density of the specimen material

E_0 is the incidence beam energy

s is the distance traveled by the electron within the specimen

S is the rate of energy loss divided by the target density. Loss of energy is denoted by the negative sign. The electron energy is reduced approximately by 1–10 eV per nanometer of distance traveled within the specimen depending on its atomic number. For every unit of length (i.e., nm) traveled, beam electrons will lose more energy in heavy elements than in light elements. However, for every unit

Fig. 3.7 Schematic diagram showing inelastic scattering where beam electron knocks out an orbital electron belonging to the specimen losing energy in the process

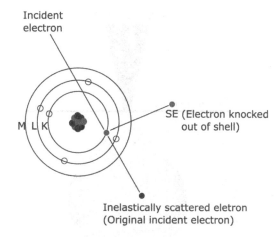

of mass (i.e., gram), light elements will extract more energy compared to heavy elements. This is due to the fact that light elements have a higher number of electrons per unit mass compared to heavy elements. This larger number of specimen electrons will give rise to a greater number of inelastic scattering events making the beam electrons lose energy more rapidly.

3.2.4 Effect of Electron Scattering

It is clear from above that when electron beam hits the specimen, the electrons do not remain confined at the surface of the specimen within a region where the electrons are converged, nor do they penetrate into the specimen in a straight line. On the contrary, elastic and inelastic scattering events make them penetrate into the depth and spread laterally across the width of the specimen forming a relatively large *interaction volume*. The density of elastic scattering events changes from the point of beam impact to the boundary of the interaction volume. It is important to realize that this is the volume of the specimen where from all imaging and the microchemical information is extracted. For instance, when an electron beam with an energy of 20 keV is converged onto the specimen surface to a very small probe diameter of few nanometers, it can penetrate 5 microns and spread 1 micron into the specimen depending on latter's density (as shown in Fig. 3.8). Therefore, the information obtained from the specimen is not restricted to the diameter of the incident probe but is gathered from a volume that is larger by hundreds to thousand times of the beam size. Due to electron scattering, the signal is generated from a large area of the specimen, thus restricting the spatial and analytical resolution of the SEM.

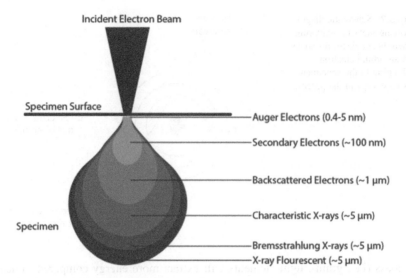

Fig. 3.8 Schematic illustrating the formation of *teardrop* interaction volume upon scattering of the electron beam within the specimen. The probe diameter of a few nanometers results in several cubic micrometers of volume where from signals are generated. This limits the spatial and analytical resolution of the SEM

3.2.5 Interaction Volume

The total volume of the specimen material that is affected by the incident electron beam is known as the interaction volume and is much larger compared to the size of the incident probe. The size of the interaction volume created depends on the specimen density, beam energy, and probe current density. Interaction volume represents the averaged behavior of a large number of electrons. Monte Carlo method is a computer simulation of electron trajectories within a specimen and aids in investigating the interaction volume created in a specific material under given beam conditions. This kind of simulation treats scattering as a statistical process and gives a visual perspective to the spatial distribution of electron scattering process. It depicts the volume of material from where various signals are collected. The simulation provides *first principles* approach using various models that take into account the collective effect of elastic and inelastic scattering to determine electron paths, deflection angles, and energy loss. The electron is treated as a discrete particle and the specimen material is taken as amorphous. Spatial distribution of secondary and backscattered electrons, as well as x-rays, can be obtained using this method [1, 8, 9].

3.2.5.1 Effect of Beam Energy on Interaction Volume

Monte Carlo electron trajectory simulations as a function of electron beam energy (E_0) in various elements are shown in Fig. 3.9. The higher the beam energy, the greater the depth and width to which the electrons can travel into the specimen as

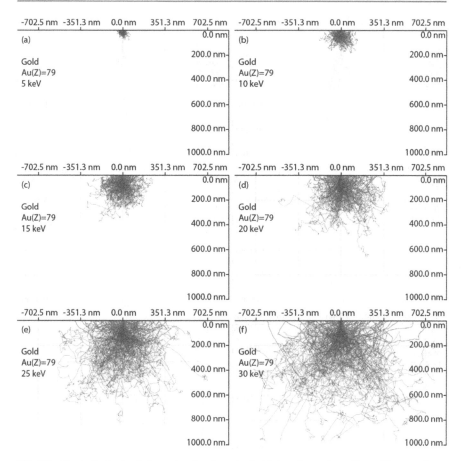

Fig. 3.9 Monte Carlo simulations showing the extent of interaction volume formed in a specimen of Au. From beam energies of 5–30 keV, depth of beam penetration increases from 60 nm to 1000 nm and width from 100 nm to 1400 nm. Note that the shape of the interaction volume does not change with beam energy

they lose energy at a lower rate which is proportional to $\frac{1}{E_0}$. Increasing beam energy also reduces its probability to scatter elastically (as a function of $\frac{1}{E_0^2}$), thus penetrating deeper into the specimen. The trajectories of the electrons near the specimen surface are straight resulting in widening of the interaction volume away from the surface.

3.2.5.2 Effect of Atomic Number on Interaction Volume

For specimens with a high atomic number, the elastic scattering is greater resulting in a deviation of the electrons from their original path more quickly and reducing the distance that they travel into the specimen as shown in Fig. 3.10a–e. The widening of the interaction volume is caused closer to the specimen surface as the probability of

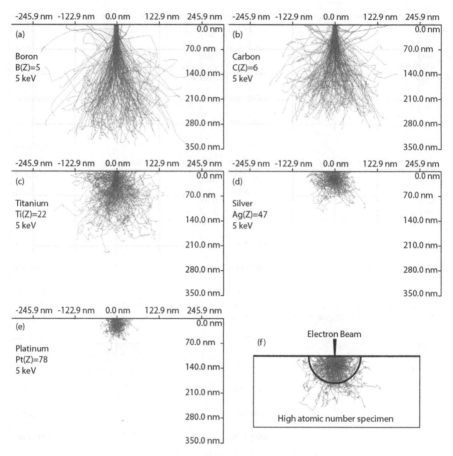

Fig. 3.10 (**a–e**) Monte Carlo simulations showing the extent of interaction volume formed within various elements at 5 keV. Depth and width of beam spread are the greatest in B ($Z = 5$) and lowest in Pt ($Z = 78$). The shape of interaction volume is considerably different for elements of low and high atomic numbers. In the elements of low Z, beam penetrates deeper before it deviates and spreads across. For elements with high Z, beam tends to deviate from its original path more quickly resulting in delocalization closer to the surface of the specimen. (**f**) Schematic representation of hemispherical interaction volume typically formed in the elements of high atomic number

elastic scattering as well as its angle increases in an element of high Z. Also, the rate of energy loss for electrons is high in the linear dimension resulting in a relatively shallow interaction volume within a high atomic number specimen. This phenomenon results in a hemispherical shape of the interaction volume (see schematic diagram in Fig. 3.10f). On the other hand, elastic scattering and the rate of energy loss per unit length are lower in low atomic number targets due to which beam electrons manage to maintain their straight trajectories for larger depths in the specimen. This results in the formation of a large interaction volume taking the form of a *teardrop* (as shown in Fig. 3.8).

3.2 Beam-Specimen Interaction

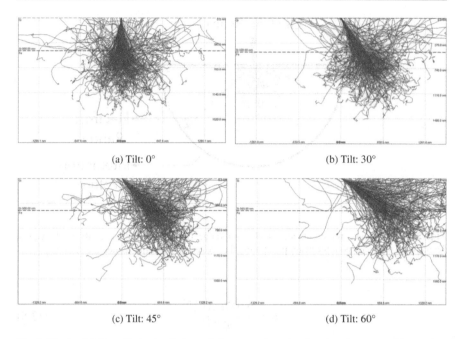

(a) Tilt: 0°   (b) Tilt: 30°

(c) Tilt: 45°   (d) Tilt: 60°

Fig. 3.11 (**a–d**) Monte Carlo simulations of electron trajectories showing the extent of interaction volume formed at various tilt angles of the electron beam. As tilt angle is increased, a large part of the interaction volume is generated close to the specimen surface which allows a larger number of electrons to escape the specimen. This also results in an asymmetry in the shape of the interaction volume

3.2.5.3 Effect of Tilt on Interaction Volume

Interaction volume also changes with the tilt of the specimen with respect to the beam. It is the maximum when the specimen is inclined 90° with respect to the beam. As the angle of inclination between the beam and specimen is narrowed, a greater number of electrons can escape the specimen surface which when tilted lies closer to the forward scattering direction (i.e., less than 90°) of electrons. This can be understood by visualizing part of the interaction volume lying outside the specimen surface, as shown in Fig. 3.11.

3.2.6 Electron Range

Electron range (or penetration depth) describes complex three-dimensional interaction volume in simple terms and can be used to provide quick rough comparisons between various materials. It is defined as the mean straight line distance of the electron from the point of entry to point of final rest in the specimen. A hemisphere is formed with a radius whose origin is at the point of beam penetration, and it contains 90–95% of the scattered electrons (see Fig. 3.12). The radius of this hemisphere is taken as the electron range given in microns as shown in the following expression [10]:

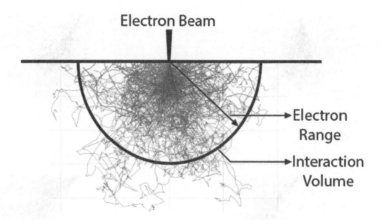

Fig. 3.12 Schematic representation of electron range (*R*) or penetration depth by drawing a hemisphere from the point of beam penetration that covers most of the scattered electrons

$$R = \frac{0.0276\, AE_0^{1.67}}{\rho Z^{0.89}} \qquad (3.8)$$

where
 R = electron range, μm
 A = atomic weight, g/mole
 E_0 = beam energy, keV
 ρ = density, g/cm^3
 Z = atomic number

This equation can be used for specimens that are considered bulk (thick), large (free of edges or boundaries), and flat. It can be seen from Eq. 3.8 that electron range or penetration depth decreases substantially at low beam energy as fewer collisions are required to bring the low-energy electrons to a stop. Additionally, the probability of inelastic scattering increases with decreasing beam energy (i.e., it is proportional to $\frac{1}{E_0}$). This results in smaller interaction volume at low beam energy. This is especially true for elements with low Z. For example, for C, Al_2O_3, and SiO_2, when the beam energy is lowered from 15 to 1 keV, the interaction volume is reduced by 5–6 orders of magnitude, and the electron range is decreased by 2 orders of magnitude. For high-Z elements, the decrease is not as severe.

For a given beam energy, the penetration depth is primarily influenced by and inversely proportional to the atomic number (and density) of specimen material. It is clear from Eq. 3.8 that the penetration depth becomes smaller as Z increases. For example at 10 keV, the beam penetrates roughly 1 μm in C while only 0.2 μm in Au. In elements of high atomic number, a large number of electrons are elastically scattered at high angles deviating them from straight trajectories resulting in small R. As a consequence, the interaction volume is smaller in elements of high Z.

Electron range can be expressed in terms of mass thickness making it independent of Z as follows:

$$\rho R = aE_0^b \qquad (3.9) \quad [11]$$

where ρ is the density, R is the electron range, E_0 is the beam energy, a is approximately 10 µg/cm², and b varies between 1.33 (at low E_0) and 1.67 (at high E_0).

Electron range of a tilted specimen decreases as the interaction volume becomes asymmetric and allows a large number of electrons to escape. Value of R will depend on the angle of tilt and can be given as:

$$R(\theta) = R(0)\cos\theta \qquad (3.10)$$

where $R(0)$ is calculated at 0° tilt

3.3 Origin of Backscattered and Secondary Electrons

3.3.1 Origin of Backscattered Electrons (BSE)

As discussed in the previous sections, beam electrons undergo multiple elastic and inelastic scattering events giving rise to a large region within the specimen called interaction volume. The extent of each type of scattering depends on the beam energy, atomic number of the target and the degree of tilt used. Elastically scattered beam electrons are deflected through large angles (>90°) and eventually find their way out of the specimen giving rise to the phenomenon of *backscattering*. These electrons are called *backscattered electrons* (BSE), and they originally belong to the incident electron beam that strikes the surface of the specimen. The surface of entrance and departure may or may not be the same. Backscattered electrons make up most of the electron signal emanating from the specimen. These electrons constitute a significant proportion of the total energy of the incident beam. With the escape of BSE from the specimen, a substantial measure of beam energy is removed from the sample. Once out in the vacuum, these electrons can be captured by a detector and used to form an image called *backscattered electron image*. The contrast exhibited by the image formed due to these electrons is called *compositional* or *atomic number (Z) contrast*.

3.3.2 Origin of Secondary Electrons (SE)

In addition to BSE, some electrons belonging to the target material are also ejected from the specimen. These electrons are generally loosely bound outer shell electrons residing close to the specimen surface and are struck by the electrons of the incident beam in an inelastic manner. Outer shell electrons are knocked out of their shells. If adequate energy is imparted to them in the process, they may find their way out of the specimen due to their close proximity to the surface. These

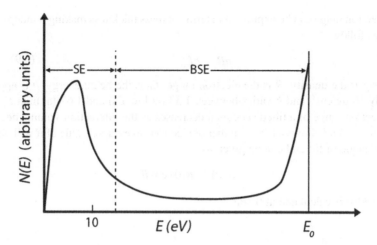

Fig. 3.13 Schematic representation of electron energy distribution where $N(E)$ is the number of electrons of energy E emitted by a specimen irradiated with incident electrons of energy E_0. Backscattered electrons dominate the energy spectrum, while low-energy secondary electrons are generated in small numbers

electrons that originally belong to the specimen material and are ejected out of the specimen are known as *secondary electrons* (SE). These can be collected by a detector to form a secondary electron image. The contrast exhibited by SE is known as *topographic contrast*.

An electron is an electron and as such, no distinction between secondary and backscattered electrons can be made. They are differentiated purely on the basis of their energy, i.e., electrons that are emitted from the specimen with less than 50 eV kinetic energy are classified as secondary electrons.

The plot in Fig. 3.13 shows the distribution of energy of all electrons that emanate from a specimen. The y-axis shows the number of ejected electrons $N(E)$, and the x-axis represents the energy of electrons emanating from specimen starting from zero to the primary beam energy incident on the specimen surface (E_0). The region to the far right of the plot represents electrons that are ejected from the specimen with no loss of energy. In addition, the greater part of the energy spectrum corresponds to electrons that are ejected from the specimen with relatively high energy. These are called backscattered electrons. The number of low-energy backscattered electrons emanating from the specimen falls close to zero as the left-hand side of the plot is approached. Beyond this point, an abrupt increase is observed in the number of electrons that emanate from the specimen. These electrons have energies of 50 eV or less at energy scale. This sudden increase in the number of electrons with low energy is attributed to the generation of secondary electrons.

3.4 Types of Contrast

Compositional and topographic contrasts are the two dominant forms of contrast used in scanning electron microscopy. These are primarily dependent on backscattered and secondary electrons which make up the most important signals used in SEM imaging. It is important to study the dependence of these signals on beam energy, atomic number, tilt, and other related factors. Most images depict a combination of contrast mechanisms. Discussion of each contrast mechanism and its applications follows in the proceeding sections.

3.4.1 Compositional or Atomic Number (Z) Contrast (Backscattered Electron Imaging)

3.4.1.1 Yield of Backscattered Electrons

The extent of backscattering in a specimen is expressed in the form of backscatter coefficient (η) which is defined as the fraction of incident electrons that leave the specimen surface, as shown below:

$$\eta = \frac{\eta_{BSE}}{\eta_B} \tag{3.11}$$

where
  η = backscatter coefficient
  η_{BSE} = number of backscattered electrons
  η_B = number of incident beam electrons

The backscatter coefficient can be measured in terms of the ratio of backscattered current moving out of the specimen to that of the incident beam current.

$$\eta = \frac{i_{BSE}}{i_B} \tag{3.12}$$

where
  η = backscatter coefficient
  i_{BSE} = backscattered electron current moving out of the specimen
  i_B = electron beam current entering the specimen

Elastic scattering occurs mostly in the forward direction that involves deflections at small angles (e.g., 5°). The incident beam tends to penetrate into the specimen in roughly the same direction it entered. Backscattering occurs only as a cumulative effect of numerous small- and some large-angle scattering events. This results in electrons changing direction to the extent that they escape as backscattered electrons from the surface they initially entered.

3.4.1.2 Energy Distribution of BSE Yield

Some beam electrons will enter the specimen, immediately deflect through large angles, and leave the specimen without any energy loss. The energy of these backscattered electrons will be equal to the energy of the incident beam. Other beam electrons will travel through the specimen until they lose almost all of their energy prior to ejecting out of specimen surface. These BSE will have essentially zero energy upon exclusion. The energy continuum of backscattered electrons, therefore, stretches from zero to incident beam energy E_0. By convention, the energy of the BSE is generally defined as 50 eV $< E_{BSE} \leq E_0$. Electrons with <50 eV energy are classified as secondary electrons. Majority of BSE emanating out of specimen undergo multiple scattering events within the specimen. Most of the backscattered electrons retain at least 50% of their energy at the time of ejection from the specimen.

In specimens with intermediate to high atomic numbers, the majority of backscattered electrons will have higher energy distribution. This is due to the fact that the degree of elastic scattering and hence backscattering increases with atomic number. On the other hand, lesser number of backscattered electrons emanating from a specimen with low atomic number will depict high energy. It can be seen in Fig. 3.13 that a broad region toward the left of the E_0 peak is covered by BSE. For elements with high Z, the majority of the BSE ejected from the specimen will exhibit energies of the order of $\geq 0.7E_0$. For the elements with low Z, the BSE energy is $\geq 0.4E_0$. Heavy elements such as Au can exhibit significant BSE yield with energy $0.9E_0$, whereas the equivalent for a light element such as C is at $(0.5-0.6)E_0$. Half of the BSE yield from Au and C will have energy of $0.84E_0$ and $0.55E_0$, respectively.

The strength of the signal output from the detector depends not only on the number of BSE but also on their energy. The contrast produced in the image, therefore, is a combination of mean Z of the specimen and the energy distribution of BSE ejected from it. Combination of high mean Z and high-energy distribution of BSE imparts a strong contrast to the BSE image of specimens with high density.

3.4.1.3 Effect of Beam Energy on BSE Yield

Backscattered electrons are generated from a depth that is roughly one-third of the penetration depth of the incident beam or electron range. The size of interaction volume formed in a given target material increases with an increase in beam energy. The electrons with high energy travel deep into the specimen. Although these electrons penetrate deeper, they still have enough energy to get deflected at large angles and leave the specimen surface. Therefore, the backscatter coefficient changes only slightly (<10%) at incident electron beam energy of 10, 20, and 30 keV as seen in Fig. 3.14a. At lower beam energy of <5 keV, backscatter coefficient increases for elements with Z < 30 and decreases for elements with Z > 30. This complicates the BSE contrast formed at low beam energies and interpretation of images needs careful consideration.

3.4.1.4 Effect of Atomic Number on BSE Yield

Backscatter coefficient η strongly depends on the atomic number of the specimen material. Target materials with high atomic number show a high degree of elastic scattering resulting in high angles of deflection and large backscattering effect. The relationship between backscattered coefficient η and atomic number Z is plotted for 10–30 keV incident beams in Fig. 3.14a.

3.4 Types of Contrast

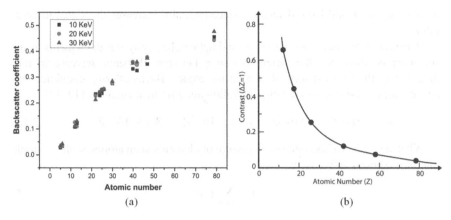

Fig. 3.14 (a) Backscatter coefficient η increases with atomic number Z. This behavior is similar at various beam energies (10–30 keV plotted in the graph). Backscatter yield is <10% for light elements and increases to around 50% for heavy elements. (b) Curve fit to the adjacent plot reveals atomic number contrast from a pair of elements that are one atomic number apart

It can be seen from the plot that the backscatter coefficient increases with atomic number. Backscatter yield is approx. 10% for light elements and increases to >50% for heavy elements. It immediately follows that two phases with different atomic numbers present within a specimen will exhibit different values of backscatter coefficient. This will result in a contrast where the phase with a high atomic number will appear relatively brighter (due to a larger number of backscatter electrons ejecting out of this phase) while phase with a low atomic number will appear relatively dark. This is called compositional or atomic number or Z contrast. It follows from the above discussion that the contrast of the phases in BSE images will depend on their chemical composition. The phases with strong contrast (i.e., appearing bright) contain heavier elements (higher Z), and the phases with weak contrast (i.e., appearing dark) constitute lighter elements (lower Z). It can be seen in Fig. 3.14a that the contrast produced between elements with a large difference in atomic numbers will be strong. For example, the contrast generated between Al ($Z = 13$) and Au ($Z = 79$) is 69% while that between Al ($Z = 13$) and Si ($Z = 14$) is mere 6.7%. It can also be seen in Fig. 3.14a that the slope is initially steep and then declines gradually beyond atomic number 40. This means that Z contrast is stronger between two low atomic number phases compared to two high atomic number phases. This relationship is expressed in a plot in Fig. 3.14b which shows decreasing contrast with increasing Z between two successive elements in the periodic table. As an example, the contrast between light B ($Z = 5$) and C ($Z = 6$) is 14% while the contrast between heavy Pt ($Z = 78$) and Au ($Z = 79$) is 0.41% only. Any contrast $\geq$10% is easily discernible in the SEM. The contrast between 1% and 10% requires effort to image, while the contrast of <1% is difficult to detect. Weak Z contrast formed between two phases with similar density requires the use of well-polished flat specimens to increase the BSE collection efficiency. In addition, use of short working distance and large detector surface area also increases the signal-to-noise ratio.

Backscattering coefficient increases and approaches unity for all elements as the specimen is tilted. So, the differences in η between elements decrease in turn diminishing the contrast formed at high tilt angles. Backscattering coefficient for an element can be calculated using the following empirical equation [12, 13]:

$$\eta = -0.0254 + 0.016Z - 1.86 \times 10^{-4}Z^2 + 8.3 \times 10^{-7}Z^3 \qquad (3.13)$$

When a target is a homogeneous mixture of elements at an atomic scale, a simple rule of mixtures can be used:

$$\eta = \sum_i C_i \eta_i \qquad (3.14)$$

where

i is an individual element
C_i is the mass fraction of element i
η_i is backscatter coefficient of element i

In addition to qualitatively identifying chemically distinct phases, Fig. 3.14a can be used to obtain quantitative chemical information from a phase. This is undertaken by measuring the BSE signal emitted from a phase, comparing it with the signal intensity from a known pure elemental standard and using the plot in Fig. 3.14a to determine the atomic number of the phase. This technique can be used to detect phases with low Z. The detected BSE signal should not contain a contribution from topography or crystallographic contrast.

A typical backscattered image of the fracture surface of an Al alloy depicting Z contrast is shown in Fig. 3.15.

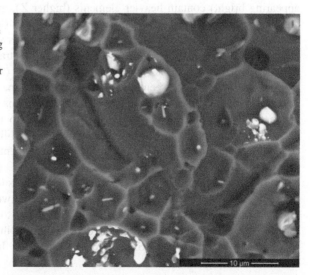

Fig. 3.15 Backscattered electron image of the fracture surface of an Al alloy showing the presence of coarse precipitates. Compositional or Z contrast makes the precipitates appear brighter indicating that they are composed of heavy elements compared to the surrounding matrix of Al alloy

3.4.1.5 Effect of Tilt on BSE Yield

Tilt angle θ is defined as the complement of the angle between the beam and the specimen surface. Backscatter coefficient changes with the degree of tilt of the specimen with respect to the beam. It is the minimum when the specimen is placed 90° with respect to the beam (tilt angle $\theta = 0°$). As the angle of inclination between the beam and specimen is narrowed (as θ increases), a greater number of electrons can escape the specimen surface as backscattered electrons. The surface of the specimen when tilted lies closer to the forward scattering direction of electrons. This means that backscattered signal from the specimen can be increased by tilting the specimen in such a way that the incidence angle of the beam with the specimen surface is small, i.e., θ is large. A monotonic increase in backscatter coefficient η with tilt angle θ can be expressed by the following equation:

$$\eta(\theta) = (1 + \cos\theta)^{\frac{-9}{\sqrt{z}}} \tag{3.15}$$

This is depicted in a plot obtained from a specimen of Fe coated with a thin (200 nm) film of Zr as shown in Fig. 3.16a. The figure also shows Monte Carlo simulations of electron trajectories for the same specimen tilted at various angles θ of (b) 0°, (c) 20°, (d) 60°, and (e) 70°. It can be seen that the interaction volume becomes increasingly asymmetric with increasing θ resulting in an increase of backscatter coefficient. It is also important to note that the energy distribution of BSE is shifted toward higher energy at high tilt angles, especially if $\theta \geq 70°$.

3.4.1.6 Effect of Crystal Structure on BSE Yield

The backscatter coefficient η is marginally sensitive to the direction of the incidence beam relative to the crystal lattice in a single-crystal specimen. If the beam penetrates along the direction of the densely packed atomic planes, the η is higher. The backscatter coefficient will be lower if the beam direction is along the low-density planes which allow the beam to penetrate deeper into the specimen before scattering, thus resulting in a lower η. The variation in η due to crystal structure is however low (i.e., in the order of 1–10% maximum). The difference in η due to crystal orientation forms the basis for *electron channeling contrast*.

3.4.1.7 Directional Dependence of BSE Yield

Upon interaction of the electron beam with the specimen, the backscattered electrons are emitted in all directions and follow straight trajectories. They follow the direction toward which they were initially ejected. However, the number of emitted BSE and hence their intensity is not the same in all directions. BSE angular distribution will follow a cosine law. Due to this reason, the collection efficiency of the detector receiving these electrons will depend on the detector's position within the specimen chamber. At zero-degree specimen tilt ($\theta = 0°$, beam normal to the specimen surface), the maximum number of backscattered electrons is emitted along the beam, i.e., normal n to the specimen surface (see Fig. 3.17). Backscatter coefficient η will be the highest along the beam direction as BSE are predominantly emitted back toward that point. At location m which is at an angle $\emptyset$ relative to the normal n, the number of emitted BSE will decrease and is given by:

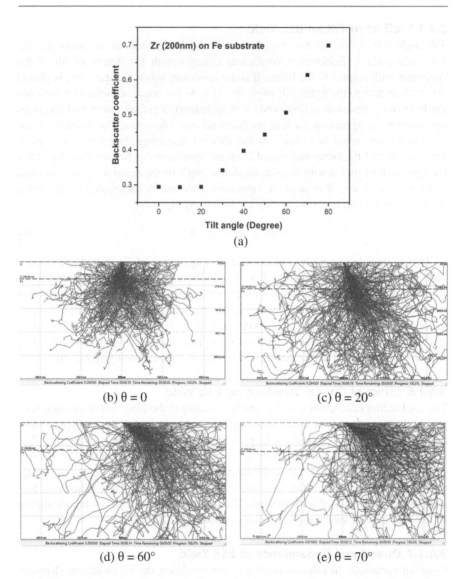

Fig. 3.16 Specimen: Fe coated with Zr. (**a**) Plot showing typical increase in backscatter coefficient as the tilt angle θ (defined as the complement of the angle between the beam and the specimen) is increased. As the incidence angle of the beam with the specimen surface is narrowed (θ is increased), a large proportion of elastically scattered electrons are able to escape the surface as the interaction volume intersects the latter. The figure also shows Monte Carlo simulations of electron trajectories at various tilt angles θ of (**b**) 0°, (**c**) 20°, (**d**) 60°, and (**e**) 70°. It can be seen that interaction volume becomes increasingly asymmetric with increasing θ resulting in an increase in backscattering [14]

Fig. 3.17 At zero specimen tilt ($\theta = 0°$), backscatter coefficient η is the highest along normal n to the specimen surface (i.e., at $\varnothing = 0$). As the value of $\varnothing$ increases, η will decrease. At $\varnothing = 45°$, η is 71% of that at n. At $\varnothing = 75°$, η is reduced to 26% of that at n

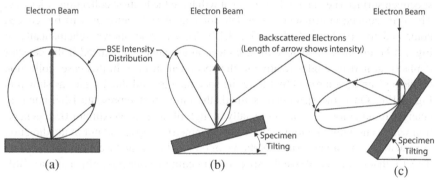

Fig. 3.18 (a) The circle represents the magnitude of backscatter coefficient η at various angles of BSE generation. Vector length within the circle corresponds to the magnitude of η at a particular angle. Large vector at a high angle of emission relative to the specimen surface indicates higher η. Small vector at a shallow angle of emission represents lower η. (**b**, **c**) As the specimen is tilted, the emission pattern changes into an elliptical shape where the highest emission corresponds to the longest vector of the asymmetric ellipse

$$\eta_m = \eta_n \cos \varnothing \qquad (3.16)$$

where

η_m = backscatter coefficient at m
η_n = backscatter coefficient at n

In Fig. 3.18a, the circle represents the magnitude of BSE intensity at various angles of emission, while lengths of the vectors drawn within the circle indicate the magnitude of BSE intensity at a particular angle. At zero-degree specimen tilt, the largest vector is obtained along the beam direction indicating maximum emission in this direction. The vector length becomes small at shallow angles relative to the specimen surface indicating smaller η. As the specimen is tilted, the circle that represents BSE intensity begins to change its shape into an ellipse. Largest vector does not correspond to the beam direction at this stage (see Fig. 3.18b). As the degree of tilt is further increased, the ellipse becomes more asymmetric with the largest vector now lying at increasingly shallow angles relative to the specimen surface (see Fig. 3.18c). This occurs because the interaction volume begins to intersect the specimen surface and more and more backscattering occurs from this surface due to the tendency of electrons to scatter in the forward direction. Maximum BSE generation will now occur along the long axis of the asymmetric ellipse and not in the direction normal to the specimen surface.

3.4.1.8 Collection Efficiency of the BSE Detector

It follows from above that the collection efficiency of the BSE detector depends on its location. If a detector is placed directly above and normal to the non-tilted specimen surface (i.e., at n in Fig. 3.17), it will have the highest collection efficiency. The BSE collection efficiency of a detector placed at location m will be reduced compared to that located at n (Fig. 3.17). This is also shown schematically in Fig. 3.19a. For non-tilted specimen, collection efficiency is reduced if the detector is placed at narrow angles relative to the specimen surface. In this case, collection efficiency can be improved by tilting the specimen toward the detector as shown in Fig. 3.19b. Once the specimen is tilted, the circle that represents BSE intensity changes to an elliptical shape especially at high tilt angles. Maximum BSE generation will occur along the long axis of the asymmetric ellipse and not in the direction normal to the specimen surface as shown in Fig. 3.19b. It can be concluded that for a detector that is located above the specimen, maximum collection efficiency for BSE

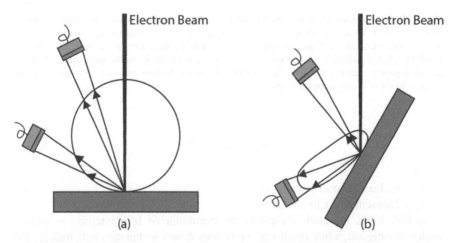

Fig. 3.19 In case of (**a**) where beam incidence is normal to the specimen surface, the maximum collection efficiency is obtained by placing the detector directly above the specimen. In case of (**b**) where the detector is placed at a small angle relative to the specimen surface, the maximum collection is obtained by tilting the specimen toward the detector

3.4 Types of Contrast

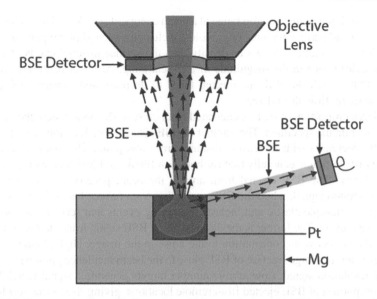

Fig. 3.20 Detector placed at a shallow angle will form an image with BSE that emanate from Pt particle and travel through Mg matrix resulting in blurring of the Pt edge that faces the detector

is obtained by keeping the specimen at zero tilt. For a detector that is placed at a small angle relative to the specimen surface, maximum collection efficiency for BSE is obtained by tilting the specimen toward the detector.

The detector placed at a shallow angle will not only receive the weaker backscattered signal but also form an image with BSE that travel a large distance through the material before they are ejected out of the specimen. This tends to degrade image resolution, especially in a multiphase material. Consider a case where the specimen constitutes of the high-Z particle (e.g., Pt) dispersed in a low-Z matrix (e.g., Mg) as shown schematically in Fig. 3.20. The BSE emanating from Pt have to pass through Mg before they are ejected and enter into the BSE detector placed at a small angle relative to the specimen surface. As a result, the edge of the Pt particle closer to the detector will lose sharpness and appear blurred in the image.

3.4.1.9 Spatial Distribution of BSE

As the atomic number of the target material increases, the depth and lateral dimensions from which the BSE signal is emanated decrease. For example, at a beam energy of 30 keV, the escape depth of BSE in Al and Au is 2.3 μm and 0.45 μm, respectively. Similarly, the lateral dimension from which the BSE signal is derived in Al and Au is 10.6 μm and 1.3 μm, respectively. The depth within the electron range from which BSE is derived depends on the atomic number of the target material. For instance, 95% of BSE are ejected from 19% of the penetration depth in Au while 95% of BSE are derived from 32% of penetration depth in C.

The depth and lateral dimensions of the volume that emit BSE signal increase with increasing incidence beam energy E_0. The BSE signal is generated from a substantial depth of the electron range at a beam energy of ≥ 10 keV. Due to this fact, any hidden subsurface features (e.g., voids, inclusions, second phase particles, etc.) with different atomic numbers or density than the surface material can be detected due to a difference in the magnitude of backscattering. At lower beam energy, the depth from which the BSE are ejected tends to decrease and subsurface features become more difficult to detect.

A large proportion of BSE is emanated from close to the point where the beam is incident upon the specimen. The number of BSE ejected from locations farther away laterally decreases with increasing distance from this point. This pattern of lateral spatial distribution is generally true for both non-tilted and tilted specimens, varying only slightly. The BSE ejected from around the beam penetration area represent high-resolution signal, while the BSE emanating from remote lateral locations have undergone multiple elastic and inelastic scattering events and serve to degrade the image resolution. The farther is the ejection of the BSE signal from the beam impact point, the greater is its contribution to the noise in the image. High atomic number targets generate a large fraction of BSE close to the beam incidence point resulting in a high-resolution signal. Low atomic number targets generate a signal which has a large proportion of BSE ejected from remote locations, giving rise to low-resolution image.

3.4.1.10 Formation of Compositional or Z Contrast with BSE

It is clear from above that the contrast in an image formed by a backscattered detector is based on the differences in the mean atomic numbers of the elements in a sample. The heavy elements appear bright and the light elements appear dark in a backscattered image. Since the intensity of the BSE signal is related to the mean atomic number of the specimen, BSE images can provide information about the distribution of chemical phases in a specimen. Differences in the chemical composition of various phases in a localized region are exhibited in a BSE image. The Z contrast C_Z obtained from two locations will depend on the difference in their backscatter coefficient which varies directly with atomic number. If the BSE detector energy response is assumed to be the same for electrons derived from both locations, the C_Z between two regions can be expressed in terms of η as follows:

$$C_Z = \frac{\eta_2 - \eta_1}{\eta_2} \qquad (3.17)$$

As shown in Fig. 3.21, two adjacent locations with the same chemical makeup (e.g., atomic number) will not exhibit any compositional contrast as they will exhibit the same backscatter coefficient. On the other hand, two locations with different Z will exhibit compositional contrast as they will yield different η, the value of which depends on their respective atomic number.

Backscattered detectors are usually annular in design placed directly above the specimen around the optic axis with a hole in the center to allow passage of electron beam. The detector is concentric with the electron beam. This design serves to maximize the solid angle of collection. BSE detectors are usually either of

3.4 Types of Contrast

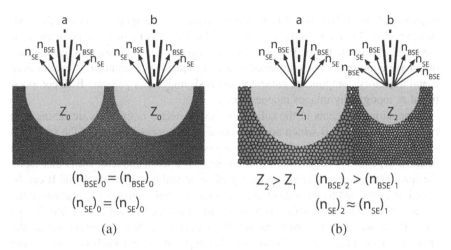

Fig. 3.21 (a) Number of SE and BSE produced from two adjacent flat locations of the same Z will be equal in number. Z contrast will not be visible. (b) Number of BSE produced from a location with high Z will be higher than those derived from low-Z location. Compositional or Z contrast will be seen with high-Z location appearing brighter than the low-Z location. The number of SE generated is taken to be the same from both adjacent locations. Adapted from [14]

solid-state semiconductor or scintillator-photomultiplier type. The solid-state device works when BSE derived from the specimen interact with Si to create electron-hole pairs (e.g., charge pulse) whose magnitude is proportional to the number and energy of the incoming BSE. The high number of pulses generates greater signal and corresponds to stronger contrast in the image. Scintillator type detector generates photons when BSE strikes its surface. The number of photons is proportional to the number of BSE interacting with the scintillator surface. Various contrast components can be selected by changing the geometry and position of the detector. Both detector types are not sensitive to low-energy electrons, thereby excluding detection of secondary electrons.

In Z contrast image, features with different densities will display different contrasts. However, features in a specimen with uniform density can also exhibit various levels of contrast depending on their orientation with respect to the location of the BSE detector. For instance, the features facing toward the detector will appear bright in the image. The features facing away from the detector will appear relatively dark even though they have the same density. This is due to the fact that BSE emitted from the specimen have straight trajectories and continue their travel in the direction they are originally emitted. A large proportion of BSE emitted from features facing the detector is more probable to enter the detector than the BSE ejected from features facing away from the detector. The signal thus produced by favorably oriented features whose surface is normal to the detector will be stronger. Apart from the atomic number Z component, this characteristic adds a directionality component to the BSE image. So the contrast in the BSE image depends not merely upon the average atomic number of the features (phases) in the specimen but also on the trajectory of the BSE emitted from those features. If the specimen is flat, the BSE trajectories will not vary and the contrast will solely depend on the Z component. If the surface is rough, the BSE trajectories will vary as per orientation of features and the image contrast will be composed of Z and trajectory components. Separation of Z and trajectory components present in the BSE image can generate BSE compositional or topography images independently.

Separation of signals in the solid-state BSE detector is made by designing it in two segments A and B which are positioned on the opposite sides of the specimen surface (see Fig. 2.37). Each of these segments can receive signals and form an image independent of each other. Each segment is a semiconductor wafer that views the specimen from a different angle. Signals received by segments A and B can be added or subtracted to separate contrast in order to highlight specific features in the images. If the signals reaching both segments are added to get a sum BSE image (A + B), it will reveal compositional differences (Z contrast) present within the specimen (see Fig. 3.22a). To obtain this type of BSE COMPO image, both segments of the detector placed above the specimen are used to collect electrons symmetrically about the beam, and the resulting images are added. In this mode, the detector has a large collection angle and the electrons are collected from all azimuthal angles. Due to this collection method, the trajectory component of the contrast is eliminated since all trajectories are collected equally irrespective of the direction in which the electrons leave the specimen surface. Atomic number component of contrast dominates in this type of image where signal becomes stronger with increasing atomic number.

If the signals are subtracted (A − B), trajectory component of the signal is accentuated to highlight surface topography (see Fig. 3.22b). This type of image is known as BSE TOPO image. If the specimen is flat, backscattered signals recorded by opposite segments of BSE detector are similar. If the surface is rough, the strength of signal reaching the segments will vary depending on its orientation with respect to the surface topography. The features of the specimen that face a particular segment of the detector appear bright in the image formed by that segment (see Fig. 3.23). Subtraction of signal produces strong BSE topographic contrast that enhances surface features of the specimen. Atomic number contrast in this type of image is eliminated since Z components received by each segment are equal in magnitude and are canceled out.

In some SEMs, individual radial segments A and B of a BSE detector can be switched on or off to control the type of contrast produced. If a scintillator-photomultiplier detector is used, the difference in BSE trajectories can be exploited to enhance contrast by physically changing detector's position. Backscattered SEM images showing compositional and topographic views of thermal barrier coating and mineral samples obtained in the manner described above are shown in Fig. 3.24a, b and c, d, respectively.

In addition to the above, BEI SHADOW mode can be used by combining signals from two BSE detectors. The first detector is placed directly above the specimen and used in COMPO mode, and the second is displaced from the end of the objective pole piece and placed at an angle above the specimen. The second

3.4 Types of Contrast 109

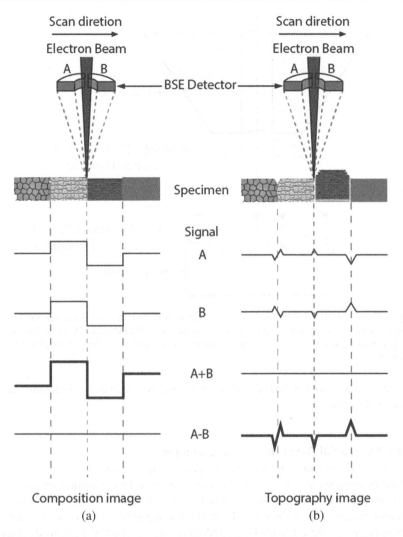

Fig. 3.22 The BSE image contains compositional and topographic information. The annular BSE detector located around the optic axis contains two segments A and B that receive signals independently of each other. (**a**) Addition of signals (A + B) gives compositional contrast, while (**b**) subtraction (A − B) gives topographic contrast [15]

detector can be a retractable BSE detector whose position can be altered. The effect of combining the signals from two detectors is to introduce "shadowing" effects in the image which serves to enhance the 3-dimensional morphology and topography of the surface features. Shadow is produced in the image since BSE have relatively straight trajectories and are blocked by the features in the specimen to reach the detector resulting in a shadow. At low magnification, BSE topographic contrast can be stronger than secondary electron topographic contrast due to sharp shadow

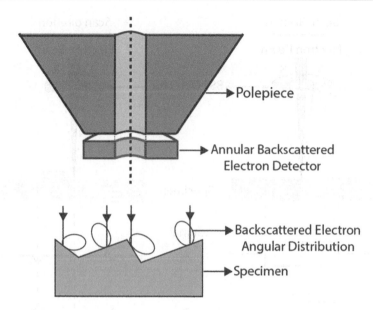

Fig. 3.23 Schematic showing angular distribution of BSE from a rough specimen surface. The features of the specimen that face a particular segment of the BSE detector will appear bright in the image formed by that segment. The same feature will appear dark in the image formed by the other segment

effects obtained with a BSE detector positioned at a narrow angle relative to the specimen surface.

3.4.1.11 Spatial Resolution of BSE Images

The spatial resolution of backscattered electron images varies between 50–100 nm for beam energies of 10–20 keV that are employed during routine imaging. The resolution is several orders of magnitude worse than that obtained in secondary electron images. This is directly related to the comparably large volume within the specimen from where the backscattered electrons are derived to form the image. Still, BSE imaging allows distinguishing between chemically distinct phases at a far better resolution than that offered by x-ray microanalysis. New technological advances such as low beam energy combined with small spot size produced by modern field emission guns can be used to reduce the BSE source volume within the specimen to a few nanometers taking it closer to the spatial resolution achieved by secondary electrons. At low beam energy of 1 keV, the information volume of SE and BSE become comparable. For high-resolution microscopy, low-loss BSE are used which are ejected from the area immediately surrounding the point of beam incidence. These electrons undergo single or lesser number of scattering events and represent high-resolution signal.

3.4 Types of Contrast

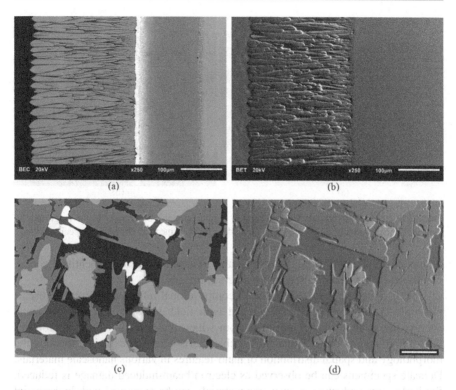

Fig. 3.24 Backscattered electron images obtained from a specimen of thermal barrier coating showing (**a**) BSE compositional contrast, COMPO mode (A + B), and (**b**) BSE topographic contrast, TOPO mode (A − B). Similarly, backscattered (**c**) compositional and (**d**) topography images obtained from a mineral sample showing multiple phases. Signals reaching both sectors of BSE detector are added to get a sum image, revealing compositional differences present within the specimen (see **a** and **c**). Signals are subtracted to highlight surface topography (see **b** and **d**)

3.4.1.12 Applications of Backscattered Electron Imaging

Backscattered electron imaging is a valuable imaging tool and provides a quick qualitative method to locate phases with different chemical compositions which then can be further investigated using elemental chemical analysis. Backscattered images reveal the distribution of phases/compounds with different compositions based on the differences in their mean atomic numbers (see Fig. 3.25a). It is particularly useful for the study of multiphase materials, catalysts, segregation at grain boundaries, and contaminants (see Fig. 3.25b). Heavy metal oxides, nitrides, or carbides appear bright in a relatively low-Z matrix such as steel. Similarly, grain or domain boundaries can be observed distinctively in a BSE image without having to etch the specimen. Likewise, BSE imaging can reveal the presence of subsurface cracks, voids, or defects. BSE imaging is especially useful in observing biological specimens. Since the BSE yield of biological specimens is low, heavy elements such as lead, silver, or gold are attached to active molecular groups present in the specimen in a process called *staining*. These locations appear bright and the rest of

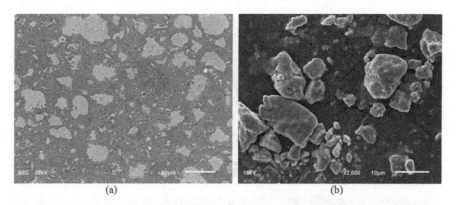

Fig. 3.25 BSE images of (**a**) polished cement sample showing differences in the magnitude of contrast formed between different phases depending on their mean Z, (**b**) catalyst specimen

the material appears dark in the BSE images. These locations are identified conveniently when compared to secondary electron images.

BSE imaging can be used to extract compositional information from nanoscale materials in the modern SEM equipped with field emission gun. At low beam energy, the volume of the specimen from which the BSE is derived approaches that of high-resolution SE_1 signal. This characteristic can be exploited to study the morphology and spatial distribution of nano-features in various nanoscale materials. Delicate specimens can be observed as electron beam-induced damage is reduced. Similarly, nonconducting or insulator materials can be examined with high spatial image resolution.

3.4.1.13 Limitations of Backscattered Electron Imaging

Contrast produced by BSE imaging depends largely on the difference in the average atomic number of various phases present within a multiphase specimen under observation. It is possible to produce two different phases with the same mean Z but having very different chemistry by mixing several elements in various proportions. The BSE images obtained from such material will not show appreciable contrast that can distinguish one phase from the other even though their chemical makeup is dissimilar [16]. Likewise, backscattering increases monotonically with atomic number, but deviation from this generally observed behavior has been reported in the literature [17]. Also, the backscattering coefficient can be influenced by the magnetic domains of a material [18]. In such cases, BSE image will fail to produce contrast that corresponds to the mean Z of the material examined. Although valid, these limitations are only encountered under exceptional circumstances and do not diminish the benefits of BSE imaging. The Z contrast formed between two phases with a comparable mean atomic number is quite weak as discussed in Sect. 3.4.1.4. This necessitates the use of well-polished flat specimens to increase the BSE collection efficiency of the detector. Likewise, short working distance and use of large detector surface area help to maintain an adequate signal-to-noise ratio.

3.4.2 Topographic Contrast (Secondary Electron Imaging)

3.4.2.1 Secondary Electron Yield

The extent of secondary electron generation in a specimen is expressed in the form of secondary electron coefficient (δ) which is defined as the ratio of the number of secondary electrons emitted from a specimen to the number of incident beam electrons that enter the specimen, as shown below:

$$\delta = \frac{n_{SE}}{n_B} \qquad (3.18)$$

where

δ = total secondary electron coefficient
n_{SE} = number of secondary electrons
n_B = number of incident beam electrons

Secondary electron coefficient can be measured in terms of the ratio of secondary electron current emitted out of the specimen to that of the incident beam current.

$$\delta = \frac{i_{SE}}{i_B} \qquad (3.19)$$

where

δ = secondary electron coefficient
i_{SE} = secondary electron current moving out of the specimen
i_B = electron beam current entering the specimen

3.4.2.2 Escape Depth of SE

Secondary electrons have low kinetic energy which enables them to leave the surface from shallow depth only. SE generated within a few nanometers of specimen surface can eject out of the specimen. SE generated away from the surface cannot escape the specimen as they lack enough energy to propagate through the solid to reach the surface and are therefore lost within the specimen. This effectively sets the escape depth for SE to a few nanometers. Therefore, the entire population of secondary electrons, generated by incident electrons at depths beyond the escape range, is not utilized for imaging.

All the random electron trajectories within the specimen are associated with the production of secondary electrons. Generation of SE relies on inelastic scattering; therefore as they pass through a specimen, their energy continually decrease. If the SE happen to reach the specimen surface, they still need enough kinetic energy (several electron volts) to overcome the potential surface barrier (work function) of the specimen. The produced secondary electrons are heavily attenuated due to inelastic scattering when they try to escape the specimen surface. The probability of the SE to leave the specimen surface is given by:

$$p(z) = 0.5 \exp\left(-\frac{z}{\lambda}\right) \quad (3.20)$$

where p is the probability, z is the depth below the surface where the SE is produced, and λ is the mean free path or the escape depth of the SE. It can be seen from Eq. 3.20 that the probability of escape of SE decreases exponentially as the depth at which they are generated increases. The maximum depth of emission is 5 times the escape depth of the SE (i.e., 5λ) [19]. The escape depth depends on the type of material. It is smaller (around 1 nm) in metals and up to 20 nm in insulators. This is due to the fact that SE generated within the specimen are inelastically scattered due to the presence of a large number of conduction electrons in metals. This scattering prevents the SE generated within greater depths of metals to escape the surface. Due to a lack of electrons in insulators, inelastic scattering of generated SE is not significant, thus allowing them to reach and escape the specimen surface from greater depths. Values quoted above are a rough estimate. In actual practice, the energy of SE will determine the mean free path so that SE with 0–50 eV range will correspond to a range of λ.

The probability of escape of secondary electrons as a function of specimen depth Z is shown in Fig. 3.26 which shows a sharp decrease in the escape probability with depth. In the range of 10–30 keV, secondary electron escape depth is around 100x smaller than the backscattered electron escape depth.

3.4.2.3 Energy Distribution of SE

The energy distribution of the electrons ejected from a specimen is spread over all energy range, i.e., from zero to the energy of the incident beam E_0. When the high-energy electron beam penetrates the specimen, it can knock out weakly bound outer shell electrons of atoms of metals, semiconductors, and insulators. There is a

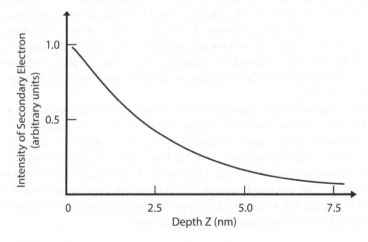

Fig. 3.26 With an increase in the depth of SE generation, the probability of SE escaping to the surface decreases [20]

3.4 Types of Contrast

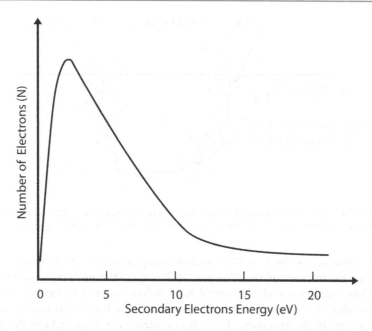

Fig. 3.27 Plot showing energy distribution of secondary electrons. Majority of the SE generated have a kinetic energy of 2–5 eV [20]

significant difference between the energy of the beam (keV) and that of the electrons belonging to a specimen (eV) in a collision event. Due to this reason, only a small portion of the beam kinetic energy is likely to be transferred to the specimen atom resulting in the ejection of low-energy electrons. High-energy electrons can also be ejected during the interaction of the beam with tightly bound electrons of the atom. However, the probability of ejection of such electrons is low. These electrons do not contribute significantly to the formation of secondary electron images due to their low numbers.

The secondary electrons that escape the specimen surface usually have a very low kinetic energy, i.e., below 50 eV. The distribution of the secondary electron energy is generally peaked in the range (2–5 eV) as demonstrated in Fig. 3.27. Although the secondary electron energy can be up to 50 eV, 90% of secondary electrons have an energy <10 eV. A small number of low-energy backscattered electrons are included in secondary electron energy range and are thus counted as secondary electrons. But the effect of this quantity is negligible.

3.4.2.4 Types of SE Signal (SE_1, SE_2, SE_3, SE_4)

The secondary electron signal obtained from a specimen is composed of two types of secondary electrons designated as SE_1 and SE_2 [21]. The secondary electrons that are produced within the narrow escape depth of the specimen (i.e., within 5λ) by the incident electron beam impinging upon the specimen surface are called SE_1 (see Fig. 3.28). These electrons make up the high-resolution image that encompasses

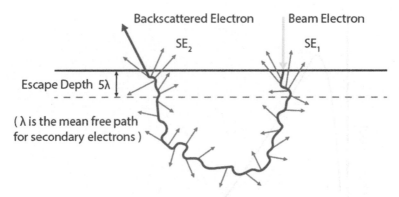

Fig. 3.28 SE_1 is generated close to the beam impact point and represents a high-resolution signal. SE_2 originates from an area away from the beam and represents a low-resolution signal

lateral spatial resolution as well as shallow sampling depth of the signal. Not all secondary electrons are produced by primary beam incident on the specimen surface. Secondary electrons are also produced in an indirect manner by the backscattered electrons that are generated within the specimen through elastic scattering and are emerging out of the specimen. These backscattered electrons inelastically scatter secondary electrons which in turn can emanate from the surface if present within the escape depth of the specimen. These secondary electrons are known as SE_2 (see Fig. 3.28). SE_1 and SE_2 are spatially separated as long as appreciable beam energy is employed. SE_1 originates close to the beam impact point while SE_2 is generated away from the probe from a region comparable to electron range of primary beam.

SE_2 is a low-resolution signal that carries different information than SE_1. SE_2 possess lateral and depth characteristics of the BSE signal that generates such signal. Change in SE_2 signal depends on the change in the backscattered signal. This could be understood by considering the fact that the extent of backscattering in a specimen with low Z is lower and it is higher in a specimen with high Z. High backscatter coefficient η in high-Z material will result in the production of the stronger SE_2 signal. This increase in SE_2 signal will produce a stronger contrast and the corresponding region will appear brighter in the image. This contrast is dependent on Z of the material and represents BSE imaging contrast rather than SE topographic contrast because the generation of SE does not change appreciably with a change in Z. Therefore, SE_2 signal is considered to signify BSE contrast.

The backscattered electrons propagate through the solid and reach the surface at much smaller angles compared to the primary beam that enters the surface at right angles. This results in greater electron path length increasing the chance of inelastic scattering of secondary electrons near the specimen surface giving rise to the stronger SE_2 signal. Additionally, backscattered electrons have lower energy compared to the primary beam and thus generate secondary electrons more efficiently. Due to this reason, the strength of the SE_2 signal is much higher than SE_1; however current density distribution of SE_1 is greater. At high magnifications (i.e., small pixel size), the resolution of SE image is thereby determined by the SE_1 signal while SE_2

Fig. 3.29 Schematic representation of the variation in the intensity of the SE signal as a function of distance from the center of the probe, i.e., spatial distribution of SE_1 and SE_2 signal. SE_1 signal is generated by the incident probe while SE_2 is produced indirectly by elastically scattered BSE away from the beam incident point. The strength of the SE_2 signal is usually higher than SE_1; however since SE_2 is generated from a much larger area, the current density distribution of SE_1 is greater

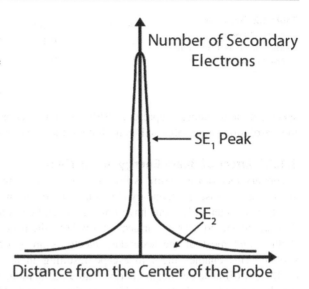

serves to contribute to the background. The signal-to-noise ratio in high-resolution images is low due to the low strength of the SE_1 signal. At mid-range magnifications, the pixel size increases (compared to that at high magnification) and approaches the range of SE_2 signal which now determines the image resolution and contrast as it constitutes the predominant part of the detected signal.

Figure 3.29 shows the intensity of the SE signal as a function of distance from the incident probe. The central peak (SE_1 signal) is generated by the incident beam electrons and thus resembles the intensity distribution profile of the probe itself [22]. The FWHM of this peak and incident probe is similar. SE_2 signal is generated farther away from the probe by beam electrons undergoing multiple scattering events. The intensity of the SE_2 signal is low but it is spread out over a greater area.

The contribution of the SE_1 signal in secondary electron emission is greater than SE_2 signal in light elements. This is due to the fact that backscattering is low in light elements due to their small atomic size. Therefore, the main component of the total SE signal is generated by the primary beam incident on the specimen surface. In heavy elements, SE_2 signal dominates since backscattering is prominent due to large atomic size.

High-energy backscattered electrons ejected from the specimen with favorable trajectories are incident upon the metallic parts of the SEM such as chamber walls, objective lens, final aperture, etc. These BSE interact with various parts of the SEM chamber to generate indirect secondary electrons which eventually find their way into the detector and form part of the SE image (see Fig. 2.32). These secondary electrons are termed SE_3 and SE_4 signal. SE_3 are generated from the walls of the specimen chamber. SE_4 signal originates from the final aperture. SE_3 and SE_4 contribute to the SE signal but will depend on the extent of backscattering in the

Table 3.2 Secondary electron emission coefficient (δ) as a function of beam energy [23]

Element	5 keV	20 keV	50 keV
Al	0.4	0.14	0.05
Au	0.7	0.2	0.1

specimen and hence represent BSE imaging contrast. These signals are low-resolution signals that will contribute to the noise in the image.

3.4.2.5 Effect of Beam Energy on SE Yield

Secondary electron emission is higher at lower incident beam energy. There is a significant increase in secondary electron emission below 5 keV. This is due to the fact that secondary electrons have low energy and can only escape the specimen if they are generated near its surface. At low keV, the penetration of incident beam is shallow and most of the secondary electrons are generated near the specimen surface. This enables them to escape resulting in a higher secondary electron coefficient (δ) at lower beam energy. Table 3.2 demonstrates experimental values for the secondary electron coefficients for Au and Al over a range of beam energies [23].

Schematic representation of the total emitted current as a function of beam energy is shown in Fig. 3.30a. As the beam energy is decreased from a higher to a lower value, total electron coefficient denoted by ($\eta + \delta$) increases to a value of unity (upper crossover energy E_2 in Fig. 3.30a). Further decrease in beam energy will result in values higher than 1 which implies that the number of electrons emitted out of the specimen due to backscattering and secondary electron emission surpasses those supplied by the beam. Increase in the production of total electron coefficient within the escape depth of the electron beam results primarily due to an increase in secondary electron coefficient δ at low electron beam energy. Further reduction in the beam energy produces a peak in total electron coefficient followed by a decrease back to unity (lower crossover energy E_1 in Fig. 3.30a) and below. At beam energy range between E_1 and E_2, the specimen will not charge as the total electron yield exceeds the beam current entering the specimen. For values of incident energy above E_2, the total yield becomes less than 1 because the beam penetrates too deep into the specimen restricting ejection of BSE and SE, resulting in charge buildup. For values of incident energy below E_1, the total yield becomes less than 1 because the primary beam lacks enough energy to create SE again resulting in negative charging.

The upper crossover energy E_2 represents a suitable beam energy point for imaging. It is in the range 0.5–2 keV for organic materials and 2–4 keV for inorganics. Insulators can produce very high total electron coefficients at low beam energies. Values of E_2 have been tabulated for various materials in Table 3.3.

At low beam energy, the SE_2 electrons are generated closer to the beam impact point decreasing the range of SE_2 signal. SE_2 spatial distribution approaches that of SE_1 under such conditions. At low beam energy, SE_1 and SE_2 signals cannot be spatially separated even at high magnifications (i.e., with small pixel size). Due to their close proximity to the beam impact point, both types of signals now serve to

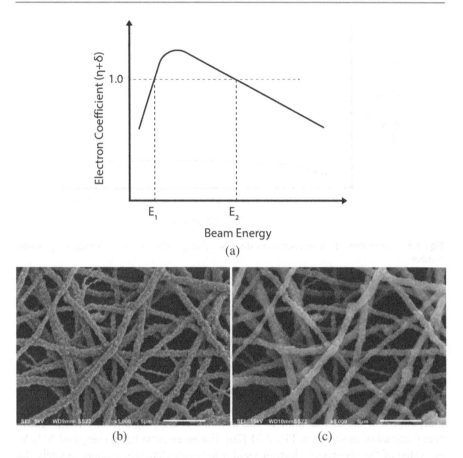

Fig. 3.30 (a) Plot illustrating the change of total emitted electron coefficient ($\eta + \delta$) with beam energy. Upper crossover energy point E_2 represents optimal beam energy to image materials. It is in the range 0.5–2 keV for organic and 2–4 keV for inorganic materials. (b) SEM image at low beam energy showing surface features. (c) Same sample: some features are lost at high beam energy

Table 3.3 Upper crossover energy E_2 for various materials

Material	E_2 (keV)	Reference
Electron resist	0.55–0.70	[24]
5% PB7/nylon	1.40	[24]
Acetal	1.65	[25]
Polyvinyl chloride	1.65	[25]
Teflon	1.82	[25]
Glass passivation	2.0	[24]
GaAs	2.6	[24]
Quartz	3.0	[24]

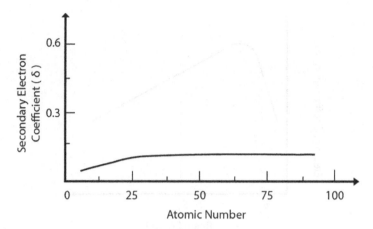

Fig. 3.31 Secondary electron coefficient does not change appreciably with increasing atomic number

make up the high-resolution image. Use of low beam energy, therefore, allows exerting better control on interaction volume and image contrast and facilitates the acquisition of high-resolution images rich in near-surface information (see Fig. 3.30b, c). The probe size, however, will be coarser at low beam energies. Aberration correction is used to overcome deleterious effects of low beam energy.

3.4.2.6 Effect of Atomic Number on SE Yield

For practical considerations, the secondary electron emission can be considered to be independent of the atomic number and does not change significantly from light to heavy elements, as shown in Fig. 3.31 [26]. For an electron beam energy of 20 keV, the value of the secondary electron yield δ for most elements is approximately 0.1 [26]. However, gold and carbon are not constrained by this value. Gold has a significantly high value of secondary electron yield of 0.2, whereas carbon has a low yield value of around 0.05. The secondary electron emission is greatly dependent on the surface condition of the specimen. Contamination at the specimen surface can pose a significant hindrance to the ejection of low-energy SE from within the target material.

3.4.2.7 Effect of Tilt on SE Yield

When the specimen is tilted at increasing angles θ, secondary electron coefficient δ increases obeying a mathematical relation that can be approximated as follows [27]:

$$\delta(\theta) = \delta_0 \sec \theta \qquad (3.21)$$

where δ_0 represents the δ value at zero tilt (i.e., when the beam is perpendicular to the specimen surface). This relationship is expressed using the schematic shown in Fig. 3.32a, b. At 45° tilt, the primary electron beam path length d increases to $d = d_0 \sec \theta$, while the escape depth of SE, d_0, remains the same. Increased path

3.4 Types of Contrast

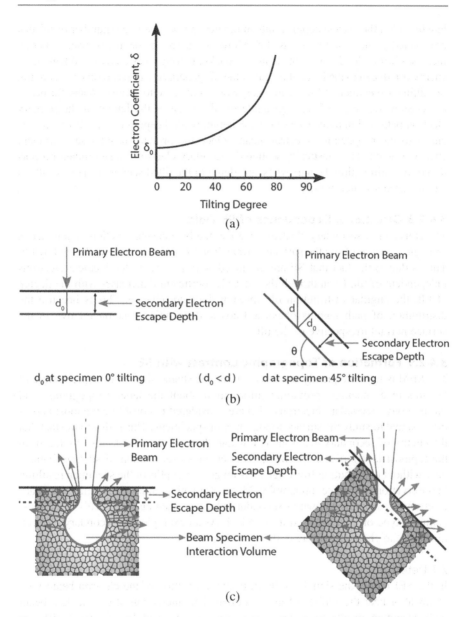

Fig. 3.32 (a) Plot showing increase in δ with tilt follows a secant relationship. (b) Specimen tilt increases the beam path length within a distance of escape depth. This results in the production and escape of a larger number of SE. (c) A large part of the interaction volume becomes close to the specimen surface when it is tilted making it more probable for the SE to escape

length within the escape depth results in the generation of a larger number of SE that are ejected giving rise to increased δ. Moreover, since the degree of backscattering increases with tilt, SE generated due to backscattering also increase. When beam energy is reduced to the level of E_2, all of the SE generated can escape even at zero tilt. So, tilting a specimen at low beam energy does not serve to increase δ any further.

As shown in Fig. 3.32c, high angle of tilt increases the length of the primary electron path within a distance of escape depth, while keeping the specimen surface closer to the propagating electron beam. Once the primary beam loses sufficient energy to generate inelastically scattered secondary electrons, a significant proportion of it is still within the narrow escape depth of the tilted specimen. This results in an increased secondary yield.

3.4.2.8 Directional Dependence of SE Yield

The emission of secondary electrons is governed by a cosine function of ϕ which is an angle measured from the surface normal. The SE yield remains unaffected by tilt. This is due to the fact that SE are produced isotropically by the incident electrons independent of tilt. Even though the total SE coefficient δ increases with the degree of tilt, the angular distribution of emission does not change. This is because the distribution of path lengths obeys a $1/\cos \phi$ distribution relation relative to the surface normal irrespective of the tilt.

3.4.2.9 Formation of Topographic Contrast with SE

The SEM is most frequently used to examine the shape, size, and surface texture of features in a specimen providing information about the latter's topography and morphology. Secondary electrons (SE) are considered to constitute the most appropriate signal to study the surface topography of specimens. This is due to the fact that the energy of secondary electrons is low and these electrons are only emitted from the topmost layers of the specimen. This enables examination of surface features only without contribution from signals from greater depths of the specimen resulting in observation of surface topography. Mode of imaging undertaken using secondary electrons is generally known as secondary electron imaging (SEI). This is the most common type of imaging used in the SEM. As an example, the secondary electron image of the charcoal sample is shown in Fig. 3.33.

E-T Detector

In the SEI mode, the signal results from the interaction of the electron beam with atoms at or near the surface of the specimen. The interaction of the incident beam with specimen results in inelastic scattering events, and low-energy (≈ 50 eV) secondary electrons are ejected from the loosely bound outermost shells of specimen atoms. Due to their low energy, electrons located within a few nanometers of the sample surface are able to escape and are used to generate topographic contrast. After their ejection from the specimen surface, the electrons are collected using the Everhart-Thornley (E-T) detector which is scintillator-photomultiplier system placed inside the SEM chamber at an angle to the specimen surface. The strength of the signal depends on the number of SE reaching the detector. This is why surface

Fig. 3.33 Secondary electron image of charcoal dipped in colloidal Au showing surface topography

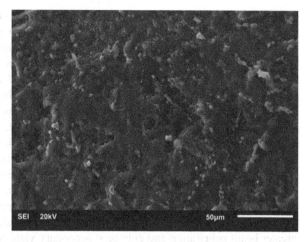

features facing toward the direction of the E-T detector generate higher contrast compared to those facing away from it. Low-energy secondary electrons are pulled toward the detector under a positive bias (+200–300 V) in the form of a wire mesh/grid/collector (Faraday cage) placed at the front end of the detector. The electrons pass through the grid and strike the scintillator surface of the detector. The secondary electrons reaching the detector possess the energy of only a few electron volts (eV) which is not adequate to generate a viable signal from the scintillator. The scintillator surface made of a plastic or crystalline material is therefore covered by a thin metal coating that is under a + 10 kV positive bias which serves to accelerate the incoming electrons onto scintillator. The electrons are converted into light (e.g., cathodoluminescence) upon striking the scintillator surface. The light then passes through a transparent pipe to reach a photomultiplier tube which converts the light into an electric signal that is amplified to levels suitable for the formation of an image. It is then converted into a digital signal and displayed on a monitor as a two-dimensional intensity distribution map. E-T detector is commonly referred to as the SE detector as the images produced are dominated by SE contrast when the collector is positively biased. Working principle of E-T detector is discussed in detail in Sect. 2.6.1.

Factors Affecting Topographic Contrast

The three-dimensional appearance of SEM images is due to differences in contrast between various structural features of the specimen when they are displayed on the viewing monitor. Contrast arises when different parts of the specimen generate varying amounts of secondary electrons when the electron beam strikes them. Areas which generate large numbers of secondary electrons will appear brighter than areas that generate fewer secondary electrons. The yield of secondary electrons by various areas may be influenced by several conditions. The orientation of the specimen topography relative to the electron beam and secondary electron detector greatly affects the yield of secondary electrons. Tilted features in a specimen yield more SE as secondary electron coefficient δ increases as a secant function of tilt θ. Yield of SE from tilted surfaces is not significantly different in various directions, so the trajectory component in the SE signal is negligible. Backscattered electrons also find their way into the E-T detector to contribute to the SE image. The Everhart-Thornley detector will receive line-of-sight BSE from features of specimen facing the detector. These BSE will form part of the SE image. In addition, tilted surfaces in a specimen will generate a higher number of BSE as backscatter coefficient η increases significantly with the angle of tilt. This will, in turn, increase the SE_2 signal. Moreover, titled features will generate BSE in a direction defined by the beam direction and surface normal of the feature. This adds a trajectory component to the SE image. Elements with high atomic number elements have a higher yield of backscattered electrons compared to elements with lower Z. This adds a number component to the topographic contrast, and high-Z elements therefore, may appear brighter in the SE image. High accelerating voltage results in lower contrast due to greater beam penetration and enhanced secondary yield from all parts of the specimen topography. Contrast can be enhanced by using lower accelerating voltage. Charge accumulation or buildup in partly coated areas of the specimen can also result in an increase in contrast. Uncoated areas tend to build up a static charge from the electron beam and cannot dissipate the charge rapidly enough. This may cause the deflection of the beam so that it strikes other areas to generate an excessive number of secondary electrons that form part of the image contrast. Likewise, naturally magnetic areas in a specimen may either deflect or attract the beam to affect the yield of SE. When crystals are oriented along certain lattice planes relative to the beam, an enhanced yield of SE may result in an increase in brightness along these lattice planes so that certain crystals will appear much brighter than others. In short, there is a large number of factors that influence the formation of contrast in SE image.

Edge Effect

Secondary electrons are emitted from all features that interact with the incident electron beam. However, the strength of the signal obtained due to secondary electrons depends greatly on the orientation of surface features of the specimen with respect to the incident beam. Greater variation in the orientation of features (i.e., higher surface roughness) results in greater variation in secondary electron signal from one location of the specimen to the other giving rise to higher topographic contrast. The number of secondary electrons emitted from specimen surface changes with the angle of incidence of the electron probe on that surface. If the incident beam is perpendicular to the surface, the interaction volume within the specimen is uniform about the beam axis resulting in the generation of a certain number of electrons. Higher incidence angle allows greater penetration of the beam into surface regions such that the escape distance toward one side of the beam decreases, and the number of secondary electrons emitted from the specimen increases. Higher secondary electron emission results in a brighter contrast. This is why thin raised areas, steep surfaces, protrusions, and edges within a specimen where interaction volume intercepts these regions tend to appear brighter compared to the broad flat surfaces. This phenomenon is called *edge effect*

3.4 Types of Contrast

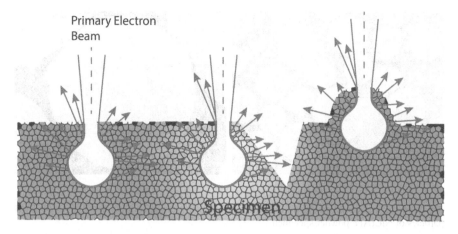

Fig. 3.34 Schematic showing the phenomenon of the edge effect. Edges, steps, and protrusions in a specimen appear bright due to generation and escape of a larger number of secondary electrons from these regions

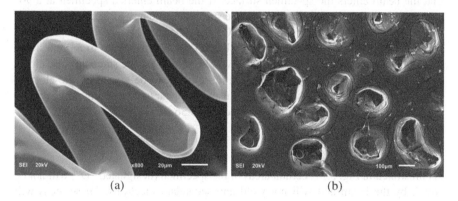

Fig. 3.35 Secondary electron image showing bright edges of (**a**) coil and (**b**) holes in carbon tab samples, due to the emission of a large number of SE

since it takes place along sharp edges in a specimen. These areas appear brighter because the secondary electrons are able to escape from all sides of the thin areas. Schematic in Fig. 3.34 shows this effect for various edge shapes.

The higher the accelerating voltage used during imaging, the greater the edge effect. This phenomenon is more pronounced when an uneven surface such as coil sample is examined in the SEM. The edges of the coil and holes in carbon tab samples appear brighter than the rest of the material as shown in Fig. 3.35a, b, respectively.

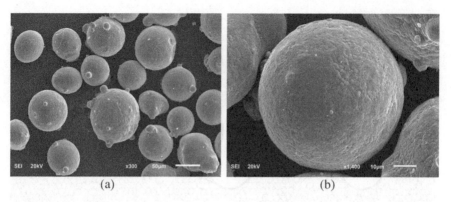

Fig. 3.36 (**a, b**) Secondary electron SEM images showing strong contrast at the edges of round particles of NiCrAlY powder specimen

Spherical Particles

As mentioned earlier, the yield of secondary electrons is also affected by the angle that the beam enters the specimen surface. If the beam enters a specimen at a 90° angle, the beam penetrates directly into the specimen, and any SE generated below a certain depth will not be able to escape. On the other hand, if the beam strikes the specimen in a grazing manner, then the beam does not penetrate to a great depth (along the surface normal), and more SE will be able to escape since they are closer to the surface. Since rounded objects are more likely to be grazed by the electron beam than would flat objects, round areas usually appear to have a bright line around them due to the enhanced yield of SE (see Fig. 3.36a, b).

Non-regular Specimens

As illustrated in Fig. 3.37, certain areas of the specimen (designated "D") will not be struck by the beam and will not yield any secondary electrons. These areas will appear dark in the image. Areas such as "I" in Fig. 3.37 will be struck by the beam, but since they face away from the detector, fewer secondary electrons will be collected and intermediate levels of brightness will be displayed in the image. Optimal yields of secondary electrons would come from areas that are struck by the beam and face the detector ("B" in Fig. 3.37). These areas would appear as highlights in the image.

Effect of Lateral Placement of E-T Detector

The contrast in the SE image partly depends on the direction where the E-T detector is placed as shown in Fig. 3.38. The features of the specimen facing the detector will appear brighter as the collection of SE from that region will be the maximum. Features facing away from the detector shall appear less bright. Organic materials will appear dark due to low SE yield.

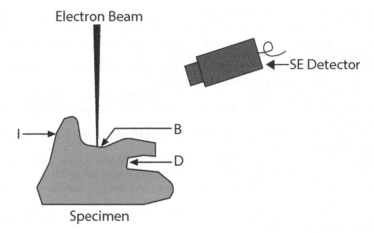

Fig. 3.37 Schematic diagram showing the effect of the irregular-shaped specimen on the SE image contrast [28]

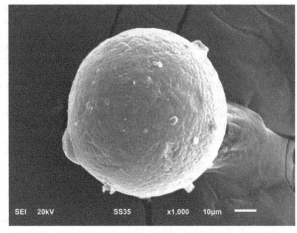

Fig. 3.38 Secondary electron SEM image of NiCrAlY powder particle placed over a C tab. E-T detector is located toward the right-hand side of the specimen. It can be seen that the surface facing the E-T detector appears bright. The C tab appears dark indicating organic material

References

1. Newbury DE, Myklebust RL (1979) Monte Carlo electron trajectory simulation of beam spreading in thin foil targets. Ultramicroscopy 2(9):391–395
2. Sim SK, Teh V (2015) Image signal-to-noise ratio estimation using adaptive slope nearest-neighbourhood model. J Microsc 260:352–362. https://doi.org/10.1111/jmi.12302
3. Frank J, Al-Ali L (1975) Signal-to-noise ratio of electron micrographs obtained by cross correlation. Nature 256:376–379. https://doi.org/10.1038/256376a0
4. Erasmus J (1982) Reduction of noise in TV rate electron microscope images by digital filtering. J Microsc 127:29–37

5. Yeap ZX, Sim KS, Tso CP (2016) Signal-to-noise ratio estimation technique for SEM Image using linear regression, Proc. of 2016 international conference on robotics, automation and sciences (ICORAS) 5–6 Nov. 2016. https://ieeexplore.ieee.org/stamp/stamp.jsp?tp=&arnumber=7872602. Date accessed 19 Sept 2018
6. Thong JT, Sim KS, Phang JC (2001) Single-image signal-to-noise ratio estimation. Scanning 23:328–336. https://doi.org/10.1002/sca.4950230506
7. Sim KS, Thong JTL, Phang JCH (2006) Effect of shot noise and secondary emission noise in scanning electron microscope images. Scanning 26:36–40. https://doi.org/10.1002/sca.4950260106
8. Henoc J, Maurice F (1991) In: Heinrich KFJ, Newbury DE (eds) Electron probe quantification. Plenum Press, New York, p 105
9. Hovington P, Drouin D, Gauvin R (1997) CASINO: A new Monte Carlo code in C language for electron beam interaction - Part I: Description of the program. Scanning 19(1):1–14
10. Kanaya K, Okayama S (1972) Penetration and energy-loss theory of electrons in solid targets. J Phys D 5:43
11. Reimer L (1998) Scanning electron microscopy: physics of image formation and microanalysis, 2nd edn. Springer, Heidelberg
12. Reuter W (1972) In Proc. 6th international congress on X-ray optics and microanalysis, Shinoda G, Kohra K, and Ichinokawa T (ed), University of Tokyo Press, Tokyo, p. 121
13. Heinrich KFJ (1966) In Proc. 4th international conference on X-ray optics and microanalysis, Castaing R, Deschamps P, and Philibert J (eds.), Hermann, Paris, p. 159
14. Goldstein JI, Newbury DE, Joy DC, Lyman C, Echlin P, Lifshin E, Sawyer L, Micheal JR (2003) Scanning electron microscopy and X-Ray microanalysis, 3rd edn. Springer, New York
15. Invitation to the SEM World JEOL Ltd. Publication, Tokyo, Japan. https://wiki.nbi.ku.dk/w/cleanroom/images/b/b5/Invitation_to_the_SEM_World.pdf. Date accessed 19 Sept 2018
16. Joy DC (1998) Scanning electron microscopy. In: Amelinckx S, van Dyck D, van Landuyt J, van Tendeloo G (eds) Handbook of microscopy-applications. VCH, Weinheim
17. Ball MD, Wilson M, Whitmarsh S (1987) In: Brown LM (ed) Electron microscopy and microanalysis 1987. Institute of Physics, London, p 185
18. Tixier J, Philibert R (1969) Effets de contraste cristallin en microscopie électronique à balayage. Micron (1):174
19. Seiler H (1967) Einige aktuelle Probleme der Sekundarelektron-emission. Z Phys 22:249–263
20. Koshikawa T, Shimizu R (1974) A Monte Carlo calculation of low-energy secondary electron emission from metal. J Phys D Appl Phys 7:1303
21. Drescher H, Reimer L, Seidel H (1970) Rückstreukoeffizient und Sekundärelektronen-Ausbeute von 10–100 keV-Elektronen und Beziehungen zur Raster-Elektronenmikroskopie. Z Agnew, Phys 29:331–336
22. Loretto MH (1984) Electron beam analysis of materials. Chapman and Hall, London
23. Reimer L, Tollkamp C (1980) Measuring the backscattering coefficient and secondary electron yield inside a scanning electron microscope. Scanning 3(1):35–39
24. Joy DC (1987) A model for calculating secondary and backscattered electron yields. J Microsc 147(1):51–64
25. Vaz OW, Krause SJ (1986) In Proc. EMSA Conference Bailey GW (ed), San Francisco Press, San Francisco, p. 676
26. Wittry DB (1966) Proc: 4th international conference on X-ray optics and microanalysis, Castaing R, Deschamps P, Philibert J (eds) Hermann, Paris, p. 168
27. Kanter H (1961) Energy dissipation and secondary electron emission in solids. Phys Rev 121(3):677–681
28. Bozolla JJ (1999) Electron microscopy, principles and techniques for biologists, 2nd edn. Jones and Bartlett Publishers, Burlington

Imaging with the SEM

4

The scanning electron microscope is routinely used to characterize wide-ranging materials due to its ease of operation and relatively straightforward sample preparation as well as due to simple image interpretation. New users can readily obtain images after little practice. However, high-resolution microscopy and examination of "difficult" samples require experience and know-how of the principles of image formation in the SEM. This chapter describes the role of various operational parameters used during microscopy in more detail. The effect of these parameters on contrast, resolution, and depth of field depicted by images is discussed. Pros and cons of microscopy conditions that have a direct bearing on the quality of images, type of information obtained, and image interpretation are elaborated. Guidelines for operation and upkeep of the SEM instrument are also summarized in this chapter.

4.1 Resolution

Theoretical resolution in a perfect optical system is given by Abbe's equation [1]:

$$d = 0.612 \frac{\lambda}{n \sin \alpha} \quad (4.1)$$

where d is the resolution, λ is the wavelength of the radiation used for imaging, n is the index of refraction of the medium between the point source and lens, and α is the half-angle of the cone subtended by the specimen plane to the objective lens in radians. The term $n \sin \alpha$ is known as numerical aperture, NA. It is not possible to focus light perfectly due to diffraction and interference effects. Diffraction changes parallel wave front that interacts with an aperture into a spherical wave front. Similarly, light gets focused not as a spot but as a set of concentric circles with

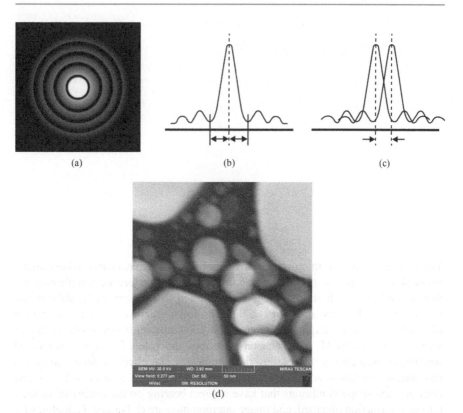

Fig. 4.1 Illustration of resolution in (**a**) airy disc and (**b**) wave front form where resolution is the distance between the first-order peak and trough. (**c**) When the distance between the first-order peaks from two points is smaller than the distance between first-order peak and trough, the points are not resolved. (**d**) An example of a high-resolution SEM image taken at a magnification of ×2,000,000. (Image courtesy of TESCAN)

diminishing intensity due to interference effect. This is known as the *airy disc* as shown in Fig. 4.1a. Wave front form is shown in Fig. 4.1b, c. An example of a high-resolution SEM image is shown in Fig. 4.1d.

The resolution is the distance between the first-order peak and the first-order trough as indicated by the width of the region bound by arrows in the waveform shown in Fig. 4.1b. If the resolution is defined as the minimum spacing between two features at which they are recognized as separate and distinct, then two features are not distinguished if the first-order peaks generated by them are separated by a distance smaller than the distance between the first-order peak and trough (see Fig. 4.1c). Lack of resolution will make them appear as a single feature. This can be considered as the resolution of an optical system that is free of lens aberrations and is limited by diffraction effects only.

4.1.1 Criteria of Spatial Resolution Limit

The ability of the SEM to distinguish fine details of a specimen is determined by its spatial resolution. Several definitions or criteria have been described for spatial resolution limit, and these are summarized in the literature [2–6]. Some of these criteria are listed below.

4.1.1.1 Rayleigh Criterion

According to the Rayleigh criterion, two point sources are resolved if the central maxima of the diffraction pattern generated by one point coincides with the first zero of the diffraction pattern generated by the second point source. Effectively, the Rayleigh's resolution limit is the distance between the first zero and the maxima of point spread function of the system. This criterion can be extended to points with non-zero sources around it. This is done by describing the resolution limit as the distance for which the intensity at the central dip in the composite image is 81% of the maximum. This criterion is based on presumed resolving abilities of the human eye.

The Rayleigh limit or the point resolution of an imaging system is given by the following equation:

$$p(r) = \frac{1}{2\pi\rho^2}\exp\left(-\frac{x^2+y^2}{2\rho^2}\right) - \frac{1}{2\pi\rho^2}\exp\left(-\frac{r^3}{2\rho^2}\right) \tag{4.2}$$

where ρ is the width of the Gaussian function, r is the absolute value of a two dimensional vector $t = (x\ y)^T$, T is for the transposition, $\rho_{p'}$ is the smallest distance over which two points can be resolved, which is given by the necessity that the value of the composite intensity distribution located at half-way between two points is 0.8 times the value of the maxima,

$$2\exp\left[-\frac{\rho_p^2}{8\rho^2}\right] = 0.8 \tag{4.3}$$

which results in

$$\rho_p \approx 2\sqrt{2}\,\rho \tag{4.4}$$

The definition proposed by the Rayleigh criterion is applicable to point-like specimen structures which are smaller than the diameter of the probe. Due to this reason, they show separation in the distribution inside the range of the order of the spot size. Resolution limits of 10–15 nm were obtained when using probe diameter of 5–7.5 nm with a LaB_6 electron gun. Resolution of 5 nm was obtained with a probe diameter of 3 nm. The distance between two points is taken to be larger than the electron probe diameter.

4.1.1.2 Sparrow Criterion

According to sparrow criterion, the smallest distance that can be resolved is when the minimum in the composite image intensity distribution just disappears. Using this criterion:

$$\rho_s = \frac{\sqrt{2}\,\rho_p}{2} \tag{4.5}$$

4.1.1.3 Schuster's Criterion

Shuster's criterion states that two point sources can be resolved when no fraction of the central diffraction pattern of one point coincides with the central diffraction pattern of the other. This criterion gives a resolution limit which is twice as large as the Rayleigh criterion.

4.1.1.4 Houston Criterion

Two point sources are resolved if the distance between the central maxima of two points is equal to the half maximum of the diffraction pattern of any one of the point sources.

4.1.1.5 Buxton Criterion

This is similar to the Rayleigh criterion, but instead of intensity, the criterion is based on amplitude diffraction pattern. Amplitude diffraction pattern is the square root of the intensity diffraction pattern. According to this criterion, two points are resolved when the components of the amplitude patterns intersect their points of inflection.

4.1.1.6 Edge Resolution

If an electron beam scans at an angle perpendicular to a sharp edge, the transmitted intensity becomes a smoothed step function.

$$I(x) = \int_{-\infty}^{x} \left[\int_{-\infty}^{\infty} j\left(\sqrt{x^2 + y^2}\right) dy \right] dx \tag{4.6}$$

The distance between points that correspond to 0.25–0.75 fraction of the maximum intensity can be defined as the edge resolution.

4.1.1.7 Radial Intensity Distribution

The radial intensity can be defined as:

$$I(r) = \int_0^r \frac{j(r)\, 2\,\pi\, r}{I_p} dr \tag{4.7}$$

With this equation, the size of the electron probe or the resolution can be defined as the diameter of the circle that contains 50–75% of the total probe current, I_p.

4.1.1.8 Maximum Spatial Frequency

The maximum spatial frequency is given as,

$$q_{max} = \frac{1}{T_{min}} \quad (4.8)$$

The spatial resolution is defined by the maximum spatial frequency given above, at which the contrast level drops so low that the periodicity T_{min} cannot be detected by the micrograph.

Other advanced theories about the resolution limit have been proposed; an in-depth review can be found in the works of Dekker and Bos [2].

According to the Rayleigh resolution criterion, the ability to distinguish between two closely spaced objects is linked to the wavelength of its illumination [3]. Since the wavelength of the electrons is much less than 1 Å, the theoretical resolution of an electron microscope is much higher than that of an optical microscope. The wavelength of electrons is given by De Broglie equation:

$$\lambda = \frac{h}{mv} \quad (4.9)$$

where λ is wavelength, h Planck's constant (6.6×10^{-27} J seconds), m mass of the particle (9.1×10^{-28}), and v velocity of the particle.

The energy of an electron is:

$$ev = \frac{1}{2} mv^2 \quad (4.10)$$

where ev is energy in electron volts ($e = 4.8 \times 10^{-10}$), m mass of the particle, and v velocity of the particle.

By substituting the values of h and m in Eq. 4.2, the wavelength λ can be expressed in terms of the accelerating voltage V as follows:

$$\lambda = \frac{1.23 \text{ nm}}{\sqrt{V}} \quad (4.11)$$

The wavelength of electrons at an accelerating voltage of 30 kV is 0.0071 nm. Above value for λ is substituted into Abbe's equation. Beam convergence angle α is small in electron microscopy; so $\sin\alpha$ is replaced by α. The value of refractive index n is taken as 1. Therefore, the equation for the resolution limit of the SEM can be given as:

$$d = \frac{0.753}{\alpha \sqrt{V}} \quad (4.12)$$

where d is the resolution in nm, α is half aperture angle, and V is accelerating voltage. At electron beam energy of 30 keV and α of 0.01 radian, the theoretical resolution limit is calculated to be 0.435 nm. At lower beam energy, it will deteriorate. At 1 keV, the limit is 2.38 nm. However, due to the presence of aberrations in electromagnetic lenses used for focusing and image formation, the attained resolution is relatively inferior. The main challenge is to be able to focus the beam into a small probe. At low accelerating voltage, electron charge repulsion tends to widen beam diameter. Use of advanced technology such as beam monochromators, beam deceleration, and immersion optics, where specimen in immersed in the magnetic field of the lens, has enhanced the attainable resolution considerably. Commercial manufacturers have claimed to attain a resolution of 0.5 nm at 30 kV and 0.9 nm at 1 kV with the SEM.

4.1.2 Imaging Parameters That Control the Spatial Resolution

The spatial resolution of the SEM is controlled by manipulating four important parameters of the electron beam, namely, probe size, beam current, convergence angle, and accelerating voltage.

4.1.2.1 Probe Size

Spatial resolution of an SEM is primarily dependent upon the probe size of the beam that falls on the specimen. Secondary electrons do not have sufficient energy to travel distances larger than 10 nm in solid specimens. Only those close to the surface are able to escape. These SE_1 electrons are stimulated from the vicinity of the beam. The spatial resolution of an image is determined by the SE_1 signal which is dependent on the beam diameter or the spot size on the specimen. The resolution of the SEM cannot be better than the diameter of the probe; however, it can be worse depending upon the level of magnification used.

It can be seen in Table 3.1 that the size of the picture element on the specimen at a magnification of 100,000× is 1 nm only. If the probe size of 1 nm is achieved in the SEM through careful use of operational parameters, it is still likely that signal delocalization will produce a much larger excitation volume within the specimen (see Sect. 3.1.2). Materials with a lower atomic number will produce large excitation volumes. Once the sampling region becomes larger than the picture element, signals from adjacent picture elements overlap, and the image may appear blurred. Thus the ability to resolve fine features is affected not only by the probe size but also by the degree of delocalization of the probe within the specimen, the magnitude of which depends on the beam energy and the atomic number of the material examined. However, it is a common observation that features are actually resolved in the SEM at much higher magnifications than mentioned above. This is due to the fact that the high-resolution SE_1 signal is superimposed on low-resolution (SE_2, SE_3, and BSE) signal when detected using E-T detector, thereby making it possible to resolve fine objects.

So, to obtain a high-resolution image, the spot size should be kept as small as possible, in order to resolve the smallest feature of the specimen. This is achieved by using a suitable choice of operating parameters.

4.1 Resolution

Table 4.1 Comparison of the electron source and final spot size of W, LaB$_6$, and field emission guns

	Thermionic, W	Thermionic, LaB$_6$	Schottky FE	Cold FE
Source diameter	15 μm	5 μm	15 nm	2.5 nm
Spot diameter	7 nm	4 nm	1 nm	1 nm

The spot size is reduced by decreasing the working distance or using the lenses in the SEM. These lenses help demagnify the beam emanating from the electron gun. It is helpful if the diameter of the beam emanating from the gun is small to start with, i.e., the source radius is small. Field emission guns have small source size and produce fine probes as seen in Table 4.1.

Small probe size while enhancing the resolution also decreases the probe current. To resolve two points of a specimen, there must be a distinct difference between the signals emanating from these points to produce contrast.

The contrast level C in a specimen in terms of critical current I_c is given by Rose criterion [7–9]:

$$I_c > \frac{4 \times 10^{-12}}{q \times F \times C^2} \tag{4.13}$$

where q is the detector efficiency and electron yield product F is the frame scan time.

4.1.2.2 Beam Current

It can be seen that for a given detection system, at a fixed scan time, there is a critical beam current required to observe a particular contrast level, beyond which it is difficult to distinguish the signal from the noise.

The spot size for a thermionic emission filament can be calculated as:

$$d = d_{\min} \left[7.92 \times 10^9 \left(\frac{I\,T}{j}\right) + 1 \right]^{\frac{3}{8}} \tag{4.14}$$

where T is the temperature and j is the current density at the surface of the filament.

$$d_{\min} = K \, \lambda^{\frac{3}{4}} \, C_s^{\frac{1}{4}} \tag{4.15}$$

From the two equations, it can be seen that to properly discern the features in an SEM, a minimum current is required which corresponds to a minimum spot size (see Sect. 2.4.4.5). However, as suggested by Eq. 2.3 for any given beam energy, smaller currents result in smaller probe sizes. For this reason, it is common practice to employ currents in the order of tens of picoamps during high-resolution imaging.

4.1.2.3 Convergence Angle of the Probe

Probe convergence angle is defined as the half-angle of the cone of electrons converging onto the specimen. The spot size can also be reduced by manipulating the convergence angle. One way is to use apertures which prevent off-axis electrons from passing through; the lens then focuses the beam into smaller spot size. This also reduces the probe current as the number of electrons allowed to pass through decrease. Another way is to manipulate the working distance; having a small working distance increases the angle of convergence, thereby decreasing the spot size and increasing the resolution. Large working distance increases lens aberrations and needs strong lenses to help focus the image. Small working distance decreases the probe size increasing the resolution, but it also reduces the depth of field in an SEM image.

4.1.2.4 Accelerating Voltage

High accelerating voltage produces high brightness and smaller spot size. The highest resolution is achieved at high beam energies. For samples with a high atomic number, the high beam energy is ideal as interaction volume within the specimen stays within acceptable limits. Similarly, for thin samples (with small interaction volume), such as nanoparticles, use of high beam energy produces high-resolution images. However, for the bulk sample with a low atomic number that exhibits high excitation volume, generation of SE_2, SE_3, and BSE signals adds to the noise component. This contributes to the lowering of the signal-to-noise ratio (SNR). However, high brightness at high beam energy serves to compensate for the lowered SNR. The inclusion of SE_2, SE_3, and BSE in the total signal serves to degrade the spatial resolution. For low Z materials, therefore, the low beam energy is preferred in order to reduce the generation of low-resolution SE_2, SE_3, and BSE signals.

Low voltage microscopy is a common technique to get high-resolution images from the SEM. The voltage used in this technique is generally in the order of 500 V to 5 kV. Use of low voltage decreases the interaction volume of the beam with the specimen (i.e., electron range decreases as a function of $E_0^{1.67}$; see Sect. 3.2.6). Due to this, the SE_2 signals that emanated from a wider volume now emanate from a smaller volume close to that of the SE_1 signal. Typical spatial distribution of SE_1 and SE_2 signal is shown in Fig. 3.29. Decreasing the beam energy serves to localize generation of the SE_2 signal. The SE_2 signals now carry more relevant information from the vicinity of the beam footprint. This enhances the resolution as now most of the SE signal carries high spatial resolution information. This also eliminates the need of separating SE_1 signal from the SE_2 signal as it carries useful information rather than the background information. Consequently, the performance of the imaging technique is improved, and better resolution and contrast are attained in SE images. This also enhances the resolution of the BSE, which now gives resolution similar to that for SE imaging.

The SE signal at low beam energies increases more rapidly than the decrease in brightness at low voltage. Due to this the signal-to-noise ratio remains good even up to voltages of 500 V. Low voltage imaging also enhances the imaging for intermediate and high atomic number specimens, because the range of signal for strongly scattering materials decreases significantly at lower beam energies.

There are certain disadvantages to using low beam energy imaging technique. Firstly, the electron source brightness which is proportional to the beam energy

4.1 Resolution

reduces linearly with accelerating voltage. Brightness is given by Eq. 2.3 where it can be seen that imaging at very low voltage might cause the beam current to fall below the threshold current for any useful contrast. This directly affects the feature visibility. At 500 eV the field emission gun has the same brightness as that of a tungsten emitter at 20 keV. The decrease in beam current is usually compensated by an increase in the SE signal. The performance of the SEM also degrades at a lower voltage due to chromatic aberration which is a result of the energy spread of the gun as can be seen from Eq. 2.10. This also causes the profile of the beam to change shape producing a broad skirt of intensity about the center of the probe, resulting in a reduction of the image contrast. These can be corrected by the use of beam monochromator which reduces the energy spread of the beam. Low-energy beams are also prone to degradation/deflection of stray electromagnetic fields, which necessitates the use of short working distances. Hence, cold field emission guns equipped with a snorkel or immersion lens are ideal for low voltage imaging. Another method of tackling the brightness problem at low accelerating voltage is the use of negative bias on conducting specimens. This slows down the incoming electrons from the beam, resulting in lower penetration; but also the number of electrons in the beam remains high, thus improving the brightness.

Another factor that can seriously impact low voltage microscopy is contamination of the specimen surface. The imaging is carried out at such low energies that signals from contamination might dominate over the actual signal from the specimen surface. Cleaning of specimen surface is therefore important which can be done with plasma beams. The rate of contamination buildup can be reduced by:

1. Avoiding high magnification.
2. Focusing and removing astigmatism in the image at an area other than that is essential for imaging.
3. Not using spot or reduced area raster mode.

In biological samples, the contamination grows due to the field-enhanced mobility of hydrocarbons across the surface of the sample. Using a thin film coating of metals can significantly reduce the growth of contaminants, as it eliminates fields from charged regions. When low atomic number specimens like biological samples are to be imaged, the lower amount of SE_1 signal makes the imaging difficult, due to low signal-to-noise ratio, which results in dark images. The signal-to-noise ratio can be improved by application of ultrathin coatings of metal. Such coatings have a thickness of 2–3 nm, to prevent distortion of surface features. Metals such as gold-platinum alloy, platinum, and tungsten pure metals are used to coat low atomic number/biological specimens. Such coatings can be easily deposited on specimens using sputter coater. At high magnifications such as 100,000×, grains are very fine and difficult to image. At high magnification, the individual particles can be difficult to distinguish from one another, due to which artifacts may arise in the images and the image might not be able to capture the actual surface detail. However, the ultrafine coating can also improve focus and make astigmatism defects easy to correct.

4.1.3 Guidelines for High-Resolution Imaging

The spatial resolution of SEM can be improved by adopting the following guidelines:

1. Sample preparation: Deficiencies in the samples prepared in haste are exposed at high magnifications. Poor polishing shows up as glaring scratch marks. Contamination at the sample surface hides surface features and leaves contamination marks in the images. Currently available commercial instruments claim a spatial resolution of 0.5 nm. At this high level of resolution, monolayer(s) of contamination becomes a limiting factor as it starts to hide structural details on the specimen surface. Such small levels of contaminants are unavoidable in the SEM chamber as currently available instruments do not operate under ultrahigh levels of vacuum. Use of thin samples in place of bulk samples can reduce the generation of low-resolution SE_2, SE_3, and BSE signals.
2. Sputtered coatings to reduce charge-up effects: The grain structure of metal coatings such as Au, Au-Pd, Pt, etc., sputtered to reduce charging effects may become visible at magnifications of roughly 100,000×. Coating thickness should be kept to a minimum by reducing the duration of the coating application. Coatings that exhibit fine grains such as those produced from Cr targets can be used. Application of thin coating also increases the secondary electron coefficient, thus contributing to the high spatial resolution.
3. Type of SEM used: Instrument equipped with advanced technology such as field emission gun, through-the-lens detector, immersion lens, beam deceleration, aberration corrector, filtering capability, high vacuum, stable anti-vibration platform, etc. serves to enable and enhance high-resolution imaging. For instance, SEM equipped with the latest lenses and detectors efficiently collects SE_1 electrons and leaves out a low-resolution signal. Entry level equipment with basic features is not used for high-resolution imaging.
4. SEM operation: SEM column should be properly aligned. Apertures should be clean and centered, and astigmatism should be removed completely. Deleterious effects of external factors such as electromagnetic interference, acoustic/electronic noise, floor vibrations, etc. become apparent at increased magnifications and need to be addressed before high-resolution imaging is undertaken [10].
5. Use of small spot size: Small electron beam diameters are responsible for high image resolution. One way to control the size of the electron probe is by varying the strength of the condenser lens (i.e., through the use of spot size). Condenser lens demagnifies the beam emanating from the electron gun. The higher strength of condenser lens results in finer probe size which can result in higher resolution, but at the same time, it lowers the probe current that may result in noisy images. Therefore, small probe size gives high spatial resolution but at the same time reduces the ability to actually distinguish features of the object under observation. For high-resolution imaging, the smallest probe size is desirable as long as enough signal-to-noise ratio is generated to get adequate contrast from the features of interest.

6. Selection of final aperture: Use of small aperture diminishes the size of the probe and also minimizes the spherical aberration. However, it can limit resolution due to diffraction effects. Intermediate aperture suitable for the beam energy and working distance employed should be selected.
7. Good signal-to-noise ratio: The objective here is to maximize the generation of a signal from the specimen. This can be achieved by using more beam current, but this also increases the spot size, so an optimal balance needs to be reached. Another way is to use a slower scan rate. Longer scan time will generate more secondary electrons from the spot being scanned. However, this may also damage the specimen. Fast scan rate can be used in combination with frame averaging.
8. Accelerating voltage: High accelerating voltage produces fine beam probes which in turn can result in high resolution. However, high beam voltage also results in large interaction volume with increased contribution from low-resolution SE_2, SE_3, and BSE signals which degrade the overall quality of images. If high resolution is to be achieved at high beam energy, then microscopy conditions are set in a way that low-resolution signal needs to be separated from the high-resolution SE and BSE signal that originates from near the surface of the specimen.

The alternative is to use low accelerating voltage (i.e., small beam penetration) where both the elastic and inelastic mean free paths rapidly decrease. This limits the interaction volume in the specimen to a point where all the signals are generated from near the surface of the specimen. Low beam energy encourages the generation of high-resolution SE_1 signal from the specimen by confining the interaction volume close to the beam impact point. Under these conditions, low-resolution (SE_2, SE_3, and BSE) signal is restricted. This eliminates the need to separate low- and high-resolution signal as SE_2 is generated from near the beam impact point and becomes part of the high-resolution signal. Surface features become prominent at low beam energy. The drawback is that low accelerating voltage decreases brightness; therefore optimum probe current to produce an acceptable signal-to-noise ratio is required. Low atomic number matrices such as polymers may suffer from low signal-to-noise ratio which is normally overcome by coating the specimen surface with a thin conductive coating. Low accelerating voltage also increases the deleterious effects of chromatic aberration.

The decision to use either high or low beam energy is made by taking into consideration the type of sample and the nature of information required from the specimen. High beam energy can be used for high-density materials where signal delocalization within the specimen is comparatively smaller. It is also suitable for thin specimens. For low Z and bulk materials and where surface features are of prime interest, low beam energy can be employed. Low voltage SEM is used at 500 V to 5 kV or less where use of high brightness field emission gun and a high-resolution lens system makes it possible to get nanometer scale resolution. Small working distance (i.e., 2 mm) is used. Spot size is 2 to 3. Field emission guns and through-the-lens (TTL) detector is normally employed for

high-resolution imaging. Also, beam deceleration technique helps to reduce the beam interaction area within the specimen.
9. Working distance: Use of shortest possible working distance will reduce lens aberrations and result in fine probe size improving resolution. The close proximity of the specimen to the column maximizes SE_1 collection by the TTL detector.
10. Type of signal used for imaging: The highest-resolution signal is SE_1 since it originates from a small area whose diameter is comparable to the probe size itself. SE_1 should be used to achieve the highest possible spatial resolution. SE_2 and SE_3 are generated away from the probe making them unsuitable for high-resolution imaging. BSE signal is produced from a region almost as large as the excitation volume generated within the sample, therefore rendering it unsuitable for resolving fine details at high magnifications. However, if BSE image is required, high-energy BSE are preferable for imaging since they undergo fewer interactions within the sample keeping the sampling volume small. Low-resolution signal can be excluded by using low beam energy and also by employing a through-the-lens (TTL) detector.

4.1.4 Factors that Limit Spatial Resolution

The factors that primarily limit the spatial resolution include probe diameter, size of excitation volume, and poor signal-to-noise ratio. Spatial distribution of SE signal within the specimen ultimately establishes the resolution of the image. Excitation volume depends on the beam energy and probe diameter as well as on the specimen density and feature topography. Attainable resolution is therefore not an instrument constant and can vary with specimen and application. Instrument manufacturers test an ideal standard sample such as small Au particles dispersed on a low Z film to measure spatial resolution. The quoted value shows the capability of the equipment and does not suggest the range of information that might be extracted from various types of samples analyzed in the same instrument. Samples that need to be imaged at low voltage or current as well as low Z samples that show poor SE yield may not achieve this level of resolution. High-resolution capability might be able to spatially resolve two features, but poor signal-to-noise ratio may not provide sufficient contrast necessary to examine the topography of the specimen. When the SEM is used in EDS or environmental mode, the resolution will be limited by the diameter of the excitation volume which can be a few tenths of a micrometer (e.g., large) depending on the density of the sample material and the electron beam energy. Use of high beam current and energy degrades image resolution. In low voltage SEM, the electron range and the escape depth of SE are of comparable size. Under such conditions, the probe size may no longer remain an indication of the measure of the resolution limit. The user, therefore, needs to determine the optimum operating parameters that can extract the required information from a sample at the optimum resolution.

4.2 Depth of Field

One of the most important advantages of SEM is the large depth of field. It is the ability of a microscope to focus different depths simultaneously such that the specimen surfaces at different distances from the lens remain in focus. The sample appears focused not only at the plane of optimum focus but also at some distance above and below it. When a specimen with large depth is observed, focusing the upper region may result in blurring of the lower region and vice versa. In such case, if the range of upper and lower features that are in focus is large, the depth of field is considered to be large [11]. SEMs have the ability to focus large depths simultaneously making it one of the most effective tools for 3-D imaging at the micro- and nano-levels. The ability of the SEM to convey three-dimensional information is largely due to its large depth of field. This feature is especially useful to image as-received rough specimens such as fracture surfaces, corrosion deposits, solids in powder form, etc.

The reason for this large depth of field is the geometry of beam optics as shown in Fig. 4.2 where the electron beam scans the surface of a rough sample at steady focus. The objective lens of SEM focuses the electron beam to a crossover at the plane of optimum focus. During the process of scanning, sample region labeled "a" coincides with the plane of optimum focus and displays the sharpest focus. At this plane, the probe diameter (or more specifically sampling region diameter) is smaller than the sample pixel size. The diameter of the probe increases both above and below this plane due to beam divergence. At some distance above and below the plane of optimum focus, if the diameter of the beam is less than 2× sample pixel size, the plane remains in focus, and any feature within this range will appear focused in the SEM image. Therefore, regions labeled "b" and "c" in Fig. 4.2 located at some distance above and below this plane, respectively, appear focused. Beyond points

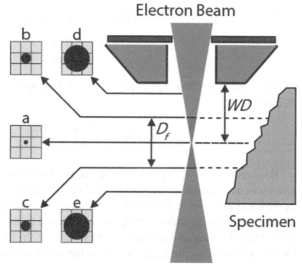

Fig. 4.2 The geometry of beam optics. The probe diameter is overlaid on a pixel at different heights of the specimen to indicate the point where the image goes out of focus. Points (**a–c**) will be in focus while points (**d–e**) will be out of focus Adapted from [12]

"b" and "c," the diameter of the probe becomes larger than 2 pixels. As a result, information from adjacent pixels overlaps and the image becomes out of focus (see Sect. 3.1.2). Regions labeled "d" and "e" will be out of focus because the probe diameter that scans these regions will be 2× pixel diameter at the selected magnification. At these regions, signals from adjacent pixels will overlap to create blurriness in the image.

The probe size within the distance labeled "D_f" in Fig. 4.2 will remain adequately small to be able to focus regions of the sample that coincide with the probe. This distance along the vertical height of a sample where all features are in sharp focus concurrently is called the *depth of field*, D_f. Features remain in focus as long as the probe diameter is <2× the pixel size at that particular magnification. Blurriness will occur at a distance of one-half of the depth of field (i.e., 1/2 D_f) both above and below the optimum focus plane where the diameter of the probe becomes too coarse to provide adequate focus. Regions of a specimen, therefore, that are scanned by coarse probe will appear blurred in the SEM image.

From the schematic in Fig. 4.3, it can be seen that:

$\tan \alpha = \frac{r}{D_f/2}$ (where α is the half-angle of convergence and r is the radius of the probe)

$$\tan \alpha = \frac{2r}{D_f}$$

$$D_f = \frac{2r}{\tan \alpha}$$

Since α is small, $\tan \alpha$ is taken as α.

$$D_f \approx \frac{2r}{\alpha}$$

Now, $r = 1 \text{ pixel} = \frac{100 \text{ μm}}{M}$ where $M =$ magnification

$$D_f \approx \frac{200 \text{ μm}}{\alpha M}$$

Now, $\alpha = \frac{R_{ap}}{WD}$ where R_{ap} is the radius of final aperture and WD is the working distance

$$D_f \approx \frac{200 \text{ μm} \times WD}{R_{ap} \times M} \quad (4.16)$$

The depth of field is dependent upon the electron beam divergence angle, which in turn is characterized by the radius of the final lens aperture and the working distance as shown in Eq. 4.16. It follows that large depth of field can be obtained by reducing the beam divergence which in turn can be achieved by using small objective aperture and a large working distance as shown in Fig. 4.3. In Fig. 4.3a, the effective focus region as defined by the depth of field is the smallest since a large-sized aperture is

4.2 Depth of Field

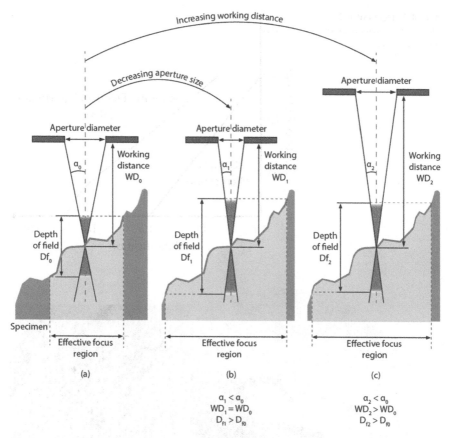

Fig. 4.3 (a) Large aperture gives small D_f. (b) Use of small aperture increases D_f and effective region in focus. (c) Increased working distance also results in larger D_f

used which creates greater beam divergence. As the aperture is changed to a smaller size (Fig. 4.3b), the beam is converged further to increase the depth of field resulting in an increase in the area of the specimen which is now in focus. The same effect can be obtained by increasing the working distance as shown in Fig. 4.3c.

Reducing magnification also increases the depth of field. Conversely, depth of field decreases with an increase in magnification (Eq. 4.16 and Fig. 4.4).

In the SEM, the divergence angle formed is very small (in milliradians) compared to that in an optical microscope. Due to this reason, the change in probe size with distance (depth) from the lens is very small. Small probe size keeps the sampling volume restricted over a large range of depth. All features in the sample along that depth will appear in focus where the diameter of sampling volume is smaller than 2× picture element of the sample. The remarkable depth of field obtained in SEM images is as important as high resolution. In fact, most of the SEM images are taken to observe the topography and morphology of specimen surface which looks so enriching due to the large depth of field. This characteristic alone is responsible for

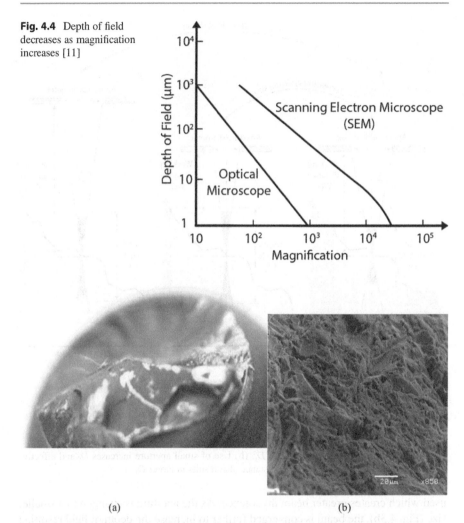

Fig. 4.4 Depth of field decreases as magnification increases [11]

Fig. 4.5 Images taken using (**a**) optical microscope and (**b**) SEM. The optical image is suitable for images at low magnifications but displays small depth of field resulting in blurring of some detail. SEM provides fully focused images of rough samples such as fracture surface

such wide usage of SEM technique across so many diverse fields of applications. The depth of field in the SEM can be between 10% and 60% of the field width depending on the selected magnification. The depth of field in the SEM is tens of times than that of the light microscope. A comparison between the two techniques is provided in the optical and SEM images shown in Fig. 4.5a, b, respectively.

The ability of the SEM to exhibit large depths of field is displayed in Fig. 4.6 where a screw sample is imaged while held in an upright position. Unfortunately, the imaging conditions that promote high depth of field concurrently reduce attainable

4.2 Depth of Field

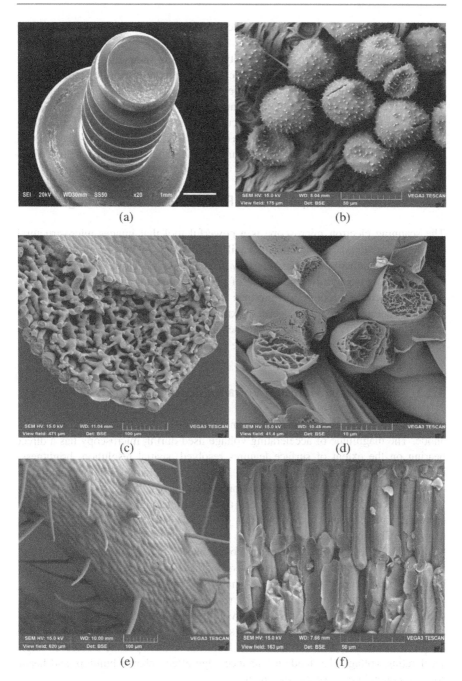

Fig. 4.6 Secondary electron SEM images showing the large depth of field capability of the SEM. (**a**) Screw sample fully focused from the top to the base. (**b–f**) Various biological samples. (**b–f** images courtesy of TESCAN)

resolution since small convergence angle α results in large focal lengths making images susceptible to aberrations.

The depth of focus refers to the ability of a lens to focus an image at varying heights relative to the image plane. In the SEM, the image is formed in an indirect manner using the signals emitted from the specimen. There is no lens that forms an image directly at any plane in the beam optics. Due to this reason, it is not appropriate to use the term "depth of focus" with regard to the SEM. The concept is more aptly defined by the phrase "depth of field" which refers to the z-range within an image where various features of a specimen remain focused within a field of view.

4.3 Influence of Operational Parameters on SEM Images

The scanning electron microscope is a powerful tool that images microstructural features of materials in significant detail. Microscopy at low magnification and imaging of conductive samples is fairly straightforward. However, during high-resolution microscopy, the knowledge of optimum imaging conditions and operational parameters is required to produce high-quality images that can reveal fine surface structure. In addition, necessary know-how helps the user to interpret SEM images and identify and differentiate between related factors that influence the results. Important imaging parameters are summarized in the following sections.

4.3.1 Effect of Accelerating Voltage (Beam Energy)

Accelerating voltage is the difference in potential between the filament and the anode. The magnitude of accelerating voltage used during microscopy has a direct bearing on the extent of surface features resolved, spatial resolution, brightness, chromatic aberration, interaction volume, edge effect, charge buildup, beam contamination, and damage and strength of analytical x-ray signal.

High accelerating voltage produces a smaller electron probe diameter, thus enabling higher spatial resolution (Eq. 4.12). It also produces brighter images (Eq. 2.4). Chromatic aberration degrades image resolution at accelerating voltages below 10 kV (Eq. 2.10). So, use of high beam energy reduces the deleterious effect of chromatic aberration. High beam energy is also required to excite x-rays from heavy elements during microchemical analysis. Apart from this, high accelerating voltage is not suitable for imaging since it results in greater beam penetration into the specimen resulting in larger excitation volume (Sect. 3.2.5.1) and generation of low-resolution signals. These signals, such as backscattered electrons, reduce image contrast and tend to hide fine features at the specimen surface. High accelerating voltage also tends to enhance edge effect, charge buildup, and beam contamination/damage of the specimen.

Operation of the SEM at low accelerating voltages (≤ 5 kV) leads to confined specimen-beam interaction restricted close to the surface producing an image with rich surface detail. This is the basis for low voltage microscopy employed for high-

4.3 Influence of Operational Parameters on SEM Images

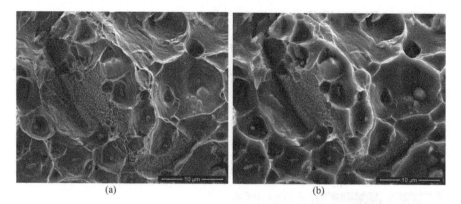

Fig. 4.7 Secondary electron SEM images of the fracture surface of Al showing typical microvoid coalescence structure. (**a**) At beam energy of 2 keV, surface details are prominent, while at (**b**) 15 keV, the surface features are not as clearly discernible, void pits are dark, and sharp fracture edges are brighter due to the enhanced edge effect

resolution imaging in field emission scanning electron microscopes. Lower accelerating voltages are suitable for examining fragile/soft samples (such as cells and polymers) and features present in low quantity and also for revealing fine surface structures since the electron beam penetration in the sample is limited. It can be seen in secondary electron SEM images shown in Fig. 4.7a, b that fine features visible at the specimen surface at 2 keV are lost when beam energy of 15 keV is used.

Another example is shown in Fig. 4.8a–c where surface features visible in a polymer sample at 5 kV are obscured when the accelerating voltage is increased to 20 kV.

The overall effect of accelerating voltage on the image quality is summarized in Fig. 4.9.

4.3.2 Effect of Probe Current/Spot Size

The current that impinges upon the specimen and results in the generation of various signals is called the probe current. For any given beam energy, smaller current results in smaller probe size (Eq. 2.16). The spatial resolution of an SEM is primarily dependent upon the probe size of the beam that falls on the specimen. The smaller the electron probe diameter the higher is the image resolution attained in the SEM. For high-resolution imaging, maximum probe current is sought for a small probe diameter by using a suitable choice of operating parameters. However, the larger the probe current the bigger is the probe diameter. The relationship between the probe current and probe diameter is depicted in a plot shown in Fig. 4.10. It is clear that the probe size increases with increasing probe current. The final probe diameter that interacts with the specimen is determined primarily by the electron source size, degree of demagnification by the electron lenses, and the extent of spherical

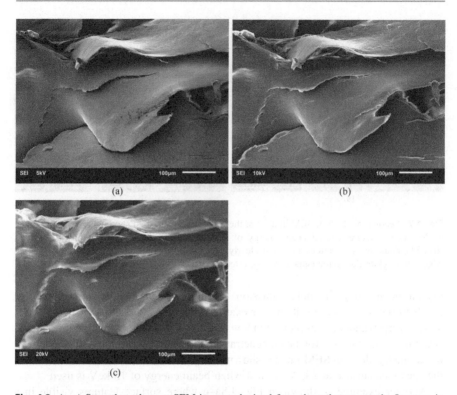

Fig. 4.8 (**a–c**) Secondary electron SEM images obtained from the polymer sample. Increase in accelerating voltage from 5 to 20 kV obscures surface features such black spots visible in (**a**)

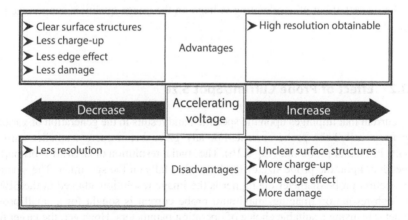

Fig. 4.9 Schematic illustrating the effect of accelerating voltage on the image quality

aberration. For thermionic emission W or LaB_6 filaments, focusing the beam into small probe results in very low beam currents. On the other hand, field emission electron guns can concentrate a large amount of current in the small probe.

4.3 Influence of Operational Parameters on SEM Images

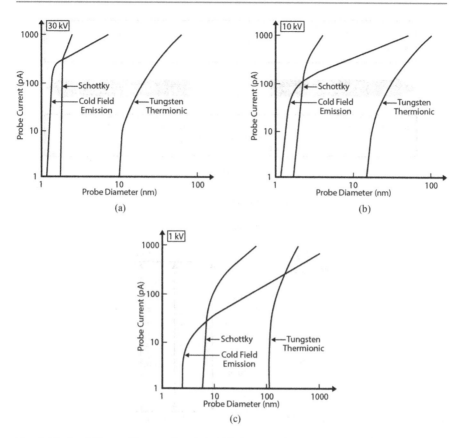

Fig. 4.10 (a–c) Plots to illustrate the relationship between the probe current and probe diameter. Probe diameter increases with probe current at all accelerating voltages from 30 to 1 kV. At higher accelerating voltage, the probe diameter is smaller. Probe size increases with decreasing accelerating voltage. Cold field emitter has the smallest probe size followed by Schottky, LaB$_6$, and W filament at all accelerating voltages and probe currents used [9]

High probe current results in smooth images but degraded image resolution. It can also induce beam damage. Low probe currents realize high image resolution while the specimen is susceptible to less beam damage. Very low probe currents give rise to grainy images which tend to hide surface details. A critical level of probe current is required to achieve an acceptable contrast in the image (Eq. 4.13). The magnitude of such current corresponds to the minimum spot size. Optimum probe current is selected based on magnification, accelerating voltage, specimen type, etc. and is usually in the order of a few picoamps. Schematic in Fig. 4.11a shows the effects of probe current on SEM images. High beam current results in greater signal strength. During imaging at low magnifications where very high spatial resolution is not required, use of large spot size is recommended. Any increase in spot size (by weakening the condenser lens) is accompanied by an increase in the beam current by a magnitude that is roughly square of the beam diameter (Eq. 2.21).

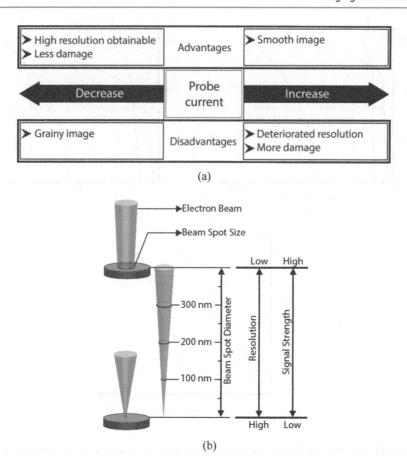

Fig. 4.11 (**a**, **b**) Schematics illustrating the effect of probe current and spot size on the image quality. Small spot size or probe current gives high spatial resolution and low signal strength. Large spot size or probe current gives high signal strength but the resolution is degraded. For imaging at low magnification, large probe diameter (≈100 nm) can be used

A large-sized beam diameter is preferred during low magnification imaging as long as the spatial resolution is not affected noticeably. At low magnifications, the size of a picture element in the sample is large and accommodates delocalization of the signal well resulting in focused images even at large beam diameters. Schematic in Fig. 4.11b illustrates the effect of spot size on the signal strength and resolution.

The effect of spot size on the image quality in the SEM is demonstrated in Fig. 4.12. It is seen that at a small spot size (small probe current), the image is noisy. With an increase in spot size (or probe current), the signal-to-noise ratio increases and the image becomes sharp.

4.3 Influence of Operational Parameters on SEM Images

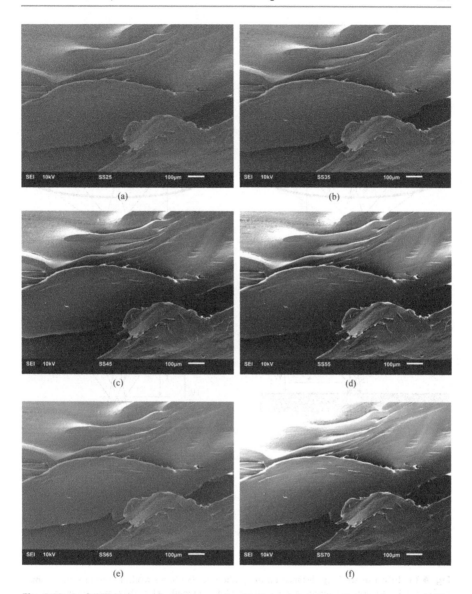

Fig. 4.12 (a–f) SEM images of polymer sample with increasing spot size (e.g., probe current). The image is noisy at small spot size. As the spot size is increased, the signal-to-noise ratio increases and the image becomes sharper. At the largest spot size, charging is observed as the polymer sample is unable to discharge the accumulated current at its surface

4.3.3 Effect of Working Distance

Working distance (WD) is the distance between the pole piece of the objective lens and the plane of best focus. Working distance is adjusted by moving the sample stage

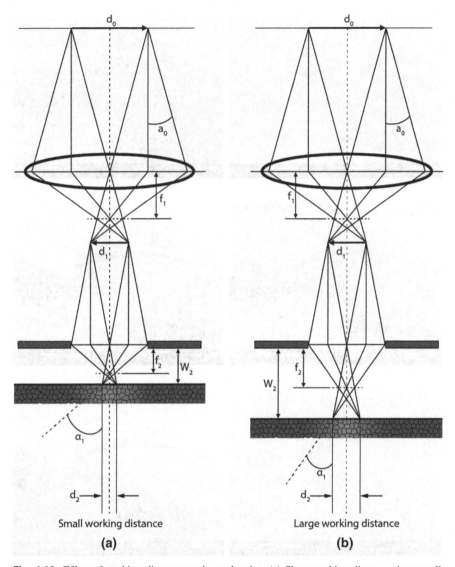

Fig. 4.13 Effect of working distance on the probe size. (**a**) Short working distance gives small probe size due to large convergence angle and small focal length, while (**b**) large working distance increases the probe size (Adapted from [9])

along the z-axis and focusing the beam on the sample surface. In order to image at a set WD, the sample is brought into optimal focus by moving the sample stage along z-axis while keeping objective lens current constant. Positioning the sample close to the lens enables high image resolution but decreases the depth of field (Eq. 4.16). The resolution is improved as the probe size becomes smaller at short WD as shown in Fig. 4.13a, b. Use of immersion lens can avail 2–5 mm WD for a specimen with

4.3 Influence of Operational Parameters on SEM Images

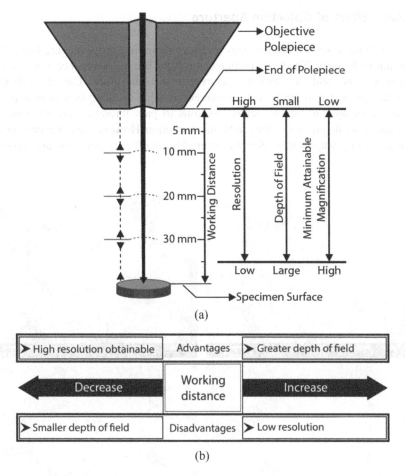

Fig. 4.14 (a, b) Schematics illustrating the effect of working distance on the image quality. Large working distance increases the depth of field and enables the specimen to be viewed at low magnification (i.e., increases field of view). Short working distance is utilized for high-resolution work

small size (<5 mm) located directly inside the lens gap. Short WD will result in small focal length which helps reduce spherical aberration.

Large WD increases the depth of field due to smaller convergence angle. It also allows the specimen to be observed at small magnifications (e.g., at 5×) encompassing the large field of view (see Fig. 4.14a, b). Large WD lowers the spatial resolution due to increased probe diameter. The signal strength at large WD decreases and the image can appear relatively noisy. Figure 4.15(a-c) shows the large depth of field achieved in the SEM. It shows the effect of working distance and aperture size on the depth of field.

4.3.4 Effect of Objective Aperture

The SEM has a set of objective lens apertures (holes in a strip) ranging from 50 to 500 μm in diameter. The final aperture controls the beam convergence angle, size of electron probe, and the amount of current in the final probe. Use of small final aperture gives rise to small beam convergence angle resulting in a large depth of field. Small aperture allows fewer electrons to pass through, thus allowing the formation of fine probe resulting in higher resolution. However, lower probe current can result in grainy images. Small aperture blocks off-axis electrons and serves to

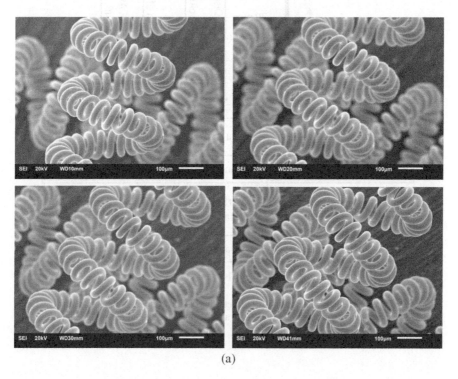

(a)

Fig. 4.15 Secondary electron SEM images of W filament retrieved from a light bulb. (**a**) Large aperture: as the WD increases from 10 to 41 mm, blurred regions of the filament come into focus. However, at WD of 41 mm, filament at the bottom surface is still not fully focused. This is due to the large size of the aperture used. (**b**) Medium aperture: as the WD increases from 10 to 41 mm, blurred regions of the filament come into focus. At WD of 41 mm, all regions of the filament are in focus. This is due to the medium-sized aperture used to take these images. (**c**) Small aperture: as the WD increases from 10 to 41 mm, blurred regions of the filament come into focus. At WD of 30 mm, all regions of the filament are already in focus. This is due to the small size of the aperture used to take these images. Also, note that the signal-to-noise ratios of images decrease from large to small apertures as the current that passes through them decrease. Edges of the filament are bright due to higher secondary electron emission for images taken with a large aperture

4.3 Influence of Operational Parameters on SEM Images

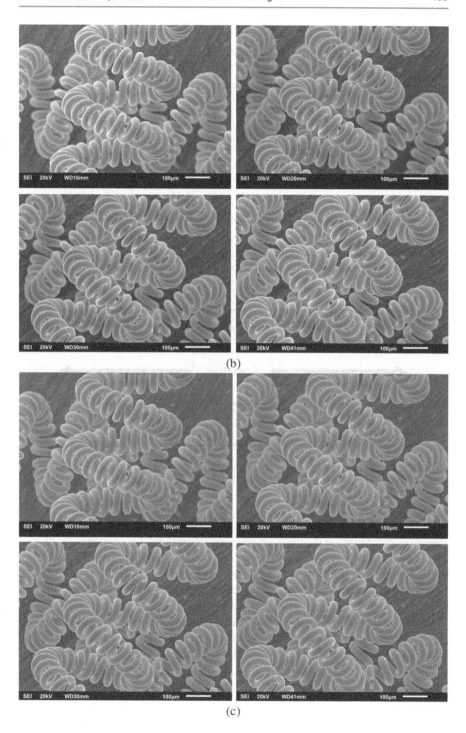

Fig. 4.15 (continued)

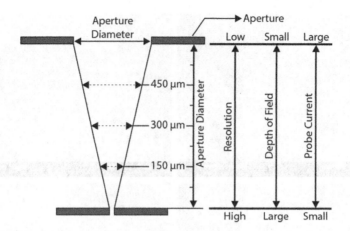

Fig. 4.16 Small final aperture results in high resolution, large depth of field and small probe current. It also reduces the effects of spherical aberration by blocking off-axis electrons

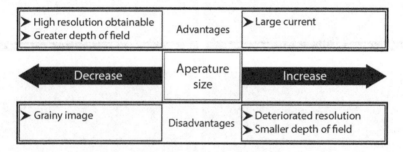

Fig. 4.17 Use of large final aperture results in large probe current which is required for x-ray microanalysis and backscattered imaging. Small aperture reduces the beam convergence angle giving large depth of field

minimize the detrimental effects of spherical aberration. Large apertures allow a larger amount of current which is required for backscattered imaging and x-ray analysis; however, it results in lower image resolution and a smaller depth of field. Schematic in Fig. 4.16 summarizes the effect of final aperture on various imaging outcomes. Appropriate aperture is selected based on the desired information keeping in mind the relative benefits of various sizes as shown in Fig. 4.17. Effect of aperture size on the quality of SEM images obtained for a steel fracture surface is shown in Fig. 4.18. It is clear that the use of small apertures helps to focus larger regions of the coil due to increased depth of field.

4.3 Influence of Operational Parameters on SEM Images

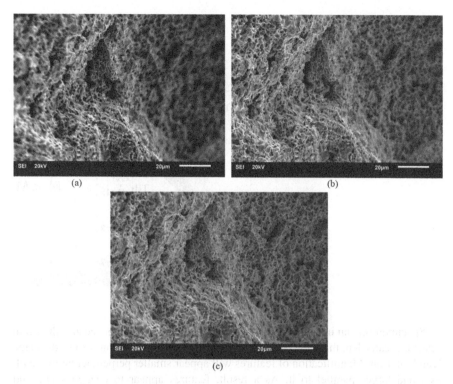

Fig. 4.18 SEM images of a steel fracture surface obtained with (**a**) large, (**b**) medium, and (**c**) small objective aperture. The image with large aperture shows the highest proportion of out-of-focus regions, while the image with small aperture shows regions at all depths focused. This shows that the depth of field increases as the size of the final aperture is decreased

4.3.5 Effect of Specimen Tilt

The specimen is sometimes tilted at certain angles in order to highlight specimen features otherwise not prominent. This can include surface topography features and side or cross-sectional views of the specimen. Tilting is also undertaken to obtain stereo micrographs (SEM images that give 3-D visual impression). Displayed magnifications are no longer valid during tilt and need to be corrected or taken at zero tilt angles when the sample lays flat perpendicular to the beam. This is apparent in the SEM image of Fig. 4.19 where a grid is tilted 45° resulting in an image which is demagnified in the horizontal direction (perpendicular to the tilt axis) by $\frac{1}{\cos 45°}$.

Fig. 4.19 Displayed magnification is not valid when the specimen is tilted as the scan length on a tilted sample (L_2) is greater than that on the untilted sample (L_1). Actual magnification of a tilted surface, therefore, will be less than the displayed number

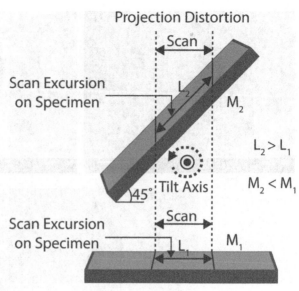

Specimen tilt can introduce distortion in an image. This occurs because the beam scans a greater length of the specimen due to tilt while projecting it onto the same length of scan. Magnification of features will appear smaller perpendicular to the tilt axis and larger parallel to it. As a result, features appear to change shape and dimensions as seen in Fig. 4.20 where grid size appears to have become substantially smaller when the specimen is imaged at 45° tilt.

Another effect of the tilt is the change in the spot size as the beam scans from the top to the bottom of the specimen surface. *Dynamic focusing* is employed to change the focal length of the lens as the beam scans over the tilted surface. This serves to keep the spot size constant along the z-axis. Tilting also results in the formation of asymmetrical interaction volume (Sect. 3.2.5.3) influencing the BSE (Sect. 3.4.1.5) and SE yield (Sect. 3.4.2.7). Also, since the location of specimen features change due to tilting, images can display *shadowing contrast* depending on the orientation of the features with respect to the detector. Detector position with respect to the specimen surface is a critical factor in producing shadowing contrast. At zero-degree specimen tilt angle, the E-T detector is located at the top of the specimen which allows for the effective collection of SE signals. However, if the specimen surface is tilted to the opposite side, the E-T detector will not be able to collect adequate SE signal since the latter have momenta in the opposite direction which results in shadowing contrast.

4.3 Influence of Operational Parameters on SEM Images 159

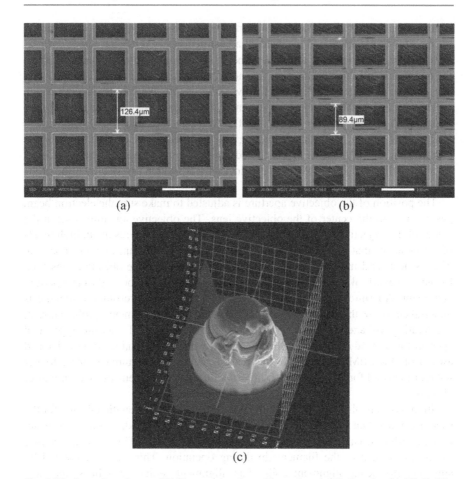

Fig. 4.20 Secondary electron SEM images showing grid sample at (**a**) 0° tilt and at (**b**) 45° tilt. The dimensions of the grid appear smaller when the sample is tilted. (**c**) SEM image showing a tilted sample. (Images courtesy T. Siong, JEOL Ltd)

4.3.6 Effect of Incorrect Column Alignment

Good image resolution can only be obtained when the microscope is properly aligned. Alignment of the microscope column enables the electron beam to fall onto the specimen surface in the most effective manner. Whenever an SEM user undertakes mechanical and/or electrical maintenance like filament change or cleaning of the column, it is important to check and adjust the SEM column alignment. During alignment, the gun, lenses, and apertures are positioned such that these are concentric about the optic axis which is an imaginary line running down the center of the microscope column. It is also not possible to remove small misalignments from within the electron column completely. Shortcomings due to imperfections in mechanical alignment pointed above are overcome by aligning the beam along the optic axis using electromagnetic coils. Alignment is obtained by means of both electrical and mechanical adjustments.

Ideally, upon generation, the electron beam should be uniformly concentric about the optic axis. For this, the filament tip should be concentric about and leveled with the aperture of Wehnelt cylinder. However, this is not likely to be achieved with flawless precision. Moreover, the position of the filament may also change slightly during SEM operation. Over the course of SEM usage, the gun will need adjustment to restore its alignment. In this procedure, gun shift and tilt are aligned to obtain the brightest image on the screen. Misaligned filament gun tip will affect the emission current passing through the Wehnelt cylinder aperture which in turn affects the probe current.

The position of the objective aperture is adjusted to make sure the electron beam passes through the center of the objective lens. The objective aperture is set at the center of the objective pole piece. Any shift from this position results in high levels of astigmatism that in turn deteriorates resolution. At optimum aperture position, there is no lateral movement of the beam as the current in the objective lens (i.e., focus) is varied. *Wobbler* control is used to oscillate the focus during aperture alignment. Aperture alignment is necessary when very high-resolution imaging is undertaken or if there is a large change in the gun alignment, probe current, accelerating voltage, or working distance. The conventional aperture type (real aperture) is located close to the SEM chamber, while the virtual aperture is located away from the SEM chamber. For this reason, the virtual aperture can serve longer without the need for cleaning which in turn reduces the requirement to align it after cleaning.

Stigmators are also aligned along the optic axis and their strength adjusted for the sharpest image that does not stretch when the focus is changed. Variation in beam current with time can introduce errors in x-ray microanalysis. This is mainly caused by the movement of the filament tip during operation. This can be corrected by adjusting the beam alignment coils. The alignment settings can be stored in a computer file available for retrieval at a later date.

4.4 Effects of Electron Beam on the Specimen Surface

4.4.1 Specimen Charging

Primary beam current i_B entering the specimen is equal to the specimen current i_{sp} flowing out of the specimen into the ground plus backscattered and secondary electron current (i_{BSE} and i_{SE}, respectively) ejecting out of the specimen as shown in the following equation:

$$i_B = i_{sp} + i_{BSE} + i_{SE} \qquad (4.17)$$

Rearrangement of the above equation gives specimen current I_{sp} as:

$$i_{sp} = i_B - i_{BSE} - i_{SE} \tag{4.18}$$

or

$$i_{sp} = i_B - \eta - \delta \tag{4.19}$$

where η and δ are the BSE and SE yield, respectively.

During the scan process, i_B remains constant while η and δ vary. At accelerating voltages typically used for imaging (>5 kV), the number of electrons leaving the specimen in the form of SE and BSE combined (i.e., total electron coefficient, $\eta + \delta$) falls significantly short of those entering it as beam current i_B. For example, pure Cu target imaged at 20 kV exhibits total electron coefficient of 0.4 only, which means that 60% of the beam electrons entering the specimen need to leave through electrical contacts to avoid accumulation within the specimen. Specimen stage is grounded for this purpose. For a metal target like Cu which is conductive, beam electrons reach the specimen stage by passing through the specimen and the specimen holder. A continuous conductive path connecting the specimen surface to the ground needs to exist for this purpose. When an uncoated insulating specimen is scanned, conductive path that serves to ground the specimen current i_{sp} does not exist. Electrons in the beam that strike the specimen surface do not find a conductive path to dissipate and thereby fail to reach the grounded specimen stage. As a result, they accumulate within the specimen in the form of a localized negative charge known as charge buildup or specimen charging. This kind of electrostatic charging increases local potential which disrupts the normal secondary electron emission from the specimen and severely degrades the imaging capability of the SEM. It could deflect the beam to another area of the specimen to generate an excessive amount of secondary electrons. Charging effect may present itself in many forms such as unusual contrast (fluctuation in image intensity such as excessive brightness/darkness in images), horizontal lines on images, beam shift and image distortion (spherical objects appear flat), etc. Whether an insulating specimen is going to acquire an electric charge will depend on the number of incident electrons impinging upon the specimen (i_B) compared to those leaving the sample ($i_{sp} + \eta + \delta$). If a balance between the incident and emitted electrons is achieved, then specimen shall not charge. If the number of incident electrons is higher than the emitted electrons, the sample shall charge. Accumulated charge at localized specimen surface could well be positive as indicated by Eq. 4.19; however, this does not pose as much difficulty as an amassed negative charge. Any positive charge created at the specimen surface is neutralized by SE emitted from the specimen and pulled back toward the surface. Various forms of charging encountered during imaging are shown in Fig. 4.21a–f.

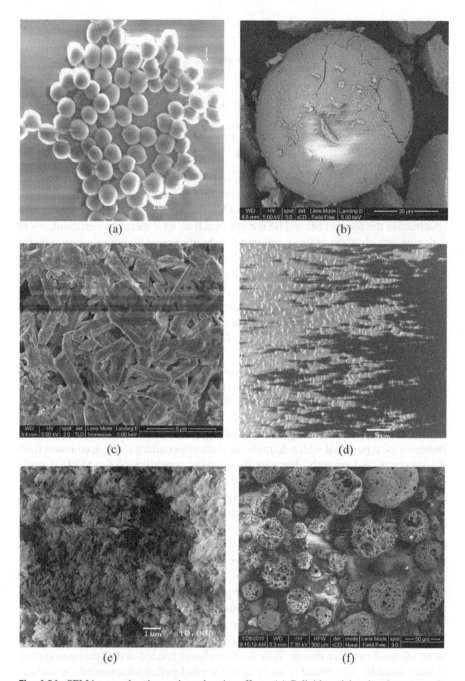

Fig. 4.21 SEM images showing various charging effects. (**a**) Colloid particles showing extremely bright regions, amplifier overloading due to saturation of signal (horizontal lines), and image distortion (spherical particles appear flat). (**b**) Catalyst particle showing bright spot at the top region. (**c**) Decreased SE collection due to changes in potential at localized regions makes some areas appear dark. (**d**) Nanoclay-polyethylene composite showing bright and dark regions due to variation in potential at the specimen surface. Dark regions represent areas where SEs are recollected due to local variation in the potential field. (**e**) Zeolite showing scan discontinuities (horizontal lines). (**f**) Carbon ash specimen showing bright regions and thin intense horizontal lines

4.4.1.1 Methods to Reduce Charge Buildup

(a) *Coating the specimen surface*

For conventional imaging in the SEM, specimen surface must be electrically conductive and electrically ground to prevent the buildup of electrostatic charge at the surface. In order to avoid charge buildup, the surface of a nonconductive specimen is generally coated with a thin conductive film such as gold, carbon, gold-palladium, tungsten, etc., prior to examination in the SEM (see Sect. 8.1.10 for details). This film has small grain size (e.g., few nanometers) and is low in thickness (few nanometers) depending on the duration of deposition. It does not interfere with the examination of the surface morphology of specimens at low to high magnifications. However, during ultrahigh-resolution imaging (~100,000×), extra care needs to be taken to ensure that the actual specimen features are being imaged and not the coating grains itself. For high-resolution imaging, a thin fine sub-nanometer coating of chromium is preferred. It is also customary to use conductive paint/tape to establish electrical contact of the coated specimen with the specimen holder and the stage.

Polymer specimens generally exhibit significant charge buildup during microscopy. While it is usually necessary to coat a polymer with conducting metal layer, it is important to keep its thickness to a minimum in order to make sure it does not mask specimen features. This is especially important during low voltage imaging where the beam penetration into the specimen is small. This could result in contrast due to the top coated layer rather than the underlying specimen surface. An example is shown in the SEM image of Fig. 4.22 where the surface of the polymer composite is buried under the thick layer of gold sputter coating deposited to eliminate charge-up. Low magnification imaging is possible with thick coatings. However, for high-resolution imaging, thin coatings need to be employed, or accelerating voltages need to be increased appreciably to allow penetration into the specimen.

It is common practice to coat specimens even if the specimen exhibits adequate conductivity. This is undertaken to increase signal strength and surface resolution, especially when samples with light elements are examined. The improvement in resolution takes place because secondary electron emission near the surface is enhanced. Instead of coating, it is more effective to stain biological samples (e.g., impregnate them with osmium or its variants) to increase the bulk conductivity of the analyzed material.

(b) *Use of low accelerating voltage, beam deceleration, and small probe current*

Effects of charging can be diminished by using low accelerating voltage/beam deceleration and probe current during imaging. These techniques restrict charge buildup by reducing the number of electrons entering the specimen, I_p. At high accelerating voltage, the beam penetration in the specimen is deep. A large proportion of SE and BSE generated cannot leave the specimen which decreases the total ($\eta + \delta$) electron yield/coefficient resulting in an accumulation of negative charge within the specimen. The decrease in accelerating voltage increases δ reaching a

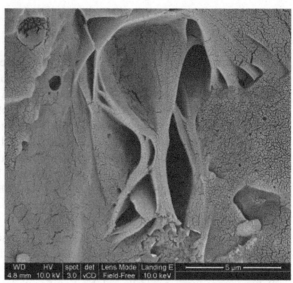

Fig. 4.22 Polypropylene specimen sputtered with Au coating for 120 s resulting in a thick coat

point where the number of electrons emitted out of the specimen due to backscattering and secondary electron emission surpasses those supplied by the beam (see Fig. 3.30a). At this point, the specimen current I_p, which maintains charge neutrality, falls to zero. At such low beam energy, the specimen will not charge. The region of zero charging exists between beam energies indicated as E_1 and E_2 in Fig. 3.30a. The optimal accelerating voltage varies with the type of material. It is in the range 0.5–2 keV for organic and 2–4 keV for inorganic materials. However, optimum beam energy where charging is reduced to zero value is found by experimentation. One disadvantage of using low beam energy is an increase in chromatic aberration effect. Therefore, field emission source is generally used for low voltage microscopy to counter chromatic aberration to a certain extent.

Semiconductor materials are prone to charging and thermal damage and are routinely examined at low kV and small probe current using field emission SEM. Samples with rough surfaces can develop complicated electric fields and interrupt charge removal. The large variation in surface topography impacts charge distribution at the surface and makes it inhomogeneous. Also, it is difficult to prevent charge buildup in pure insulator materials by means of these techniques only.

(c) *Use of samples with small dimensions*

SEM chamber can accommodate large specimens. However, it is preferred to use specimens with small dimensions as long as it adequately represents the material or component under investigation. Small-sized specimens are free of extra material and can be well prepared to avoid charge buildup.

4.4 Effects of Electron Beam on the Specimen Surface

(d) *Proper specimen mounting*

Carbon tape is widely used as a mounting material for samples in the SEM. Carbon tape may outgas in the SEM chamber and create an environment conducive to charge buildup in the specimen. On the other hand, it also serves to enhance the conductivity of the specimen. These contradictory effects dictate the use of a small quantity of tape only and keep microscopy sessions as short as possible. In addition, the sample should be secured to the underlying tape with carbon or silver tape to ensure proper conductivity. Charge buildup in powder samples is uneven and may make some portions of the specimens look very bright and others dark. Powder samples should be distributed evenly in the form of a single thin layer on SEM stub, and excess material should be blown off.

(e) *Tilting of the sample*

As discussed in Sects. 3.4.1.5 and 3.4.2.7, tilting a specimen changes the BSE and SE yield. This phenomenon can be used to advantage to reduce the charging effect by tilting the specimen to an angle that enhances emission of electrons from the surface.

(f) *Use of BSE detector*

Emission pattern of energetic BSE is not disturbed by the change in local potential at specimen surface. BSE emitted from the specimen have adequate energy not to be attracted back to the localized positive potential regions created at the specimen surface due to charging. Therefore, imaging with the BSE detector can eliminate the effects of low-intensity charging.

(g) *Fast scan and frame averaging*

Images are obtained at fast (TV rate) scan to reduce the dwell time of the beam at any pixel to reduce charge accumulation. This, however, results in noisy images. A series of images or frames of a single field of view are taken, and their pixel intensity is averaged to increase the signal-to-noise ratio and thereby eliminate the effects of charging.

(h) *Use of low vacuum or environmental SEM*

In a low vacuum or environmental SEM, gas or water molecules are injected above the specimen surface. These are ionized by the electron beam on its way to the specimen producing a mass of positive charge. If a specimen starts accumulating a negative charge at its surface, this positive mass is attracted toward it and serves to neutralize the charge buildup. This technique is discussed in more detail in Sect. 5.2.

(i) *Energy filtering*

Charging is primarily produced by lowest energy secondary electrons. Electron energy filters (such as E × B filter and r-filter) separate electrons based on their energy and suppress the role of low-energy SE which serves to reduce charging effects. The working principle of energy filters is discussed in Sect. 5.1.

4.4.2 Surface Contamination

If the surface of a specimen is scanned for long durations, it may cause loss of sharpness in the image with an accompanying dark rectangular smudge at its surface as shown in Fig. 4.23. This mark is caused by carbon deposition which occurs due to the interaction of the electron beam with residual gas molecules present in the vicinity of the specimen surface. Usually, this residual gas is volatile hydrocarbon molecules that are ionized by the electron beam and deposited on the specimen surface as nonvolatile carbon. This phenomenon is known as specimen *contamination* which occurs at the point of beam impact. This thin film of carbon is deposited on the area that is scanned with the beam for a considerable amount of time and can be observed by zooming out during live imaging. This contaminant layer serves to obscure and blur the surface details of the specimen with an accompanying darkening of the scanned area.

Despite the presence of vacuum in the specimen chamber of the SEM, some degree of gas molecules is present in the environment, which results in specimen contamination. Source of these contaminants could be the hydrocarbons introduced by the specimen itself due to outgassing, the organic material used to prepare/mount specimens, instrument surfaces or grease, backpressure from rotary oil pump used to evacuate the SEM chamber, etc. These hydrocarbons are broken down into its

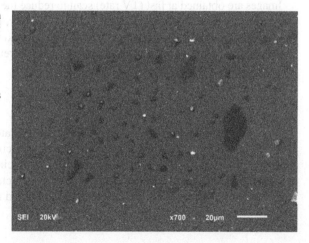

Fig. 4.23 Secondary electron SEM image of Ni-based alloy specimen showing contamination effect after long exposure to beam scan. The scanned area contains large dark spots and also loses sharpness due to contamination buildup

4.4 Effects of Electron Beam on the Specimen Surface

constituent materials and while nitrogen and oxygen are pumped out by the vacuum system; carbon deposits on the specimen surface.

Residual hydrocarbon molecules can also be present on various components of the SEM column such as apertures. Beam interaction with these residuals can produce contaminants on these component surfaces that can result in beam instability. Contamination can become a serious issue, while imaging at very low accelerating voltages and probe currents as the electron beam is not energetic enough to penetrate the deposited contaminant layer. In this case, the contaminants may be imaged instead of the underlying specimen surface. Moreover, low-energy x-rays emanating from the specimen may be absorbed in the contamination layer and introduce error in the EDS microanalysis.

Contamination from the instrument is reduced by employing dry pumps or installing a vapor trap in the pump backing line that can control hydrocarbon contamination originating from vacuum pumps. In addition, cold traps can be employed to seize contaminants, and the SEM chamber is purged with dry nitrogen gas during specimen exchange. Contamination from the specimens can be reduced by proper handling (e.g., use of gloves and completely dry specimens) and use of minimum amount of adhesive tapes or conductive paints. Size of outgassing biological or hydrocarbon volatile specimens that need to be imaged should also be kept to a minimum. Embedding agents and resins used for sample preparation should be carefully selected as some might give off a high amount of gas. Also, since organic gas is given off when the resin surface is irradiated with an electron probe, use of the smallest possible surface area is recommended for imaging, or surface is to be coated with a conductive material.

4.4.3 Beam Damage

During electron beam-specimen interaction, heat is generated due to ionization at the irradiated spot as energy is transferred from the beam to the specimen. The magnitude of heat generated or level of temperature achieved depends on the accelerating voltage, probe current, time of exposure, specimen area, and the ability of a specimen to dissipate heat. Beam damage can occur due to ionization and subsequent chemical reaction at the specimen surface due to the incident beam. The extent of the damage varies with the nature of the specimen material. Conductive specimens such as metals and alloys can dissipate heat effectively and therefore are more resistant to beam damage. Polymers and biological specimens, on the other hand, are poor conductors of heat and thereby more prone to beam damage.

Radiation damage in organic materials occurs as a result of inelastic scattering which disturbs the valence electron configuration and introduces permanent changes to the chemical bonds of the solid. The effects of damage in these materials may result in specimen heating, structural damage, mass loss, reduction in crystallinity, and contamination. Susceptibility to this type of damage makes it all the more difficult to undertake high-resolution microscopy and microchemical analysis of polymeric, biological, and life science specimens [13].

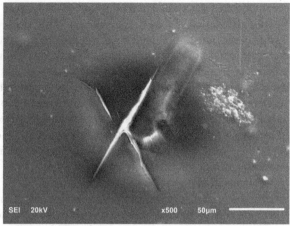

Fig. 4.24 Secondary electron SEM image showing beam radiation damage induced in a polyethylene specimen during microscopy

Some polymers are more sensitive to the radiation damage than others. Aliphatic and amorphous compounds are more prone to damage than aromatic and crystalline compounds. Presence of oxygen in materials results in peroxide formation enhancing damage. Radiation damage exhibits itself in the form of cracks, bubbles, holes, depressions, and dimensional changes. An example of the damage caused by the beam is shown in a polyethylene sample in Fig. 4.24.

Beam damage is an irreversible process. It can occur fairly quickly and sometimes is difficult to judge whether a feature is part of a specimen or a consequence of beam damage. During imaging of sensitive specimens, certain steps can be taken to contain beam damage, i.e., use low accelerating voltage, decrease probe current, reduce exposure time, use low magnifications/large scan areas, and apply conductive coatings such as of gold, carbon, etc., at the specimen surface to improve thermal conductivity.

The phenomenon of radiation damage is utilized in electron beam lithography in the manufacturing of integrated circuits. The electron beam is scanned over the surface of a thin polymer film to introduce *controlled damage* at the surface in the form of a pattern. The pattern is then exposed to a mild etch, such that the regions damaged by the electron beam react at a different rate than the unexposed regions. The pattern is thus used as a mask for subsequent deposition.

4.5 Influence of External Factors on SEM Imaging

External factors that originate from the environment and poor maintenance of the SEM can influence the quality of SEM images as discussed below.

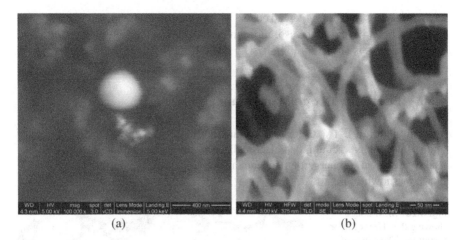

Fig. 4.25 High magnification secondary electron SEM images showing image distortion introduced due to electromagnetic interference effect. The edges of (**a**) Ca-YSZ particle and (**b**) carbon nanotubes appear jagged in the horizontal direction due to EMI

4.5.1 Electromagnetic Interference

Image distortion can be produced due to the presence of external electromagnetic interference (EMI) effects in the area where the SEM is operated. This interference is caused by electrical equipment located in the vicinity of the SEM. These could be transformers, distribution boards, high tension cables, lights, improper electrical grounding, and other lab equipment. The source of this equipment needs to be identified, and if their removal is not possible, their effects should be canceled by installing EMI cancelers. This equipment eliminates or minimizes EMI by applying magnetic screening/shielding or applying a field of similar magnitude in the opposite direction that serves to cancel the stray field. SEM images distorted due to electromagnetic interference are shown in Fig. 4.25a, b. It could be seen that the edge of imaged feature exhibits sharp spikes. Such an effect is more visible at higher magnification. Also, the lower the beam energy the greater is the interference effect. Use of high accelerating voltage and short working distance can reduce the effects of EMI.

4.5.2 Floor Vibrations

The location where the scanning electron microscope is installed has to meet certain specifications regarding mechanical vibrations and stray magnetic fields. Vibrations can arise due to mechanical vacuum pumps, motors, etc., and also if the microscope is installed on higher floor levels in a building. This is why most microscopes are equipped with anti-vibration mounts/table or soft extension springs and installed in the basement or ground floor of the building. Image distortion produced due to floor vibrations is similar to that produced by EMI such as features exhibiting jagged edges. These distortions are more visible at large magnifications during high-resolution imaging. SEM image showing the effect of floor vibrations is shown in Fig. 4.26.

4.5.3 Poor Microscope Maintenance

The scintillator of E-T detector and HT tank are supplied with very high voltages of up to 10 kV. Poor vacuum, bad electrical connections, contamination, and dust can lead to electrical discharge that appears as horizontal lines and a bright spot in the image as shown in Fig. 4.27.

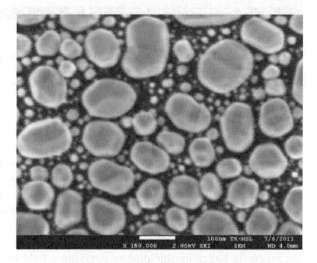

Fig. 4.26 SEM image showing the effect of floor vibrations. The edges of particles show jagged appearance

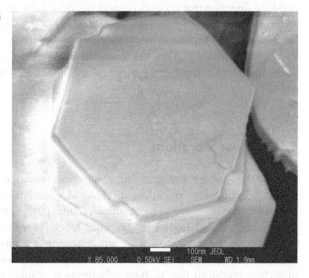

Fig. 4.27 Secondary electron SEM image showing the probable effect of electrical discharge brought about due to poor microscope maintenance

4.6 Summary of Operating Conditions and Their Effects

Effects of different SEM operating conditions on image quality are summarized in the following table.

Operating conditions	Effects
Misaligned microscope	Images lack sharpness and focus
Coated specimen	Eliminates charge buildup Better contrast Coating grains visible at magnifications > ×100,000 Not suitable for low voltage microscopy
Uncoated specimen	Specimen charging Low contrast Suitable for low voltage microscopy
High beam energy	High resolution Good for bulk surfaces Required for EDS microchemical analysis Lacks surface details due to high beam penetration and large interaction volume Specimen prone to charging Specimen prone to beam damage Specimen prone to contamination Pronounced edge effects
Low beam energy	Low resolution Good for surface morphology examination due to low beam penetration and small interaction volume Not adequate to measure the chemistry of heavy elements Less charge buildup Fewer edge effects Less specimen damage
Large probe current (large spot size)	Broad beam probe Strong signal Low resolution Smooth image Increased surface damage Required for microchemical analysis
Small probe current (small spot size)	Small beam probe High resolution Grainy image Less surface damage Weak signal Not adequate for microchemical analysis
Large working distance (WD)	Large field of view Small minimum magnification High depth of field Low resolution
Small working distance (WD)	Small field of view Large minimum magnification Small depth of field High resolution

(continued)

Large objective aperture	Low resolution Small depth of field Smooth image Weak contrast Suited to microchemical analysis
Small objective aperture	High resolution High depth of field Grainy image Strong contrast Not suited for microchemical analysis
Secondary electron imaging	Suitable for surface morphology examination High resolution Prone to specimen charging
Backscattered electron imaging	Suitable for contrast based on the composition Low resolution Less prone to charging
Energy dispersive x-ray detector (EDS)	Qualitative and quantitative microchemical analysis Spot (point analysis), area analysis, line profile, 2D area profile
Low vacuum mode	Backscattered electron images Reduces charge buildup Low resolution Low contrast
Low voltage imaging	Backscattered electron imaging Suitable for examination of surface features Reduces charge buildup
STEM imaging technique	Used for detection of electrons transmitted through the sample
Beam deceleration	Reduces charge buildup Makes use of high accelerating voltage with low penetration depths Reduces chromatic aberration Surface study due to low interaction volume

4.7 SEM Operation

Electron gun generates an electron beam with an accelerating voltage that can range from 500 V to 30 kV. The beam is focused into a fine probe of approx. 1 nm to 10 nm by electromagnetic condenser lenses located within the column. The fine electron probe is then rastered over specimen surface in a rectangular area by scan coils. The sample sits in the SEM chamber. The electron beam penetrates into the sample in the form of a teardrop/hemisphere extending from 100 nm to 5 μm depending on accelerating voltage and sample density. This interaction produces a variety of signals including secondary and backscattered electrons and x-rays which are collected and used to produce images as well as to determine the elemental composition of the specimen material. Images are digitally processed, displayed on computer screens, and saved on hard drives.

4.7 SEM Operation

This section focuses on the practical aspect of the technique. It includes a stepwise guide to the use of SEM with an aim to get useful images. The way the controls and software user interface is laid out differs from one model of the microscope to the other. Nomenclature used may also vary depending on the manufacturer. It is not the intention of this chapter to describe specific details for use of instrumentation and software of a particular model. This information can be found in the relevant user manual of the microscope. The aim here is to explain practical steps to undertake scanning electron microscopy irrespective of the model in use. This can serve as a source of guidance to a new or casual user.

4.7.1 Sample Handling

4.7.1.1 Sample Size
The sample should be of an appropriate size to fit in the SEM chamber. The sample is generally mounted on a holder whose size can vary from 10 mm to 30 mm. Generally, various sizes of holders are available for use with a particular microscope. The holders are placed on the specimen stage located within the SEM chamber. Still bigger specimens can be accommodated since the dimensions of the stage can generally be in the order of 150 mm to 250 mm. Figure 2.28 shows pictures of various types and sizes of holders available for mounting samples. Specimen holders are exposed to vacuum in the SEM chamber. Therefore, gloves are used to handle specimens and specimen holders to minimize contamination that might cause problems during imaging.

4.7.1.2 Sample Preparation
Polished specimens of metals and alloys prepared using metallographic sample preparation techniques are mounted in epoxy or Bakelite in appropriate sizes of mounts to fit into the available holders. Loose powders are placed on C tabs or Cu/Al adhesive tapes that are attached to Al stubs which are in turn inserted into specimen holders for examination in the SEM. As-received specimens such as broken metal pieces are held down on the holder with the help of adhesive tape. Conductive tapes, paints, and tabs are available to dissipate current and reduce accumulation of electrostatic charge on the specimen surface during observation in the SEM. Metal, alloy, ceramic, and glass samples do not require any preparation except for coating. It is normal practice to coat samples in order to reduce charging effects during imaging. Nonconductive samples like polymers, rocks, glass, etc. need to be coated; however, conductive samples like metals and alloys are also coated to get good imaging results.

Usually, the SEM is operated in a high vacuum which necessitates the use of dry specimens. If the specimens are wet (e.g., rocks, soils, corrosion deposits), they can be dried in an oven or by simply leaving them out in the air for an appropriate length of time. Polymeric samples charge significantly and therefore need to be coated prior to the examination. In some cases, it is desirable to dip polymer samples in liquid nitrogen to make it brittle and then smash it to reveal fresh fracture surface. This procedure allows imaging of certain features otherwise not visible in as-processed surfaces. Samples and sample holders are stored in dry and dust-free environments to minimize contamination of the SEM chamber.

Some samples (e.g., biological samples or tissues) might change their shape or structure as a result of drying. These specimens are subjected to techniques such as freeze drying or critical point drying. They are dried slowly in a controlled fashion in order to secure fine details of their structure. These sample preparation techniques are discussed in more detail in Chap. 8. Cryo-SEM has been used to examine wet specimens such as plants, oily rocks, etc., in a high vacuum environment. Currently, variable pressure SEM instruments are available to examine wet specimens without any preparation.

4.7.2 Sample Insertion

Specimen should be of a correct size to fit in the SEM chamber. It is held or mounted on an appropriate stub or holder. It is usually coated with gold or carbon to improve its conductivity. A conductive path between the specimen and stub/holder is ensured to dissipate electron current and prevent the buildup of excessive charge. The microscope should be in a ready state. Electron beam should be off. In microscopes where the chamber is purged with dry N_2 gas during sample exchange, the gas supply is turned on. Sample insertion procedure is started by pressing the appropriate control button on the SEM console or clicking the button in the computer software interface. Once the specimen chamber is appropriately vented with air or N_2 gas, the chamber door is opened, and the specimen which is already held in a holder is placed onto the specimen stage. Some microscopes use specimen exchange airlock (load lock) system to keep the vacuum within the chamber intact during specimen exchange. Once the specimen is placed onto the stage, the door is closed, and the chamber is evacuated by pressing or clicking the appropriate button. In order to reduce the level of contamination in the chamber and keep good vacuum, any components exposed to the inside of the chamber including the holders, stubs, and specimens should be handled with lint- and powder-free gloves. For high vacuum operation, the specimen should not outgas or get damaged. The specimen is brought under the objective lens by clicking appropriate buttons in the software program. Some microscopes require aligning of the specimen stage. The specimen is brought up to the correct working distance (WD). This could be 10, 5, 2 mm, etc., depending on the type of imaging required. High-resolution imaging is undertaken at short WD, while large WD is used for high depths of field and low magnification microscopy. Movement of specimen stage can be controlled manually as well as through the software. Evacuation normally takes 1–2 min during which time the electron beam cannot be switched on.

4.7.3 Image Acquisition

Once the vacuum is in the ready state, the HT button is turned on. For a microscope equipped with tungsten filament, the filament heating knob is slowly turned clockwise to gradually increase the current in order to heat the filament. This is done to the point where the screen reaches its maximum brightness; after this point, the brightness starts to decrease. This is the point of maximum saturation. Using a filament beyond this point will drastically reduce the service lifetime of the filament which can last up to 100 h of usage or more. Some users keep filament knob set at the saturation point. In this case, only the HT needs to be turned on, and the filament reaches the set saturation point by itself. In modern field emission microscopes, only the HT needs to be turned on by clicking the appropriate button in the software. Once the HT is on and the filament is saturated, a secondary electron image of the specimen should appear on the screen.

Magnification is kept low (e.g., 100× or so) and scan rate is set to a fast raster scan. Brightness and contrast are adjusted. Contrast is set at minimum value and brightness is adjusted to show a slight change in intensity to the screen. Contrast is then increased to get a reasonable image on the screen. The auto brightness contrast feature can be used to get an appropriate image. Magnification is increased to 1,000× or so and focus adjusted. If the sample starts to charge up or show beam damage, a faster scan speed is used.

Appropriate spot size is selected. Spot size dictates the amount of current in the beam. The smaller the spot, the lower is the beam current. Smaller spot size will reveal finer details in the specimen but the image will get noisier. So, a balance needs to be achieved for a current setting that reveals as much as detail of the specimen without rendering the image too noisy. A noisy image is improved by lowering the scan speed. Spot size is controlled through a condenser lens which is the first lens beneath the electron gun. Spot size is the actual area on the specimen where the beam is focused. Focused beam area and the beam current both increase with increasing spot size. Smallest spot size is selected for ultrahigh-resolution imaging (e.g., >200,000×), intermediate size is used for standard imaging, while bigger spot size is required for microchemical EDS analysis, cathodoluminescence, EBSD, etc. Larger than required spot size gives out of focus images, while smaller size produces grainy images due to low signal strength. Good focus and astigmatism correction are indicative of optimum spot size.

The focus is controlled through the objective lens which is the last lens in the SEM column. Microscopes usually have coarse and fine control knob for adjusting the focus. Usually, a feature of interest with distinct edges on a specimen is used for focusing. Appropriate scan rate (about 0.1 μs to 3 μs dwell time) is selected. Different areas of interest can be examined using x and y stage controls operated through manual knobs provided on the door of the SEM chamber or through the software using a handheld device such as a mouse. The specimen can also be rotated using rotation controls in order to align particular features in the specimen.

The next step is to remove astigmatism. In order to check for astigmatism, the image is magnified to 10,000×, and the focus knob is turned to positions of under-

and overfocus. If the image stretches to opposite directions 90° apart during this operation, the image is deemed to be astigmatic. To remove astigmatism, the image is set midway between under- and overfocus, and one of the knobs for astigmatism correction is used to sharpen the image as much as possible. The image is refocused again followed by adjustment through the second knob provided for astigmatism control. This procedure can be repeated to get a sharp image. Astigmatism needs to be corrected when there is a change in imaging conditions, objective lens aperture, or after specimen exchange. Astigmatism in the image is usually better visible at higher magnifications (3,000× or more). It is not possible to carry out astigmatism correction fully if the objective lens aperture is dirty or if the magnification is too high for the beam spot size in use or if the sample is charging.

Magnification is modified to the proper level to take an image. Brightness and contrast are also adjusted. High brightness and low contrast produce soft images, while high contrast and low brightness produce sharp images. Image quality is enhanced by adjusting contrast, brightness, magnification, and focus with an aim to maximize the image quality. Care is taken not to scan the area of interest for too long in order to avoid contaminating or damaging the sample before the final image is taken. The usual practice is to move away from the feature of interest onto the adjacent area using x and y stage controls and focus until the image is sharp. Focusing is performed at a higher magnification than the one at which the image is taken. For example, for an image required at 10,000×, focusing is performed at 30,000× or so. An image is taken by pressing the appropriate button on the control or clicking the button in the computer software. Modern microscopes provide filtering functions which improve image quality by averaging two or more frames. It can be used to decrease the high noise level generated during fast scans. Different frames can be added into a single averaged frame. However, it is necessary to ensure that the specimen isn't charging and the beam is stable for this function to produce good results. Images can be saved in many formats including TIF, BITMAP, JPEG, GIF, PNG, etc.

Advanced microscopes have more than one lens including immersion or semi-immersion lens for high-resolution microscopy. The specimen is usually placed very close to this lens at a short working distance so that the specimen is immersed in the high magnetic field created by this lens. The correct lens mode needs to be selected, and the specimen (if magnetic) is to be held securely in its position to avoid being pulled by the field.

4.7.4 Microscope Alignment

It is necessary to align the microscope column after each filament change. The SEM needs to be properly aligned during microscopy to get optimum imaging. The purpose of alignment is to get the gun, lenses, and apertures concentric about the optic axis which can be considered as an imaginary line passing through the center of the SEM column. Gun alignment procedure may vary depending on the microscope model. A general guideline to align a microscope is as follows:

A conductive specimen is placed at 10 mm working distance and focused at a magnification of 10,000× with an accelerating voltage of 30 kV. Objective aperture used is large, and condenser lens strength (beam current/spot size) is relatively high (i.e., large spot size).

Firstly, the objective aperture is aligned by activating the focus wobbler which starts to change the focus of the objective lens automatically from over- to underfocus positions. Due to aperture misalignment, periodic change of focus results in the translation of the image. Aperture misalignment is corrected by adjusting X and Y knobs provided on the column near the aperture. Once the aperture is aligned, image translation diminishes and the image appears to wobble in one position.

Secondly, the gun tilt is aligned by adjusting the X and Y controls provided on the SEM console. The brightest image on the screen is obtained at the correct alignment. The same procedure is adopted to correct the gun shift.

Thirdly, stigmators need to be aligned along the optic axis. For each of the X and Y stigmators, image movement or stretching is minimized by using X and Y controls. The focus is adjusted every time a stigmator control is used. The adjustment should result in a sharp image which should not stretch or elongate when the focus is changed.

Checklist for acquiring good quality images can be summarized as follows:

- The microscope is aligned with correctly mounted and properly cleaned filament assembly.
- The specimen is prepared and mounted on the holder. It is preferably coated and electrically ground to specimen holder.
- Proper objective aperture size (typical range 30–100 μm) is selected, i.e., for high-resolution microscopy (30 μm) and for general imaging and EDS (40–50 μm).
- The appropriate accelerating voltage is selected.
- Optimum working distance is selected.
- Appropriate probe current is selected.
- The specimen is focused.
- Astigmatism is removed.
- Brightness and contrast are set at an optimal level.

4.7.5 Maintenance of the SEM

Maintenance of the SEM is essential to keep it at an optimum operable condition and to realize its maximum useful service lifetime. Both preventive and corrective maintenance on a regular basis are important. Most of the complicated maintenance and regular servicing of the SEM is carried out by qualified service engineers. The tasks required for an operator to undertake for the upkeep of instrument are kept to a minimum. Reliable instrumentation used in the microscope ensures long uptimes and renders frequent servicing unlikely. Usually, service engineers are contracted to pay 6-monthly visits for regular servicing and maintenance of the instrument. Some of the maintenance activities are summarized as follows:

All parts exposed to the electron beam are kept clean and highly polished. The aim is to free them from dirt, scratches, or any media which can charge-up and degrade the image. During operation of the SEM, some contamination builds up in the column and chamber. These are cleaned and polished. Components such as removable detectors are also cleaned. Lint-free cloth with a small amount of soft scrub is used for cleaning. A cotton swab or toothpicks can be used for inner and small holes, respectively. Lint-free nylon or latex surgical gloves are worn during the cleaning operation. Cleaned parts (not detectors) are washed with deionized or distilled water in an ultrasonic bath to remove any contamination or polishing residue. It is again cleaned with alcohol or isopropanol. Threaded parts are not polished as they are not exposed to the beam. They can trap cleaning material and become a source of contamination.

Specimen stage is inspected periodically and cleaned of any residue samples that might have fallen during specimen exchange. Small vacuum cleaner or small bursts of dry nitrogen gas are used for this purpose. Abrasives and solvents are not used to clean stage components. Care is taken not to cause harm to the pole piece or detectors within the chamber.

Sample holders are cleaned using a lint- free cloth and mild abrasive cleaner. They can be rinsed in tap water and ultrasonically cleaned in distilled water or alcohol. Parts should be washed separately. The Water chiller is checked on a regular basis to make sure there are no leaks and water temperature and pressure is within prescribed limits.

Components such as emitter, anode aperture, standard apertures, extractor aperture, pre-vacuum pump, etc. need to be serviced at regular intervals or changed when required. O-ring seals should be replenished. Bake out of the SEM column should be performed at regular intervals (after several months) to keep vacuum at an optimum level. Circuit breakers should be checked. Protective covers should be kept in good condition. Safety labels should be legible. Hard drives of the computer should be defragmented and cleaned. The computer system should be protected by an antivirus system.

A logbook that contains the complete record and history of any maintenance done on the equipment should be kept by the custodian of the equipment or lab. An up-to-date inventory of the spares should be managed.

4.8 Safety Requirements

4.8.1 Radiation Safety

The SEM produces ionizing radiation (x-rays) when high voltage electron beam strikes the specimen surface or any walls of the SEM column or chamber. High energy BSE emanating from the specimen can also produce x-rays when they

4.8 Safety Requirements

interact with the SEM components. Exposure to x-rays can produce permanent damage to the human body such as skin burn, pigment alteration, dermatitis, and tumor. Hence, safety regulations need to be enacted and safe practices are to be followed to minimize radiation hazards and ensure personal safety.

The SEM manufacturers provide proper shielding to prevent any radiation leakage. Radiation safety tests should be conducted at the time of purchase and upon installation. The joints between different sections of the SEM column and the points where apertures are located at the column are the sensitive points. Interlocks of the machine should be checked and the ports within the SEM chamber need to be secured against any x-ray leakage.

Radiation should be checked at high beam voltage and current and with all apertures removed. Radiation leak check should be conducted at least every 1–2 years. Radiation level should be comparable to the background. Radiation limit of 5 µSv/r is considered safe. Operators/users should be educated and made aware of the radiation hazards associated with the equipment. The warning label should be put at the door of the room where the SEM is located. A similar label should be posted on the microscope itself clearly stating that it is a radiation generating equipment.

4.8.2 Safe Handling of the SEM and Related Equipment

The SEM should be operated by authorized trained personnel only. Operating procedure should be prepared and made available to all interested personnel. Proper training sessions should be organized. Start-up and shutdown procedures should be summarized. The microscope usage should be password controlled. The logbook should be available to accurately record personnel and usage data. User operational manual of the SEM should be at hand. Safety devices should not be allowed to be tempered with. Select personnel should have the clearance to override interlocks or warning devices. Rules for electrical safety should be followed to avoid high voltage shocks from equipment such as sputter coaters. Electrodes in vacuum evaporators should be handled carefully to avoid burns. Eye protection should be worn, and direct observation of the heated bright filament should be avoided to prevent eye damage. Pressurized gas cylinders are to be handled with established safe work practices.

4.8.3 Emergency

Record of all SEM machines in an organization should be kept complete with serial number, model, manufacturer, date of installation, and contact information. Standard emergency procedures should be in place. Emergency contacts should be available. First aid kits should be kept stocked.

References

1. Abbe E (1873) Über einen neuen Beleuchtungsapparat am Mikroskop [About a New Illumination Apparatus to the Microscope]. Archiv für mikroskopische Anatomie (in German). Bonn, Germany: Verlag von Max Cohen & Sohn. 9: 469–480 doi:https://doi.org/10.1007/bf02956177
2. Den Dekker AJ, Van Den Bos A (1997) Resolution: A Survey. J Opt Soc Am A 14(3):547–557
3. Reimer L (1998) Scanning electron microscopy: physics of image formation and microanalysis. Springer, New York
4. Rayleigh L (1874) On the manufacture and theory of diffraction gratings. Phil Mag 47(310):81–93
5. Yao N, Wang ZL (2012) Handbook of microscopy for nanotechnology. https://doi.org/10.1073/pnas.0703993104
6. Rose A (1948) The sensitivity performance of the human eye on an absolute scale. J Opt Soc Am 38:196–208
7. Rose A (1973) Vision: human and electronic. Plenum, New York, NY
8. Goodhew PJ, Humphreys JF, Beanland R (2001) Electron microscopy and analysis, 3rd edn. Taylor and Francis, New York
9. Goldstein JI, Newbury DE, Joy DC, Lyman C, Echlin P, Lifshin E, Sawyer L, Micheal JR (2003) Scanning electron microscopy and X-Ray microanalysis, 3rd edn. Springer, New York
10. Bozzola J, Russell L (1992) Electron microscopy, 2nd edn. Jones and Bartlett Publishers, Massachusetts, USA
11. Scanning Electron Microscope A to Z (2009) JEOL Ltd, p. 8. https://www.jeol.co.jp/en/applications/pdf/sm/sem_atoz_all.pdf
12. Hafner B (2007) Scanning electron microscopy primer. University of Minnesota, Twin Cities, pp 1–29 http://www.charfac.umn.edu/sem_primer.pdf
13. Egerton RF, Li P, Malac M (2004) Radiation damage in the TEM and SEM. Micron 35(6):399–409 https://doi.org/10.1016/j.micron.2004.02.003

Specialized SEM Techniques

5

This chapter describes various imaging techniques used in the SEM. Some of these techniques require specialized equipment, devices, or detectors, while others are simply accomplished by manipulating standard operational parameters available with the SEM. The techniques discussed in this chapter include imaging at low voltage and low vacuum, focused ion beam (FIB), STEM-in-SEM, electron backscatter diffraction (EBSD), electron beam lithography, electron beam-induced deposition (EBID), and cathodoluminescence.

5.1 Imaging at Low Voltage

The two types of secondary electron signals obtained from a specimen can be designated as SE_1 and SE_2. The secondary electrons SE_1 are produced within the narrow escape depth of the specimen and are localized within a few nanometers of the impinging electron beam. These electrons make up the high-resolution image as they correspond to local fine features. Secondary electrons SE_2 are produced by the backscattered electrons that inelastically scatter secondary electrons which in turn emanate from the specimen surface. These secondary electrons carry a low-resolution signal. Similarly, those backscattered electrons that emanate from the immediate vicinity of the incident beam and have lost minimal of incident energy form high-resolution backscattered signal. A high-resolution image is obtained by separating the high-resolution SE and BSE signals from low-resolution SE and BSE signals, respectively. One way to achieve that is to undertake imaging at low accelerating voltages ranging from 0.1 to 5 kV. At high accelerating voltage (i.e., 15–30 kV), the signal is derived from larger depths of the specimen which tends to obscure surface features. As the beam energy is lowered, the specimen interaction volume decreases sharply resulting in a high-resolution high contrast SE and BSE signal emanating from fine features close to the specimen surface giving rise to images with greater surface detail. Due to the small interaction volume generated at low kV, both SE and BSE signals produced are of a high spatial resolution giving rise to stronger image contrast. Production of low-resolution signals such as SE_2, SE_3, and BSE farther away from the probe is eliminated. Effects of specimen charging and edge brightness are also reduced at low beam energies. This imaging technique is also suitable for beam-sensitive specimens as it minimizes radiation damage.

However, use of low voltage during imaging is not free of challenges (also see Sect. 4.1.2.4). Disadvantages of this technique include decreased gun brightness, increased chromatic aberration (see Eq. 2.10), increased diffraction contribution at the aperture (see Eq. 2.11), and contamination buildup relative to the low depths from which the signals are generated. Low-energy beams are also susceptible to electromagnetic interference effects. If the beam current and gun brightness are kept constant, operation at low kV results in a significantly larger spot size resulting in decreased resolution. The brightness of gun source decreases due to lowered accelerating voltage; however, signal-to-noise ratio remains solid down to 500 V due to an increased SE signal. Use of high brightness source with low energy spread, and an immersion lens typically used in the modern field emission microscopes helps to maintain reasonable image contrast and provide surface-sensitive information. Cold field emitters are also least affected from *Boersch effect* [1] (e.g., defocusing at crossover) due to low beam current employed in this type of gun. It is advisable to employ short working distance during imaging at low voltages to mitigate the effects of lens aberrations and any extraneous electromagnetic field present in the work environment. Images taken at low beam energy appear *flatter* (less 3-D like) and translucent (less solid). Compositional (Z) contrast is also less evident. The usual practice is to undertake BSE imaging at low voltages to avoid charging effects, although the images tend to be slightly noisy. The rate of contamination buildup can be reduced by avoiding high magnification, by focusing and removing astigmatism in an area other than that used for imaging, and by not using spot or reduced area raster mode. Clean specimen chamber with high-quality vacuum system, stable and vibration-free platform and proper shielding from electromagnetic influences has enabled imaging at a few tens of volts. Example of images taken at three different accelerating voltages is shown in Fig. 4.8a–c.

5.1.1 Electron Energy Filtering

Various advanced techniques have been developed for low voltage imaging [2, 3]. One approach is to use energy filtering of the signal that enters the in-lens or through-the-lens (TTL) detector present in the SEM that is equipped with field emission gun and an immersion lens. The field emission source and advanced optics serve to achieve a probe at nanometer scale, while the filter works to separate the low-energy SE from the high-energy SE and BSE. The final image can be selected to compose of mainly SE or BSE or combination of both depending on the detected signal. Two types of filters are used in commercial SEMs. One employs a control electrode with an E × B filter (developed by Hitachi High-Technologies) [4, 5], and the other is known as r-filter (JEOL Ltd.) [6].

5.1.1.1 E × B Filter
The control electrode is located within the objective lens in the SEM column below the upper detector. Conversion electrode and E × B filter (Wien filter) are positioned in the column above the control electrode (see Fig. 5.1a). E × B filter has electrostatic and magnetic field crossed at right angles to the trajectory of SE. When the control electrode is positively biased, low- and high-energy SE enter the detector and line-of-sight BSE strike the conversion electrode to emit SE which then enter the detector. When the control electrode is negative (see Fig. 5.1b), low-energy SE can be rejected, or its detection can be controlled in combination with BSE by regulating the extent of negative bias on the control electrode. Low-energy SE are primarily responsible for charging effects. The ability to filter low energy from the high energy signal enables better control during low voltage imaging.

BSE to SE conversion occurs at a large solid angle of collection. The SE yield for each backscattered electron that strikes the conversion electrode actually increases at low accelerating voltages, making this technique quite useful for low voltage imaging. The SE signal generated in this manner contains information about the BSE emanating from the specimen.

5.1.1.2 r-Filter
In this technique, a cylindrical conversion electrode is placed within the objective lens as shown in Fig. 5.2. The voltage applied to the electrode produces an electric field which deflects the electrons of certain energy and eliminates them through collision with cylinder walls. In this manner, the signal to the detector can be continuously controlled by regulating the voltage applied to the electrode. This serves to "filter" the signal on the basis of energy and can be used to create SE or BSE image or a combination of both. Figure 5.3a, b shows images taken using r-filter technology.

5.1.2 Detector Technology

5.1.2.1 Energy Selective Backscatter (EsB) Detector (Made by Zeiss)
Energy filter can be made part of the detector. Zeiss has manufactured an in-lens scintillator detector with a filter attached to its front end for its GEMINI SEM column. It is called EsB (energy selective backscatter) detector and is located above the in-lens SE detector [7, 8]. It is used to create BSE images. The filtering grid fixed at the front of the detector is set at 0 to −3 keV potential that repels low-energy SE and allows only BSE to pass through to the detector.

5.1.2.2 Upper Electron Detector, UED (Made by JEOL Ltd)
A set of two detectors can be used in the column. The filter attached to the upper detector is set at a potential, for example, −300 V. Electrons with kinetic energy

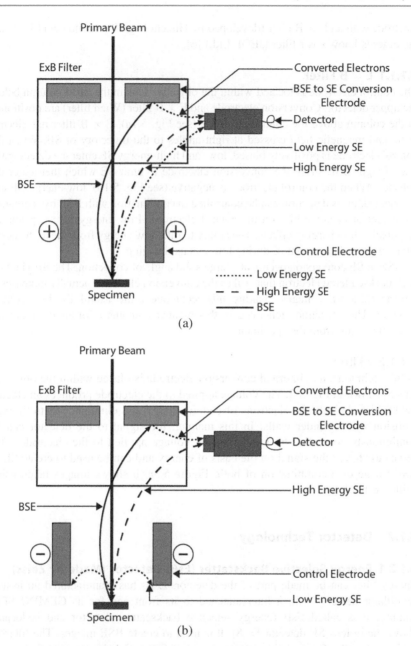

Fig. 5.1 Schematic showing energy-filtering techniques that consist of a control electrode and an E × B filter that has electrostatic and magnetic fields perpendicular to the incoming SE signal. (**a**) When the control electrode is positively biased, low and high-energy SE enter the detector. Line-of-sight BSE strike the conversion electrode to generate SE which are also directed into the detector. (**b**) When the control electrode is negative, low-energy SE are rejected, while high-energy SE and BSE are detected. The magnitude of negative bias on the control electrode can be regulated to determine the energy range of the electrons that are detected

5.1 Imaging at Low Voltage

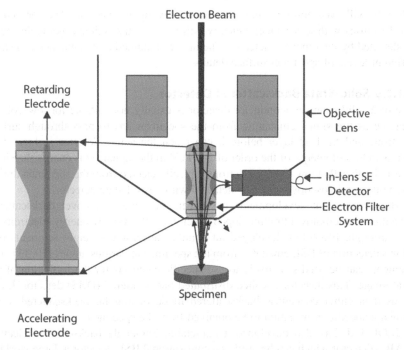

Fig. 5.2 Schematic showing r-filter technology where cylindrical conversion electrode placed within the objective lens deflects the electrons within a selected energy range and eliminates them through collision with cylinder walls. The signal is thus filtered based on electron energy and is used to create SE or BSE image or a combination of both

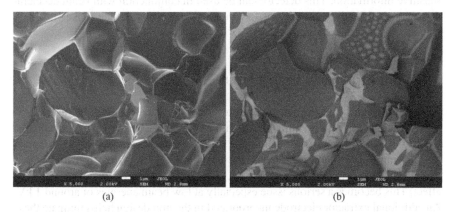

Fig. 5.3 SEM images of fractured Al alloy surface obtained at an accelerating voltage of 2 kV and a working distance of 2.8 mm, showing the capability of r-filter developed for energy filtering. (**a**) SE image formed predominantly by SE showing clear surface details and (**b**) BSE image formed using primarily BSE showing the presence of a distinct phase (light gray contrast) within Al grains. Note the surface details within Al grains are relatively obscured in this image

>300 V will pass through and be detected by upper electron detector (UED) [9]. Electrons with lesser kinetic energy will not pass through and get deflected to be detected by the other SE detector. The images obtained in this manner will show different levels of specimen surface details.

5.1.2.3 Solid-State Backscattered Detector

Use of a solid-state semiconductor detector is usually not suitable for low voltage imaging as the signal emanating from the specimen has to pass through surface electrode and Si dead layer before it can reach the active detector surface. This results in loss of energy of the order of 2–5 keV in the signal. This energy threshold clearly implies that low-energy electrons typically ejected from the specimen during low voltage microscopy cannot be detected with a solid-state detector.

However, recent developments have seen the advent of novel backscattered detectors that employ ultrathin doped layer that allows low-energy electrons to pass through. FEI® has developed an annular detector with eight segments that allows detection of BSE emanated from the specimen at various angles [10, 11]. The segments can be used selectively to use specific electrons for the formation of the final image. This detector is called distributed backscattered (DBS) detector. It can be used in either concentric backscattered mode or angular backscattered mode allowing specific information to be obtained from the specimen.

JEOL Ltd. has also developed a retractable low-angle backscattered electron (LABe) detector which can be used as a conventional BSE detector at large working distances and intermediate accelerating voltages to collect high-angle BSE [12, 13]. It can also be employed at a very short working distance where it collects low-angle BSE at low accelerating voltage. In the latter case, it provides surface-sensitive information. This detector can be used in conjunction with beam deceleration with final landing energies of few hundred volts only, which makes it highly suitable for imaging charging samples.

5.1.3 Electron Beam Deceleration

As mentioned earlier, there are some disadvantages associated with the use of low accelerating voltage during imaging. Unlike at high beam energy, space charge within the SEM column is not negligible during low voltage operation. Gun emission current is low, energy spread of the beam is large, and crossover diameter is increased due to Coulomb interaction between the electrons known as the *Boersch* effect. This becomes a serious issue especially at low beam energies of around 1 kV. An additional extractor electrode incorporated in the gun design helps mitigate these effects. Another advancement to overcome the drawbacks of low voltage imaging has been the introduction of *beam deceleration* [14]. In this technique, the electron beam is kept at high energy as it passes through the SEM column. Once it exits the final lens, the beam is decelerated before it strikes the specimen surface. By maintaining the beam at high energy during its movement through the column and lenses, large energy spread, *Boersch* effect, and chromatic aberrations are avoided.

5.1 Imaging at Low Voltage

The beam lands on the specimen surface with lesser energy which serves to reduce beam penetration and interaction volume. Beam deceleration technique manages to inhibit specimen charge-up by reducing landing energy significantly. With this technique, greater flexibility in the selection of beam voltages becomes available. It enables detection of electrons scattered at shallow depths emphasizing its surface features. It improves microscope resolution and contrast at low accelerating voltages. Beam deceleration is a relatively simple technique that can be incorporated within the existing electron sources and columns eliminating the need for a separate SEM system.

Beam deceleration is accomplished by applying a negative bias (up to -4 kV) to the stage which sets up an electric field between the specimen and the detector, acting as an additional electrostatic lens working to retard the beam accelerating voltage immediately before it hits the specimen. The energy with which the beam lands onto the specimen surface is known as *landing energy* and is equal to accelerating voltage minus stage bias. The landing energy can be controlled by varying the electron gun voltage and stage bias to achieve the optimum imaging quality. Since the beam is confined to a small surface area on the specimen, the effect of stray magnetic fields on imaging is also curbed. The electric field generated on the specimen surface due to stage bias tends to counter small electric fields that may otherwise exist at the sample surface under usual imaging conditions. This serves to minimize effects such as streaking and any possible disruptions to the trajectories of electrons emitted from the specimen. In addition, emitted secondary electrons are accelerated during beam deceleration, which increases signal collection efficiency. The specimen, however, needs to be flat to be able to remain unaffected by the strong electric field created at its surface. For rough, tilted, or composite samples consisting of conductive and insulating material, complex electric fields generated at the specimen surface may render the use of beam deceleration technique less viable.

Beam deceleration can be used with both backscattered and secondary electron detectors. However, backscattered detectors are more suitable, while the standard E-T detector is less efficient. SEM images shown in Fig. 5.4a, b reveal the benefit of beam deceleration technique. It can be seen that surface details are clearly visible without any charging effects in Fig. 5.4b. Figure 5.4c–f show SE and BSE images of uncoated nonconductive toner cartridge particle and paper samples obtained at various landing energies of 300–2000 eV. It can be seen that surface details are visible without significant charge-up.

5.1.4 Recent Developments

Aberration correctors have been developed for SEM, and this technology has seen great improvements in the last decade. Multipole aberration correctors are used to minimize chromatic and spherical lens aberrations in microscopes equipped with cold field emission guns [15–20]. Despite remarkable technological advancement, their use is presently cumbersome and puts limitations on the depth of field and size and shape of the specimen that can be examined. Nevertheless, aberration correction

Fig. 5.4 Use of beam deceleration technique enables imaging of nonconductive materials at high magnification without any significant charge-up. High magnification secondary electron image at (**a**) 500 V and (**b**) 300 V. The latter image shows more surface details. (**c**) Backscattered SEM image of toner cartridge sample. Landing energy is 300 V. (**d**, **e**) Backscattered SEM images of paper at landing energies of 500 and 2,000 eV, respectively. (**f**) Secondary electron image of paper at landing energy of 2 keV

technology is expected to improve steadily in the future and overcome these drawbacks.

An advanced design suitable for low voltage imaging consists of an electron source that is immersed within the electromagnetic field of low aberration condenser lens and can produce 5 nm spot size with 5 nA current at an accelerating voltage of 3 kV. Present-day SEMs have demonstrated resolutions of 1.4 nm at 1 kV and 5 nm at 0.1 kV. Another important development in the field of low voltage imaging has been the introduction of a monochromator for field emission SEM. Such FE-SEM reduces the energy spread of Schottky field emission source to <0.2 eV, thus diminishing the influence of chromatic aberration. This makes imaging at low kV possible, without having to forego high current capabilities of Schottky field emitter [2, 3].

5.1.5 Applications

FE-SEM can be used to observe nanomaterials such as nanoparticles, nano-wires, and nanotubes by utilizing a STEM-in-SEM technique. This method reduces knock-on damage, compared to imaging with conventional high voltage TEM or STEM. Low kV FE-SEM can also be used to image layers of graphene deposited on Ni substrate. The lower rate of thermal damage at low kV also makes this technique suitable for beam-sensitive materials. By using a combination of reduced beam current and low kV, FE-SEM can be used to image semiconductor materials with high resolution and minimum damage.

5.2 Imaging at Low Vacuum

5.2.1 Introduction

In the normal working mode, specimen chamber is usually under high vacuum (e.g., 10^{-3} Pa or less). The high vacuum within the SEM column and specimen chamber ensures smooth travel of electron beam from the gun source to the specimen without getting scattered due to any residual gas molecules. Without an appreciable vacuum, beam electrons will be scattered by air molecules present in the chamber. The high vacuum also prevents oxidation damage to the electron gun. While these vacuum conditions are suitable for imaging dry specimens, they are unfavorable for examining wet and dirty samples. Such specimens serve to degrade vacuum or contaminate the specimen chamber under normal vacuum. In addition, an uncoated and nonconductive specimen can acquire electric charge due to impingement of the electron beam at its surface. A specimen is thought to be charged up if the number of incident beam electrons impinging upon the specimen is more than the specimen current i_{sp} flowing out and secondary and backscattered electrons emitted from the specimen. Due to its nonconductive nature, the specimen is unable to discharge the negative charge. This can cause contrast variation, beam instability, and image distortion. In order to overcome these problems, SEMs capable of imaging damp, dirty, or insulating samples under low/degraded vacuum were invented. Depending on the manufacturer, these microscopes go by different names such as low vacuum SEM, variable pressure SEM, environmental SEM, nature SEM, etc.

5.2.2 Brief History

Development of such apparatus started before conventional electron microscopes were even commercialized. An arrangement for imaging liquids in conventional SEM was proposed by Thornley back in 1960 [21]. This was achieved by placing the sample between two carbon films, which prevent the liquid from evaporation, thus minimizing the contamination of vacuum present in the column and in the chamber. Further advancements in this field allowed researchers to vary the pressure in the chamber to simulate the natural environment without degrading the vacuum present in the chamber itself. The specimen chamber was converted into a substage by isolating it from the electron column with the help of pressure differential apertures [22]. Isolation of specimen chamber from the column allowed the researchers to achieve 1,000 times greater pressure as compared to the column while allowing electron gun to function and access the chamber normally. Achieving high pressure in the chamber and high vacuum in the column simultaneously is made possible by isolating both components from each other and providing a pumping system to each component. This isolation is undertaken with the help of small differential aperture or pressure-limiting aperture (PLA) placed below the objective lens, which serves to separate the vacuum in the SEM column from that in the specimen chamber (see Fig. 5.5a–c). The vacuum in the column remains high while that in the chamber can be degraded typically by 1,000×. Wet or nonconductive specimens that charge up can be imaged using SEM equipped with a low vacuum mode where the vacuum is intentionally degraded to a pressure of several hundred Pascal. Environmental SEM can operate with pressures as high as 3,000 Pa and at a high relative humidity (up to 100%) within the chamber.

5.2.3 Working Principle

In low vacuum mode, gas or water vapor is injected into the specimen chamber around the specimen surface area. High-energy electron beam penetrates the water vapor with some scatter and interacts with the specimen surface. Secondary and backscattered electrons emanating from the specimen strike the water molecules and produce secondary electrons which in turn produce more secondary electrons upon interaction with the surrounding water molecules. Water molecules are changed into positive ions as a result of this interaction with incident beam and secondary/backscattered electrons emerging from the specimen. Positive bias applied to a detector accelerates secondary electrons toward the detector, while positive ions are pushed toward the negatively charged areas on the specimen. Further ionization

5.2 Imaging at Low Vacuum

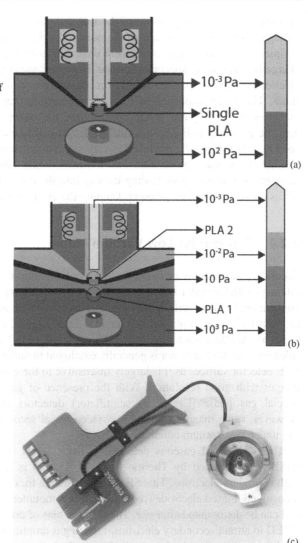

Fig. 5.5 The use of PLA allows maintaining a high vacuum within the column, while the vacuum in the chamber is degraded. (**a**) Schematic showing the use of pressure-limiting aperture (PLA) in SEM using low vacuum mode. (**b**) In another design, a combination of two PLA is used to degrade vacuum in the SEM chamber by an order of magnitude. (**c**) Photograph of PLA (seen protruding out) attached to low vacuum detector

events are produced due to this accelerated movement of electrons and ions. The water vapor thus serves to produce positive ions and also increases the number of secondary electrons resulting in gas amplification. Water vapor is a preferred choice for this application as it ionizes into positive ions easily. Generation of positive ions and their movement toward the negatively charged areas of the sample neutralize the negative charge accumulated at the specimen surface. This allows for charge-free imaging of nonconductive specimens without having to coat them with a conductive material. Another reason to use low vacuum can be to image moist or wet specimens such as clay or biological material. These samples can dry up under normal high vacuum and lose their features. Factors that affect the quality of imaging at low vacuum include vapor pressure of the injected gas, working distance, accelerating voltage, spot size, and the area of the nonconductive surface.

Injection of water vapor in the SEM chamber serves the purpose of dissipating the charge buildup at the specimen surface. However, the components within the SEM electron column needs to be saved from water vapor and kept at high vacuum at all times. This is accomplished by inserting pressure-limiting aperture (PLA) beneath the objective pole piece at the point where the beam enters the chamber. This aperture is simply a disc with a small hole (shaped like a circular conical cylinder, see Fig. 5.5c). Since the pinhole has very small dimension, it is able to separate the two levels of vacuum without *interdiffusion* while allowing the beam to pass through. Any water vapor finding its way into the column is pumped out using oil diffusion or turbomolecular pump keeping the electron gun safe.

5.2.4 Detector for Low Vacuum Mode

Secondary electron image is usually not available in low vacuum mode because SE interacts with the water vapors immediately upon emitting from the specimen. Also, conventional E-T detector relies on a high bias (+10 kV) applied upon the scintillator to enable electron-to-photon conversion. Such high bias can easily ionize water vapor and arc the detector to ground. Thus, the use of E-T detector for SE imaging is ruled out, and BSE detector is generally employed to undertake BSE imaging. Use of BSE detector suffices as it is largely insensitive to the charging effect and produces images with good resolution. With the presence of gas particles in the chamber, special gas phase (large area scintillator) detectors are required for imaging. However, some manufacturers do provide special secondary electron detectors for use under low vacuum conditions.

The specialized gaseous detector, called gaseous secondary electron detector (GSED), (developed by Thermo Fisher Scientific) is used to undertake imaging with secondary electrons. This SE detector suitable for operation in a low vacuum is a positively biased electrode (see Fig. 5.6). It is mounted on the objective pole piece and can be dismounted after use. The positive bias of up to +600 V is applied on the GSED to attract secondary electrons. Due to gas amplification, the current collected by the detector is hundreds or even thousands times greater than the original signal. The detector is also placed closer (a few mm) to the specimen compared to an E-T detector, thus collecting SE efficiently. Positive bias on GSED also drives positively ionized water molecules toward the specimen to effectively neutralize the accumulated negative charge at specimen surface.

The advantage of using GSED in low vacuum mode is that specimen can be imaged in its original state without coating with gold or palladium. Images are free of distortions or contrast variations that commonly occur due to charging effects. Pits and voids at the specimen surface show better contrast as lack of charging allows electrons to eject out of the depths of the deep holes. GSED is not sensitive to light and therefore can be used to image specimens at high temperatures. Also, wet specimens can be examined.

5.2 Imaging at Low Vacuum

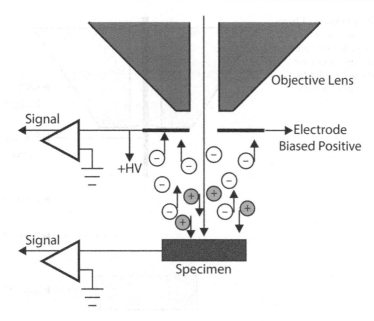

Fig. 5.6 Schematic of gaseous secondary electron detector (GSED) with a central bore

A conical pressure-limiting aperture is provided at the center of the detector to sustain low vacuum in the chamber while maintaining a high vacuum in the electron column. The smaller the bore size in the center of the GSED, the higher is the pressure that can be maintained in the specimen chamber. For instance, one-half of a millimeter can support a pressure of 1.3 kPa. Large field-of-view GSED with 1 mm bore size is also available. This type of GSED can support pressures of 750 Pa. Large bore size enables a maximum field of view of approx. 1 mm in low vacuum mode, and the minimum attainable magnification is reduced by a factor of 2 to around 125× compared to a GSED with 0.5 mm bore size.

5.2.5 Gas Path Length

The average distance traveled by an electron without colliding with an air molecule is termed mean free path (MFP). The average number (n) of scattering events per electron depends on the total scattering cross section of the gas molecule for electrons (σ_g), the pressure of the gas (p_g), and the length of the path that an electron travels in gas (termed gas path length, GPL or simply L) and is given by:

$$n = \frac{\sigma_g \, p_g \, L}{kT} \qquad (5.1) \quad [23]$$

where k and T are the Boltzmann constant and the temperature, respectively. Under normal vacuum (10^{-3} MPa) at which the SEM operates, MFP is extremely large (several kilometers), and the electrons can travel the short distance through the

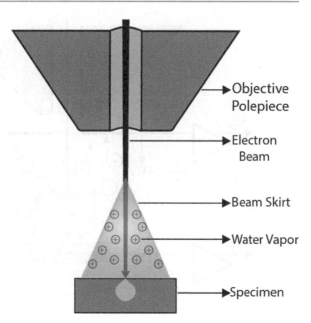

Fig. 5.7 Formation of beam skirt (electron beam scattering) due to the presence of water vapor above the specimen surface during imaging at low vacuum

column and the chamber without scattering. If gas pressure or the gas path length in the chamber is increased, the number of scattering events also increases. For instance, at 100 MPa (possible gas pressure used in low vacuum mode), the MFP reduces to around 10 mm. This is the typical working distance used in the SEM. Therefore, electron beam on its way to the specimen gets scattered at low vacuum. The scattered electrons move away from the focused beam and strike the specimen surface at a point away from the probe. This results in broadening of the electron beam which takes on a *skirtlike* form (Fig. 5.7). The skirt radius becomes large as the scattering increases and can be estimated with the following equation:

$$r_s = \frac{364\ Z}{E} \left(\frac{p}{T}\right)^{1/2} L^{3/2} \qquad (5.2) \quad [23]$$

where r_s is skirt radius (m), Z is atomic number, E is beam energy (eV), p is gas pressure (Pa), T is temperature (K), and L is gas path length (m). This radius can be as large as 100 μm at an accelerating voltage of 10 kV, water vapor pressure of 10^3 Pa, and gas path length of 5 mm.

Schematics in Fig. 5.8a–c illustrate the electron beam scattering pattern with or without gas molecules present above the specimen surface. It can be seen in Fig. 5.8a that electron beam scattering in conventional SEM is rare and electrons travel straight to the specimen without any deflection. In the presence of high pressure and imaging gas inside the sample chamber, two cases are possible. The first case is shown in Fig. 5.8b where the beam will scatter, but the central part will remain in focus. This phenomenon is known as *oligo scattering*. The second case is illustrated in Fig. 5.8c wherein the presence of high pressure, the beam will scatter completely giving rise to *plural scattering*.

5.2 Imaging at Low Vacuum

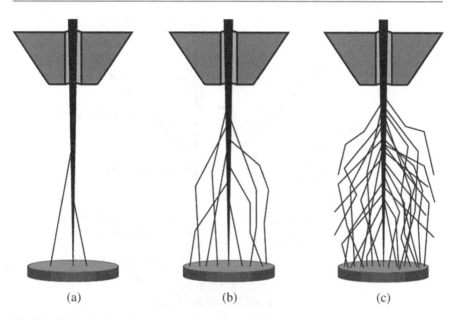

Fig. 5.8 Schematic showing scattering of the electron beam with increasing gas pressure from (**a**) to (**c**). An average number of collisions is defined as m in the figure. At $m = 0.05$, 95% of electrons do not have any collision [24]. (**a**) Minimal scattering; scatter < 5%; $m < 0.05$. (**b**) Partial scattering; 5–95% scatter; $0.05 \leq m \geq 3.0$. (**c**) Complete scattering; scatter > 95%; $m > 3.0$

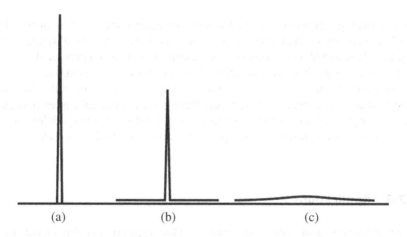

Fig. 5.9 (**a**–**c**) Schematics showing beam intensity profile corresponding to the three beam scattering cases shown in Fig. 5.8. (**a**) Minimal scattering regime; (**b**) Partial scattering regime; (**c**) Complete scattering regime

Figure 5.9a–c shows the beam intensity profile resulting from all three abovementioned cases, respectively. For the beam scattering cases shown in (a) and (b), there will not be much difference in the resolution of the image. The skirt electrons result in the generation of a signal from the area of interest and

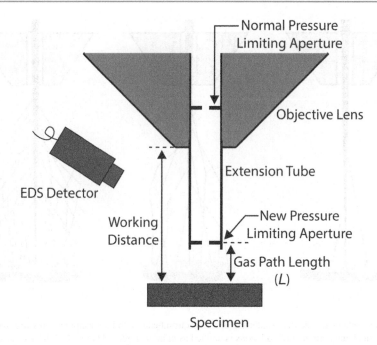

Fig. 5.10 Use of an extension tube mounted on the objective pole piece serves to reduce the gas path length (L) of electrons and results in less electron scattering in low vacuum mode

its surroundings. However, a small amount of beam current is lost in the skirt, resulting in a lower signal from the point of beam impact. In case of (c), the beam is primarily scattered and a signal of appreciable strength is not generated.

This scattering effect can be reduced by employing an extension tube with pressure-limiting aperture mounted at the end, as shown in Fig. 5.10. This long tube is fitted to the objective pole piece. Electrons enter this tube after emanating from the objective lens assembly. In this manner, the distance (gas path length) that the electrons have to travel in gas vapor is reduced, resulting in less scatter.

5.2.6 Applications

Types of specimens suitable for imaging using low vacuum capability include moist biological samples that shrink and change structure if dried. Likewise, insulating specimens such as polymers can also be imaged in a low vacuum without coating which occasionally tends to hide specimen features. Coatings may also interfere with microchemical analysis results. Similarly, wet colloids or oil-bearing rock samples can be examined in an as-received condition. Figure 5.11a, b show backscattered SEM images of a sample at high and low vacuum, respectively.

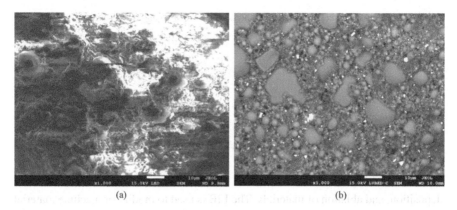

Fig. 5.11 Backscattered electron images of a sample at (**a**) high vacuum and at (**b**) low vacuum. Charge-up present at the specimen surface under high vacuum conditions is mitigated under low vacuum

5.2.7 Latest Developments

Electron beam-gas interactions limit the imaging resolution of the microscope. However, improvements are being continuously made in this regard. At present, field emission microscopes also offer low vacuum capabilities with improved spatial resolution. Pressure-limiting aperture is small (few hundred microns in diameter) allowing for greater pressure differences between the column and chamber. The vacuum in the chamber is easily controlled via a leak valve operated through computer software, and the type of gas used can be selected based on requirements. Since the PLA is placed close to the specimen, the distance the electron beam has to travel through the gas is shorter than the working distance employed during conventional microscopy.

5.3 Focused Ion Beam (FIB)

5.3.1 Introduction

The focused ion beam (FIB) is an instrument that uses positively charged heavy ions (instead of electrons) to raster the specimen surface. Use of ion source turns FIB into a versatile instrument. When the focused ion beam interacts with the surface of a material, it results in the generation of secondary ions, secondary electrons, and neutral atoms. Information from secondary electrons and secondary ions help in the formation of an image in the same manner as that in the SEM. The resolution of the FIB image can be as high as 5 nm.

Ions are heavier than electrons and carry a greater momentum. Use of heavy ions makes it easier to remove material from the specimen. Therefore, FIB is used for sputtering, etching, or micromachining of materials. It is also useful for milling,

Table 5.1 Comparison between different characteristics of FIB and SEM sources

	FIB	SEM
Particle size	0.2 nm	0.00001 nm
Charge	+1	−1
Beam energy	Up to 30 keV	Up to 30 keV
Beam current	pA to nA	pA to µA
Penetration depth in Fe	20 nm (30 keV) 4 nm (2 keV)	1800 nm (2 keV) 25 nm (2 keV)
Generation of secondary electrons per 100 particles at 20 kV	100–200	50–75

deposition, and ablation of materials. The FIB is used to modify or machine material surface on a micro- and nanoscale due to its ability to sputter materials with its positively charged heavy ions. Features that are milled can be as small as 10–15 nm in dimensions. One atom layer of a material can be etched without disrupting the layer underneath. Material removal and deposition can be controlled to a nanometer scale. Different gases can be injected into the system near the surface of the specimen to deposit required materials.

Imaging capability enables it to carry out these operations on specific sites selected by the user. It can be used to characterize and fabricate semiconductor materials and also prepare thin film sections for examination in a transmission electron microscope. Since ions are positive, large, and heavier compared to electrons and react only with outer shell electrons of the specimens, they exhibit high interaction probability and low penetration depth in specimens. Ions can also be used as dopants since they can be trapped easily due to their large mass. Comparison of ions and electron characteristics is given in Table 5.1. The spatial resolution of images generated using SEM is greater than those of the FIB. Focused ion beam uses a column similar to the one used for the SEM. It is widely used in semiconductor industry for failure analysis, circuit writing, thin sample preparation, etc.

A schematic of FIB system is shown in Fig. 5.12.

High current density removes the atoms or molecules from the surface of the specimen called sputtering. Using high current density, micromachining or milling is also achieved as the ions carry a large amount of energy and momentum. Small current density does not sputter a greater amount of material, but it still generates secondary electrons which are then used to get an image as in the SEM.

An advantage of focused ion beam over scanning electron is the characterization of a nonconducting sample which can be imaged using positive primary ion beam. An electron flood gun with low energy is used to neutralize the charge on the surface of the specimen generated by the focused ion beam, and the positive secondary ions are collected by the detector to get an image without destroying the specimen. The FIB is also equipped with gas injectors to produce specific chemical reactions at the surface during ion beam-assisted deposition or etching process. Due to adequate resolution attained in modern FIB instruments, use of a separate SEM may not be required for imaging. However, instruments are available that are equipped with two columns, one for FIB and the other for SEM (see Fig. 5.13). In these instruments,

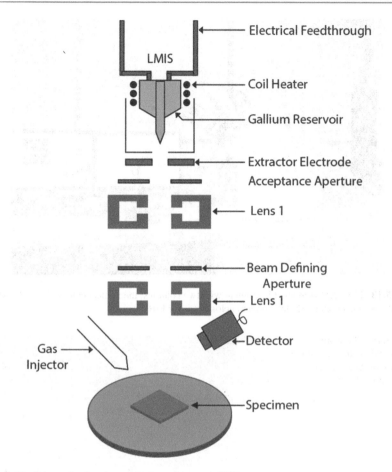

Fig. 5.12 Schematic showing various components of FIB

specimens can be prepared with FIB and can be examined using high-resolution SEM. Thin transparent TEM foil is prepared using FIB column and is imaged using STEM detector located within the same combined equipment. Availability of such equipment has made it possible to undertake high-resolution 3-D microscopy and nano-tomography.

The FIB has become a popular instrument for materials science and semiconductor industry applications. It is used for defect analysis, circuit board preparation, and repair in the semiconductor industry. It is widely used to prepare cross-sectional TEM samples in research labs. Specimens with thin areas at specific locations such as grain boundaries, cracks, pits, etc. can be produced. The ability of FIB can be seen in the example shown in Fig. 5.14 where word "KFUPM" has been milled on an aluminum alloy surface.

Focused ion beam is widely used for TEM sample preparation. The steps involved in such a preparation are shown in Fig. 5.15.

Fig. 5.13 FIB-SEM combined instrument with two columns; the vertical column is for the SEM, and the inclined column (hidden under enclosure) is for FIB

Fig. 5.14 The word "KFUPM" is milled onto the surface of Al alloy using focused ion beam

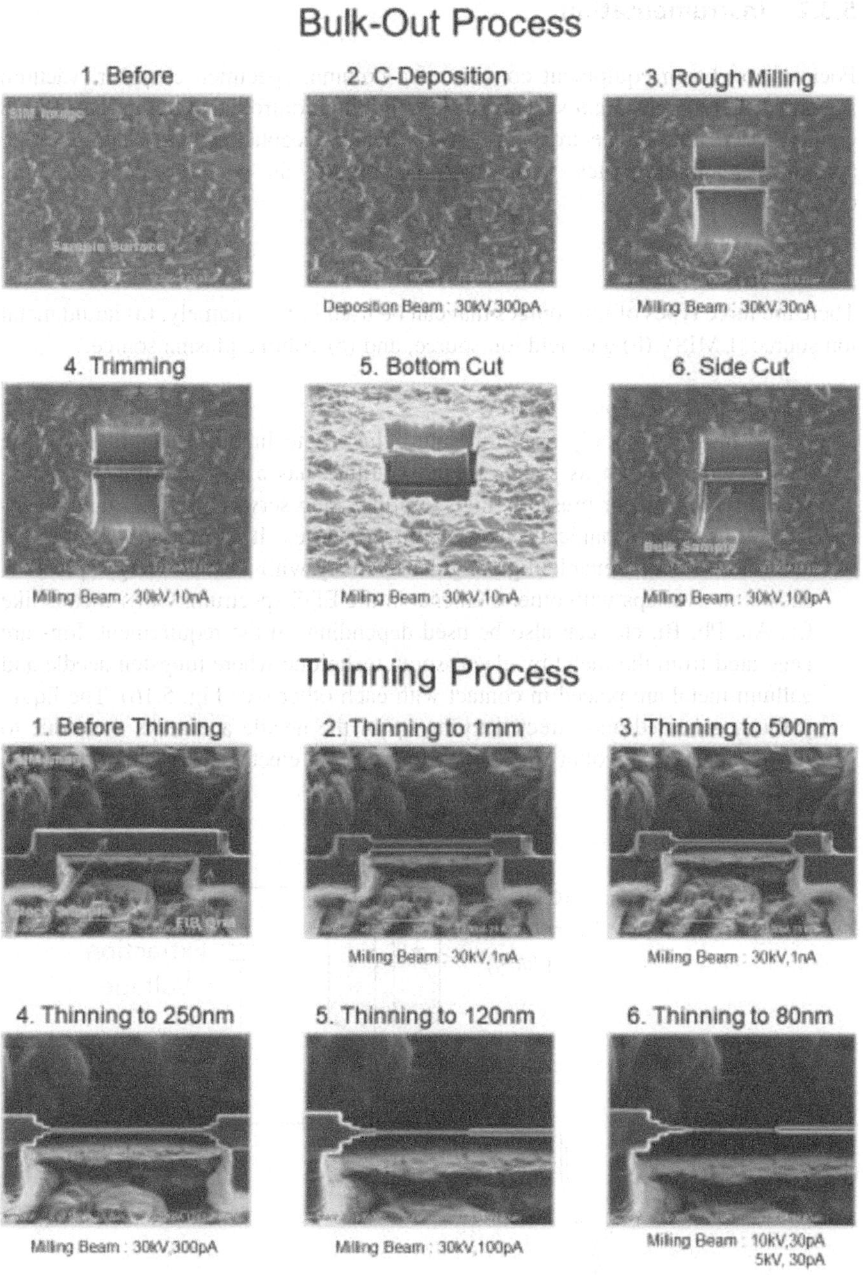

Fig. 5.15 Thin foil (lamella) sample preparation for transmission electron microscope (TEM) using a focused ion beam instrument. Complex cross-sectional samples can be made within hours. At high beam currents, a large amount of material is removed which allows very fast site-specific milling of specimens compared to that done using argon ion milling

5.3.2 Instrumentation

Focused ion beam equipment consists of a column, specimen chamber, vacuum system, detectors, gas input system, and a computer-controlled system. The column contains ion source, electrostatic lenses, beam acceptance and beam-defining apertures, blanking plates, a steering quadrupole, an octupole deflector, and detectors.

5.3.2.1 Ion Sources
There are three types of ion sources that can be used in FIB, namely, (a) liquid metal ion source (LMIS), (b) gas field ion source, and (c) volume plasma source.

(a) *Liquid Metal Ion Source (LMIS)*

The most commonly used ion source in FIB technique is liquid metal ion source with gallium as source metal. Gallium has a low melting point, low volatility, low vapor pressure, high stability, long service lifetimes, and excellent electrical, mechanical, and vacuum properties. It exhibits good emission characteristics that enable high angular intensity with a small energy spread. It shows no overlaps with other elements in the EDS spectrum. Other metals like Cs, Au, Pb, Bi, etc. can also be used depending on the requirement. Ions are generated from the metal by electrospray technique where tungsten needle and gallium metal are placed in contact with each other (see Fig. 5.16). The liquid gallium is heated and directed to the tip of the needle and stays there due to surface tension. A voltage is applied to create an electric field which results in

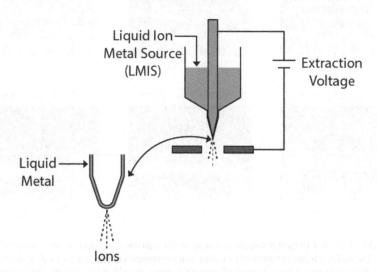

Fig. 5.16 Liquid metal ion source used in the FIB technique to generate ions. Schematic shows the formation of ions at the Taylor cone tip and their movement toward the electrostatic lenses [25]

Taylor cone formation which is the shape the liquid metal takes at the tip of the tungsten needle. The radius of the cone tip is about 2 nm. When the voltage is further increased, it results in the ejection of liquid metal from the cone tip in a thin stream. This technique is known as *electrospray* technique. The electric field acting on the cone tip is more than 1×10^8 V/cm that ionizes the liquid metal, and field ejects the ions. Ions are accelerated to 1–50 kV toward the electrostatic lens. This method produces high current density focused ion beam which has a very small spot size in the range of a few nanometers.

(b) *Gas Field Ion Source (GFIS)*

The setup of gas field ion source is similar to the liquid metal ion source. The only difference is that the liquid metal is replaced with a condensed gas. Usually, noble gases are used. Most common gases used are H_2, He, Ar, N, etc. The tungsten needle is kept at cryogenic temperatures. The gas is inserted and condensed at the needle tip. The cryogenic temperature of the needle helps in the condensation of the gas. Then an electric voltage is applied across the needle and the other electrode which results in the ionization of the gas in the same manner as in the liquid metal ion source. The current density can be enhanced by increasing the electric field at the protrusion of condensed gas. The smaller the size of the condensed gas cone tip, the greater will be the electric field and hence the higher the current density. By making the size of cone tip smaller, the electric field lines can be forced to get almost parallel to each other resulting in a very focused and thin ion beam which increases the brightness of the image [25]. Due to the short service lifetimes of around 160 h, this source has not found commercial use.

(c) *Volume Plasma Sources (VPS)*

Volume plasma sources are used for metal deposition, lithography, and ion milling machines. The desired species to be implanted is introduced in the form of gas and is bombarded with electrons to create the plasma, and an electric field is applied to accelerate the ions toward the substrate. Volume plasma ion sources generate a high current density although the ions are not emitted from a single point as in LMIS and GFIS. A brief comparison of ion sources is included in Table 5.2.

Table 5.2 Comparison of different ion sources [25]

Ion source	Ion species	Virtual source size (nm)	Energy spread (eV)	Brightness (A/cm^2 sr)	Angular brightness (μA/sr)
Liquid metal	Ga$^+$	50	>4	3×10^6	50
Gas field	H$^+$, H^{2+}, He$^+$	0.5	~1	5×10^9	35

5.3.2.2 Lens System

Unlike scanning electron microscope, focused ion beam instrument uses electrostatic lenses which focus the ions near to the lens. Upon generation, ions are accelerated in the form of a beam under the influence of an applied voltage toward the beam acceptance aperture and enter the condenser lens. Steering quadrupole of the condenser lens aligns the beam so it can pass through the center of the beam defining aperture. Steering quadrupole of the objective lens adjusts the trajectory of the beam with the optical axis of the objective lens. The blanking apertures present between the condenser and the objective lenses protect against constant milling of the specimen by the ion beam. The octupole deflectors present below the objective lens provide astigmatic correction.

5.3.2.3 Stage

The stage is fitted in the specimen chamber and holds the sample while it is being bombarded with the ion beam. One can maneuver the stage in X and Y axes either by rotation or by traversing. The stage can also be tilted as per requirement.

5.3.2.4 Detector

The detector consists of a glass array having millions of tiny channel electron multipliers forming a microchannel plate (MCP). Charge species formed after ion beam interaction with the specimen are attracted to the detector resulting in the formation of an image or in the identification of elements. Scintillators are also part of the detection system which are made from a material that converts electrons or other radiations into a photon. Detectors are just mounted at some angle above the specimen to get secondary particles generated from ion-solid interactions.

5.3.3 Ion-Solid Interactions

When a primary ion strikes the target material or solid specimen, it transfers its energy and momentum to the solid, and a number of phenomena take place, including sputtering, backscattering, ion reflection, electron emission, electromagnetic radiation emission, specimen damage, ion emission etc. The primary ion after losing all its energy comes to rest and gets deposited in the solid. During this period, from striking until deposition, it strikes a number of atoms resulting in a collision cascade. Since the primary ion carries energy and momentum, two types of ion-solid interactions can take place: elastic interactions and inelastic interactions. During the inelastic interaction, striking of a primary ion with the specimen ionizes some of the neutral atoms in it resulting in the generation of some electromagnetic radiation and emission of electrons, whereas during elastic interaction between primary ions and the specimen, primary ions transfer their energy to the specimen atoms or molecules in the form of translational energy resulting in the knocking out of atoms from the specimen or causing displacement from the initial position. This leads to damage of the specimen. The probability of inelastic scattering between a specimen and electron beam is less as compared to that of between a specimen and ions.

The energy needed to displace an atom from its original site is known as displacement energy. It is a critical value of energy. When a high-energy ion strikes a solid, it may transfer its energy as translational energy to a solid atom. If this translational energy is greater than the critical displacement energy, the ion will knock out the solid atom from its site generating a defect known as an interstitial-vacancy point defect. The primary high-energy ion may have some energy left after this collision. If this is the case, the primary ion will still move forward striking more solid atoms in its path displacing them from their original site and resulting in a cascade of collisions until it loses all its energy and gets embedded in the solid. If the interaction only happens near the surface of the solid, the recoiling atom has a chance to get out of the solid after the collision, leading to sputtering. Here it should be understood that the displacement energy is always higher than the binding energy between the two atoms which indicates that the collisions are nonadiabatic. After the primary ion has stopped, the result is the emission of electromagnetic radiation, some particles (electrons, ions, etc.), lattice defects, heat and incorporated primary ion in the solid. Monte Carlo calculations are suitable for simulation of collision cascade due to ion-solid interaction. In inelastic collisions, most of the energy of the primary ion is lost due to solid heating or vibrations of solid atoms rather than displacement.

5.3.4 Ion Imaging

Ion beam scans the surface of the specimen like the electron beam in scanning electron microscope resulting in the generation of electrons, ions, and electromagnetic radiations. In case of scanning electron microscope, the resulting electrons are called secondary electrons which are generated when electron beam hits the sample. In focused ion beam microscopy technique, the resulting electrons are often known as ion-induced secondary electrons and have low energies. For each 5–30 keV Ga ion hitting the surface of the sample, around ten ion-induced secondary electrons with an energy of 10 eV each are generated. The surface of the sample may have been oxidized before the ion beam strikes it; therefore, the electron yield ejecting out of the sample might be low. But as the ion beam strikes the surface, it results in the sputtering of the oxidized layer, resulting in the removal of a large number of electrons. Thus electron yield changes with time and is more when the surface is clean.

Electron beams are more focused than ion beams; therefore, the resolution due to the ion beam is lesser than that of the electron beam. But the focused ion beam provides a greater channeling contrast as compared to that provided by the SEM. When crystal orientation is such that the primary ion channels through the atoms in the sample, there are fewer interactions of a primary ion with the atoms, and the induced secondary ion yield is less. But when the orientation is such that primary ion channels through the atoms but interacts with more atoms near the surface, more induced secondary electrons escape from the surface of the sample resulting in a higher contrast. Atomic mass has a direct relationship with the contrast of the image. The heavier the atomic mass of the sample, the greater is the probability of generation of secondary induced electrons as compared to that of the sample with lower atomic weight. Surface geometry also affects the contrast of the image.

5.4 STEM-in-SEM

Traditionally, transmission electron microscopes (TEM) have been used to study the microstructure of materials at high accelerating voltages of 120–300 kV. In this type of microscope, the electron beam is transmitted through the specimen to form an image. The TEM can be equipped with a scanning mode where the beam is scanned and transmitted through the specimen at the same time. This type of equipment is known as scanning transmission electron microscopes (STEM). In another variation, a specialized TEM that can only be used in the scanning transmission mode is called "dedicated" STEM. However, advances in technology have enabled the use of SEM in a scanning transmission mode, providing a cost-effective alternative. This is undertaken by mounting a STEM detector within the SEM and using ultrathin specimens that are transparent to electrons at 30 kV. Resulting STEM-in-SEM is used to undertake scanning transmission electron microscopy in addition to the conventional study of the surface topography of materials. STEM detector is considered an important tool that serves to enhance the characterization capability of the SEM without having to invest heavily in the acquisition of the more expensive transmission electron microscope. Use of STEM detector allows imaging of inner structural details of materials such as alloys, coatings, carbon nanotubes, nanopowders, catalysts, etc.

5.4.1 Working Principle

Scanning transmission electron microscope (STEM) detector is mounted below the sample stage and is used to collect electron signal transmitted through the specimen. The stage is altered to allow the continued progression of the electron beam and also to create space for the placement of STEM detector. The specimen has to be thin enough to transmit an electron beam with an accelerating voltage typically used in the SEM. A small probe is scanned across the sample surface, and the detector collects the signal after the beam is scattered and transmitted through the specimen. Image formation is similar to a conventional SEM where signal obtained from a specific location of the specimen during the scan is processed through the detector and displayed on a corresponding location on a monitor that is scanned in sync with the beam scan. The strength of signal obtained from a pixel of the specimen determines the intensity of the corresponding pixel on the monitor during the synchronized scan. Fine powders and nanotubes can be examined in an as-received condition, while metallic specimens are prepared using standard TEM sample preparation techniques such as electro-jet polishing and ion beam milling.

A STEM detector can be a scintillator/photomultiplier or solid-state type detector as shown in Fig. 5.17a. This type of detector is used to form both bright and annular

5.4 STEM-in-SEM

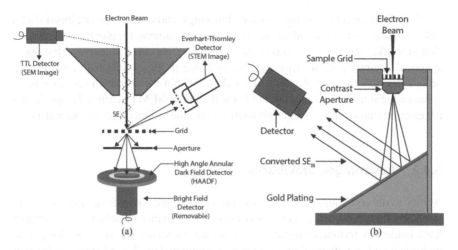

Fig. 5.17 (a) Schematic representation of STEM-in-SEM arrangement where the electron beam is scanned over and transmitted through an ultrathin specimen. Electrons scattered at low angles are collected by a bright field (BF) detector, and those scattered at high angles are intercepted by high-angle annular dark field (HAADF) detector. (b) Schematic representation of less expensive "STEM converter" arrangement where the beam transmitted through the specimen is scattered by slanting Au-coated Si surface. Secondary electrons generated due to beam interaction are collected by a conventional E-T detector placed in the specimen chamber

dark field images. The bright field image is formed by collecting electrons that are scattered at small angles and are centered on the optic axis of the microscope while passing through the specimen (e.g., *direct beam*). The *incoherent* dark field image is formed by (off-axis) electrons scattered at high angles and shows the atomic number and mass-thickness contrast. The degree of contrast shown by different elements varies depending on their atomic number. The detector that collects the strongly scattered electrons to form a high STEM image contrast is called high-angle annular dark field (HAADF) detector. The size of the detector is fixed, but the collection angle of the detector can be varied by moving the detector away or close to the specimen. Use of a multi-segment detector allows collection of signals by each segment independent of each other which then can be added or subtracted to form an image.

An inexpensive alternative to the scintillator or SSD device is to have a "STEM converter" as shown in Fig. 5.17b. In this arrangement, the beam transmitted through the specimen passes through an underlying small aperture (few mm in diameter) and is scattered by Au-coated mirror surface placed at an angle. The impact generates secondary electrons that are collected by a conventional E-T detector available in the specimen chamber. Aperture size can be changed to vary contrast. Smaller-sized aperture will produce stronger contrast but a smaller field of view. This type of detector is less expensive but forms images with low signal-to-noise ratio and is used for bright field imaging only. Through-the-lens (TTL) detector placed in the column above the specimen can be used to capture the signal scattered upward.

The STEM detector is usually installed in a high-end microscope equipped with a field emission gun to take advantage of its high brightness. Employment of immersion or snorkel objective lens allows the use of short working distance which reduces spherical aberration in the lens and creates a fine probe. Additionally, use of thin section restricts the size of interaction volume within the specimen resulting in an enhanced spatial resolution of 0.6 nm in the STEM-in-SEM [3]. Since the specimens are electron transparent, the contrast formed is similar to that in a conventional TEM.

5.4.2 Advantages/Drawbacks

Ability to observe multiple specimens, automated stage navigation, efficient analysis, ease of use, and cost-effectiveness along with enhanced resolution and contrast have made this technique popular with life and materials scientists working with biological and polymeric thin sections and nanopowders. Use of low accelerating voltage (typically 30 kV) in STEM-in-SEM reduces the probability of beam damage and provides enhanced contrast for the low atomic number and low-density materials. Bright field and dark field images can be recorded simultaneously during a single scan. Owing to a lack of imaging lens below the specimen, the solid angle of collection of transmitted electrons is large resulting in substantial signal-to-noise ratio. Owing to thin electron-transparent samples, signals originating from depths of the sample are eliminated. The transmitted electrons carry information about the internal structure of the material under examination. However, low electron beam energy used in this technique requires the specimen to be adequately thin to be electron transparent. Due to a different configuration than standard STEM equipment, image interpretation can be relatively complicated.

5.4.3 Applications

Imaging of submicron sized or nanopowders presents a challenge in the conventional SEM, especially if the powder material is composed of light elements. Electron beam penetrates the fine powder grains and is scattered. The volume of scattering produced within the material, known as the interaction volume, is bigger than the grain size of the powder itself. Consequently, the beam is scattered off the substrate material that is used to hold the powder specimen. Electrons scattered from the substrate contribute to the noise in the signal and diminish the image quality. In nano-sized particles, the noise dominates the signal, and contrast from the powder becomes too low to resolve its features. Use of a STEM detector can overcome this problem. The fine powder is placed over a copper grid coated with a holey carbon film (3 mm diameter, routinely used for examining TEM specimens) to reduce scattering from the substrate. The electron beam is transmitted through the powder specimen and collected by the STEM detector to form an image as shown in Fig. 5.18.

5.4 STEM-in-SEM

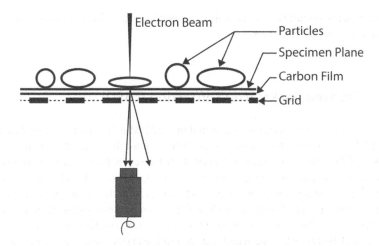

Fig. 5.18 Schematic representation of imaging of nanopowders in STEM-in-SEM. The specimen is placed over a thin carbon film to reduce extraneous scattering from the specimen holder. Use of thin holder reduces noise in the signal

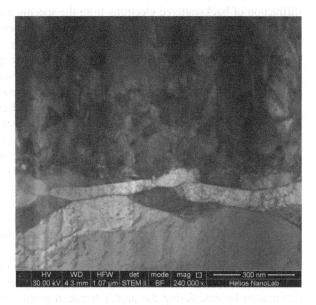

Fig. 5.19 High-angle annular dark field (HAADF) STEM image showing Ni coating on steel substrate in cross-sectional view. Grain boundaries, coating-substrate interface, and dislocations present in the substrate underneath the interface are clearly visible in one plane

Thin specimens for STEM imaging are prepared using electropolishing, argon ion milling, or focused gallium ion beam methods. Due to the thinness of the specimen, the transmitted image lacks contribution from low-resolution backscattered or SE_2 signals, thereby resulting in improved resolution and image contrast. Normally, STEM microscopy is conducted with the highest accelerating voltage available in the microscope to enable maximum transmission and brightness in the image. An example of a bright field STEM image of a cross-sectional specimen of Ni deposited on steel substrate is shown in Fig. 5.19. The image clearly shows grains of Ni along with its grain boundaries, coating-substrate interface, and dislocations present in the substrate underneath the interface.

5.5 Electron Backscatter Diffraction (EBSD)

Usually, phases in a material are identified in the SEM using microchemical analysis with EDS technique. This kind of analysis cannot be termed conclusive. For instance, TiO_2 may have different crystal structures with the same composition. Other phases such as iron oxides may exist as FeO, Fe_2O_3, or Fe_3O_4, and it may not be possible to distinguish them solely on chemistry. Another example is the identification of austenite and ferrite in steels which is not possible using EDS due to low carbon contents. This limitation in the SEM can be overcome by electron backscatter diffraction (EBSD) technique which can determine crystal structure as well as grain orientation (texture) of materials. Material surface is specially prepared for this analysis, and specimen is tilted at approx. 70° from the normal position. High accelerating voltage such as 20 kV is used to strike the specimen surface resulting diffraction of backscattered electrons from the specimen planes that are imaged as straight lines. Different orientation of planes at a specimen surface produces various sets of intersecting straight lines called Kikuchi pattern. The relationship between these lines determines the correlation between atomic planes. For instance, the distance between lines represents angles between crystals. Indexing of Kikuchi patterns is done using dedicated software which is used to calculate structural information and measure grain orientation. Specialized equipment such as EBSD detector with a phosphor screen and CCD camera, as well as computer software to control acquisition and processing of EBSD data, are required to undertake such an analysis. EBSD detectors in the modern field emission SEMs are able to examine materials with fine grains down to 100 nm in dimensions.

Availability of EBSD in SEM greatly enhances the latter's power as an analytical tool. With this addition, the SEM can not only be used to examine surface morphology and determine chemistry but also obtain crystallographic information from the material. Ability to determine crystal structure greatly improves the ability to identify unknown phases. Crystallographic information that can be gathered using this technique includes crystal spacing, crystal symmetry, and the angle between planes. Low- and high-angle boundaries between crystals can be determined by calculating the angles between grains. This technique has found application in the study of recrystallization and grain growth. EBSD can acquire crystallographic data from a crystalline specimen and correlate it to its microstructure. It is used to identify phases and study phase fractions/distribution, grain size/shape, aspect ratios, strain, material texture, crystal symmetry/orientation, defects, and grain boundaries. Traditionally, such type of structural analysis had been conducted using transmission electron microscopy.

5.5.1 Brief History

The exploration of diffraction which lays the foundation of present-day EBSD can be linked back to the year 1928 when an electron beam with an energy of 50 keV, and an angle of incidence of 6° produced from a gas discharge was directed on to a cleavage face of calcite by Seishi Kikuchi [26]. The patterns which emanated as a result of diffraction were captured onto photographic plates positioned 6.4 cm in front and back of the crystal. The patterns were described as "...black and white lines in pairs due to multiple scattering and selective reflection."

Nine years later, in 1937, the same phenomenon was observed by Boersch [27], and he was able to produce some excellent patterns on photographic film. Both transmission and backscattered diffraction patterns from cleaved and polished surfaces of various materials including NaCl, KCl, quartz, mica, diamond, Cu, and Fe were reported by him. In 1954, Alam, Blackman, and Pashley [28] used a cylindrical specimen chamber and a film camera to produce high-angle diffraction patterns from cleaved LiF, KI, NaCl, and PbS_2 crystals.

The coming of commercial SEM in 1965 paved the way for marked progress during the years from 1969 to 1979 where three prominent discoveries came to light. These included selected area channeling patterns (SACP) by Joy et al. at Oxford [29], Kossel diffraction by Biggin and Dingley at Bristol [30], and electron backscatter patterns (EBSP) by Venables and Harland at Sussex [31]. A TV camera and a phosphorous screen were employed for the first time to record the EBSP patterns.

5.5.2 Working Principle

In a crystalline specimen, atoms are positioned in a regular periodic three-dimensional arrangement called lattice. In this technique, electrons from the primary beam (10–30 kV accelerating voltage, 1–50 nA incident current) in an SEM strike the surface of a tilted (70°) and highly polished flat strain-free crystalline specimen and are scattered forming an interaction volume within the specimen. Backscattered electrons spread in all directions within the interaction volume. This can be thought of as a divergent source of electrons present within the specimen close to the surface. Part of these high-energy backscattered electrons is incident on sets of parallel lattice planes present within the crystal and are scattered in a manner that satisfies Bragg's equation which is written as:

$$n\lambda = 2d \sin \theta \qquad (5.3)$$

where n is the order of diffraction, λ is the electron wavelength, d is the lattice plane spacing, and θ is the Bragg angle of diffraction. This type of scattering is termed as electron diffraction. This is constructive interference of electron waves where the difference in path length traveled within a single lattice plane between two waves is a multiple of λ and the incident and emergent angles of the wave are equal. Upon scattering within the specimen, the electrons spread in all directions, and for each set

212 5 Specialized SEM Techniques

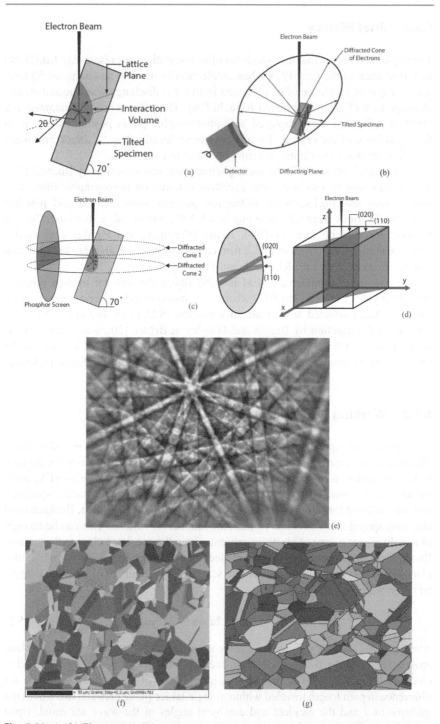

Fig. 5.20 (a, b) Electron scattering acts as a divergent source of electrons within a specimen. These electrons are incident upon a lattice plane satisfying Bragg's equation. Resulting diffraction forms a pair of cones at the front and back end of the plane. (c) Diffracted rays are formed along the

5.5 Electron Backscatter Diffraction (EBSD)

of lattice planes for which the above Bragg condition is fulfilled, the diffracted beams emerge out of the specimen in all directions in the form of a cone (see Fig. 5.20a, b). Diffracted beams lie on the surface of this cone whose axis is normal to the diffracting lattice plane. In fact, two cones are formed for each set of lattice planes, one at the front and the second at the rear of the lattice plane, as seen in Fig. 5.20c. These cones intersect the phosphor screen as two dark lines bordering a bright band. Bragg reflections emanating from various planes present within a specimen give rise to a network of a pair of sharp lines with bright bands intersecting each other at various angles. This network is called Kikuchi pattern and consists of a set of parallel lines crossing each other at different angles. Every set of two parallel lines represent a family of parallel planes with a specific value of d-spacing. One line represents the positive and the other line represents the negative plane, and the distance between the two lines is inversely proportional to the d-spacing for that specific plane.

Kikuchi pattern can be used to determine crystal orientation, angles between lattice planes, bend contours, electron channeling patterns, and fringe visibility maps. Projection of lattice planes in a Kikuchi map is shown schematically in Fig. 5.20d. A typical Kikuchi pattern is shown in Fig. 5.20e. The width of a Kikuchi band is determined by Bragg conditions and the distance between the specimen and the phosphor screen. The surface area of a specimen can be scanned to obtain an EBSD map from each scanned point in that area (see Fig. 5.20e). EBSD grain maps of stainless steel and Al are shown in Fig. 5.20f, g, respectively.

5.5.3 Experimental Setup

The experimental setup used for EBSD in an SEM is shown in Fig. 5.21a. The specimen is tilted to 70° using a pre-tilt holder or the SEM stage. At high tilt angles, near-surface material is excited, and the total interaction volume formed close to the surface is very large compared to the interaction volume deep within the material. Formation of large interaction volume close to the specimen surface allows easy escape of electrons from within the specimen and thus increases the ratio of diffraction component to the yield of backscattered electrons. Without the use of high tilt, the proportion of diffracted electrons in the overall electron yield may be too low to be detected, and adequate contrast may not be produced in the pattern. The detector for EBSD is attached to one of the free ports in the specimen chamber. Its front end consists of a fluorescent phosphor screen as shown in Fig. 5.21a. The screen converts electrons emanating from diffracting planes of the specimen into light which passes through a lead glass window located after the phosphor screen. Lead glass serves to isolate the detector assembly from the evacuated SEM chamber

Fig. 5.20 (continued) surface of cones and strike the phosphor screen resulting in a pair of lines for each set of lattice planes. (**d**) Schematic illustration of electron beam interaction with the lattice planes giving rise to Kikuchi bands on a phosphor screen (**e**) Kikuchi pattern obtained from a bcc iron. EBSD grain map of (**f**) stainless steel and (**g**) Al

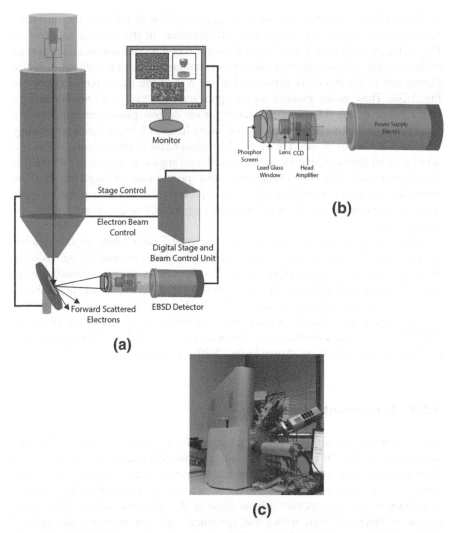

Fig. 5.21 (a) Standard experimental setup for electron backscatter diffraction (EBSD) in SEM showing a crystalline specimen tilted 70° toward the phosphor screen. The diffracted pattern generated due to specimen-beam interaction intersects the screen that converts electrons into light which is then imaged using a CCD camera. The Kikuchi patterns and mapped images are displayed onto the computer screen. Forward scattered electrons can be detected by placing an optional diode detector at the lower end of the EBSD detector. (b) Illustration of an EBSD detector which includes a phosphor screen, lead glass window, lens, CCD camera, amplifier, and associated electronics in a single compact assembly. (c) An EBSD detector is seen fixed to the specimen chamber of an SEM

while allowing light to pass through it. The light travels through a lens and onto the surface of a sensitive CCD camera which detects it and converts it into an image. The pattern formed on the phosphor screen is visualized and imaged with the help of this camera. The signal is then amplified and fed to the computer for display onto the monitor. Schematic and photograph of a compact design EBSD detector are shown in Fig. 5.21b, c, respectively. Diffracted electrons form only a small proportion of the total number of electrons scattered from within the specimen that strikes the phosphor screen. In other words, Kikuchi pattern is superimposed on a background which needs to be removed to make visualization possible. Materials with a high average atomic number (i.e., with high electron scattering factors) produce a relatively larger number of diffracted electrons resulting in higher contrast in the electron backscatter patterns (EBSP).

Electron backscatter diffraction detector is usually coupled with a Schottky field-emission electron microscope which provides high stable beam current and good mechanical stability. These characteristics are important since the acquisition of high-resolution large orientation EBSD maps can take several hours. EBSD is an extremely surface-sensitive technique with information acquired from depths of only a few to tens of nanometers. Surfaces are generally prepared using electropolishing or ion milling to remove any deformation.

5.5.4 Applications

Types of materials examined with EBSD include metals, alloys, minerals, ceramics, thin films, solar cells, intermetallics, and semiconductors. EBSD is used to identify and determine the distribution of intermetallic phases, secondary phase particles, precipitates, and minerals in a wide variety of materials. Each EBSD pattern is unique, and its characteristics are governed by lattice parameters of the crystal under examination, positioning of that specific crystal in 3-D space, the wavelength of the incident electron beam (which is proportional to the acceleration voltage), and the distance between the sample to the EBSD detector. Angles between bands within a Kikuchi pattern are measured to distinguish between different crystal structures with varying unit cell dimensions and interplanar angles. Phases that possess the same crystal structure but different lattice parameters can be differentiated by using more complicated routines that also measure bandwidth in Kikuchi patterns.

Dedicated software programs are used to index the Kikuchi patterns and display EBSD maps. One method of indexing is to provide crystal structure information and microscope operating conditions to the computer. The computer program measures the position of the Kikuchi lines and calculates the angles between them to compare the data to the provided crystal structure. In this manner, the crystallographic orientation of the specimen is determined. In another method, EBSD patterns obtained from various points are indexed by the software program and compared to the stored values of crystallographic lattice planes in the database (i.e., crystallographic indices of lattice planes, lattice spacing d, the interplanar angles, and intensity of lattice planes) to report the best fit for identified phase and its orientation. Kinematical electron diffraction model is used to determine the best fit. Study of grain size, orientation, and morphology is another common application of EBSD. Distribution of high- and low-angle grain boundaries and twin

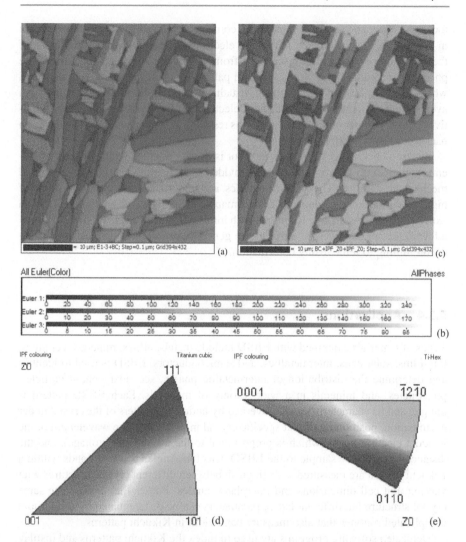

Fig. 5.22 (a) Euler map obtained from the Ti alloy specimen. Colors in this map represent the specific orientation of phases as shown in the orientation color key in (b). (c) Example of IPF coloring of the same specimen. Legend of IPF map for (d) cubic and (e) hexagonal Ti

boundaries can be investigated. Calculation of misorientation between each pixel reveals the sub-grain structure of a material.

EBSD can also be used to check if a material has a specific orientation (i.e., texture). In this method, crystal orientation measurements are made at multiple points within a phase, and the information is combined to conclude if the phase has texture. The orientation data can be displayed in a "Euler" map (see Fig. 5.22a, b). Generation of such maps is useful since they make it easy to visualize the distribution of texture within a specimen. Inverse pole figure (IPF) maps (see Fig. 5.22c–e) are also used

5.5 Electron Backscatter Diffraction (EBSD)

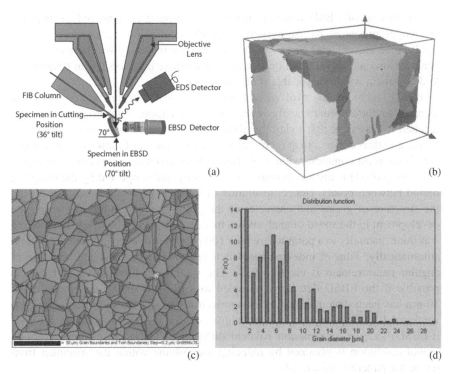

Fig. 5.23 (**a**) Experimental setup depicting the combined use of EBSD and FIB in an SEM. (**b**) 3-D EBSD orientation map of annealed copper wire. (**c**) Grain boundaries from a stainless steel sample determined by EBSD. It highlights the grain boundaries (black) and twin boundaries (red) distribution in the steel. (**d**) Grain size distribution of stainless steel sample

to show 3-D texture on a 2-D plane by converting crystallographic directions into points.

EBSD and EDS can be integrated to acquire simultaneous crystallographic and compositional data, respectively, from each analyzed point in a synchronized manner. Microchemical analysis results obtained using EDS can be used to shortlist candidate phases based on composition. The shortened list is then used to index the EBSD pattern. For a material with known phases, the distribution and fraction of phases can be determined using EBSD. Combined EDS and EBSD technique is helpful during mapping to separate phases that possess similar crystal structure.

EBSD combined with focused ion beam (FIB) instrument (see Fig. 5.23a) has enabled scientists to acquire 3-D microstructural information from specimen volume. In this automated process, milling is used to expose fresh surface within a volume to acquire an EBSD map. The series of maps obtained at increasing depths of the specimen are later combined to generate a three-dimensional microstructure of the analyzed specimen volume (see Fig. 5.23b). Examples of grain boundary structure and grain size distribution obtained using EBSD are shown in Fig. 5.23c, d, respectively.

The design of EBSD detectors has been continuously improved to meet the challenges presented by nanomaterials which require low accelerating voltage, small probe current, and short working distance during analysis. These conditions generate low-intensity diffraction signals, and sensitive detectors had to be designed to detect them within reasonable acquisition time.

EBSD technique has evolved in terms of automated accuracy along with the increase in the speed of analysis and data acquisition. This improvement has reached to an extent that today the ability of the incident beam to scan multiple points and thus create a map representing the orientation of grains of the scanned area is the most common method of evaluating material microstructure. An orientation map (OM) is distinct because of its locality and scope and by the sampling period between points. Hence, resolution of OM can be accustomed to exposing the structure of grains and character of the grain boundary. Over the years, the development in the speed of analysis has made a shift from a point where indexing was done manually to a point where now 100 patterns could be indexed per second automatically. Time of indexing/analysis is few tens of seconds, and precision of angular measurement is close to 0.5°. The spatial resolution of 20–100 nm is possible if the EBSD detector is coupled to a FESEM. Improved resolution of 10 nm has been demonstrated by using electron-transparent specimens with conventional EBSD hardware. This technique is referred to as transmission EBSD (t-EBSD) [32] or transmission Kikuchi diffraction (TKD) [33]. Improvement in spatial resolution is obtained by reducing the volume within the specimen from where the pattern is generated.

5.6 Electron Beam Lithography

5.6.1 Introduction

Semiconductor device fabrication is based on a sequence of photographic and chemical processes to manufacture structures at the micro- and nanoscale. The photographic process is known as lithography, a word that originates from the Greeks words *lithos*, meaning "stone," and the word *grapho*, meaning "to write" [34]. Accurately translated as "writing on stone," the technique of lithographic printing, from the late eighteenth century, used a flat stone slab onto which fat or oil was applied to split the slab into hydrophobic and hydrophilic regions. The ink applied to the slab would obey to only the hydrophilic regions, and when the paper was contacted with the slab, the ink would transfer to the paper making a copy of the hydrophilic regions on the slab. In lithography of semiconductor device fabrication, the "stone" is known as a mask and contains the pattern, and the "paper" used for printing the pattern is called substrate.

The substrate can be made from any material that can be formed into a flat plate, for example, Si, GaAs, and Quartz, among others, are widely used. The substrate material is often selected due to its electrical or thermal properties which may allow specific types of semiconductor devices to be realized. To write on these substrates, many lithographic-based techniques could be used. The most common techniques are imprint lithography, optical lithography, and electron beam lithography. The illumination source is used in optical lithography; this source is projected through the mask that contains the wanted pattern. Imprint lithography is based on the molding process, in which a deformable layer placed on the substrate surface could be structured directly. In electron beam lithography, an accelerated electron beam is used on the substrate surface placed on a movable stage. That stage moves the sample carefully so that the electron beam can scan across the whole substrate surface to trace the desired pattern.

None of these processes directly affect the surface of the substrate, so to allow the pattern to be transferred into the substrate, a thin layer of *resist* is used. In the case of electron beam lithography, the resist layer is sensitive to photons or electrons. This sensitivity results in a change in chemical properties of the resist at the points where it has been exposed to the radiation source, resulting in a change in dissolution rate in a given solvent which allows the pattern to be formed in the resist layer. Electron beam lithography is the gold standard in terms of being able to produce the highest-resolution features; however, its extremely low throughput precludes its use in large-scale industrial manufacturing. The high resolution and flexibility of electron beam lithography have ensured that it has a particular role in research and development applications. A good overview of the technique is written by Nabity et al. [35].

5.6.2 Experimental Set-Up

Electron beam lithography has been used since the 1950s and can be performed on a wide spectrum of hardware from converted scanning electron microscopes, through to sophisticated commercial tools engineered expressly to provide the ultimate performance. A schematic diagram of the electron beam lithography system is shown in Fig. 5.24. The key systems are the cathode (cold or Schottky field emitter) where electrons are generated, the electromagnetic lenses and scan coils which focus the electron beam and provide a method to control the deflection of the beam across the substrate, and X-Y-Z stage on which the substrate coated with resist is mounted, along with the vacuum, electronic control, and power supply systems. Dedicated E-beam control unit is provided, which contains a pattern generation system to convert computer-aided designed patterns into control signals for the beam deflection and blanking system. Converted SEM may have an external pattern generator system. More advanced tools may also contain a substrate handling system allowing multiple substrates to be automatically loaded and unloaded from the machine for exposure in a sequential or "batch" fashion. The entire machine may be mounted on a plinth and/or contain a vibration isolation table, which stabilizes the system reducing effects of seismic activity and isolating the equipment from environmental sources of mechanical vibration.

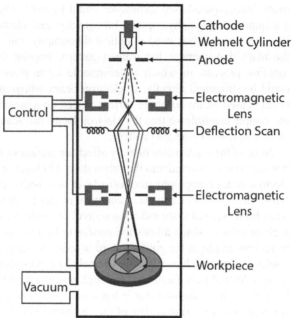

Fig. 5.24 Schematic representation of an electron beam lithography system [36]

5.6.3 Classification of E-beam Lithography Systems

Electron beam lithography systems can be classified on the basis of the beam shape and beam deflection method. An important distinction between different types of electron beam lithography tools is that of shaped-beam versus Gaussian-beam. With a Gaussian-beam system, the electron beam is focused down to have as small a diameter as the beam current will allow and the cross-sectional intensity profile of such a beam can be approximated by a Gaussian function. Shaped-beam systems form the electron beam into a wide beam of uniform intensity. This wide beam is then passed through a series of interchangeable apertures which allow the beam to be directly formed into trapezium shapes. In this way, while with a Gaussian-beam system, each shape is formed from a series of point exposures, in a shaped-beam system, each shape can be exposed as one "shot" or exposure. As a result, throughput is vastly superior with shaped-beam systems when large patterns are exposed. However, this comes at the cost of ultimate resolution. Shaped-beam systems have found significant use in advanced manufacturing facilities, but their relative expense compared with Gaussian-beam systems and their slightly lower-resolution capabilities mean that Gaussian-beam systems are of more interest in a university research setting [37–39].

5.6.4 Working Principle

5.6.4.1 Beam Deflection and Blanking

In electron beam lithography, two strategies are used to control the beam deflection and blanking: raster scanning and vector scanning. With raster scanning, the beam is scanned in a series of parallel scan lines across the complete subfield, and the beam blanker is switched on and off to control which parts are written. In a vector scan system, the beam is deflected to the start of each shape to be written and then scanned across that shape, before being deflected to the start of the next shape. It means that the beam is on for a much greater proportion of the writing time since it is only blanked when moving between shapes. Both tools are ultimately limited by the brightness of the electron source.

5.6.4.2 Pattern Design and Electron Beam Resist

The whole aim of lithography is to form a pre-defined pattern on a physical object. To create the pattern in the first place, it is extremely common to use a form of computer-aided design (CAD) to create an electronic file containing the required arrangement of shapes. All conventional forms of lithography rely on the use of a layer of resist, which is a layer of material which can be selectively removed so that it protects some areas of the substrate and exposes other areas to subsequent processing, for example, etching. In electron beam lithography, the resist is formed from a compound that undergoes chemical changes when exposed to energetic electrons. There are two forms that these changes can take, and this determines the type of resist, positive or negative [40]. When electrons interact with a "positive" resist, they cause scissions in the polymer chains that make up the resist. This makes the exposed regions more soluble in a chemical solution known as a developer. With negative resist, the incident electrons cause the molecular chains to cross-link [41]. This makes the exposed regions less soluble in the developer. The illustration in Fig. 5.25 shows the final resist profile for both positive and negative resists.

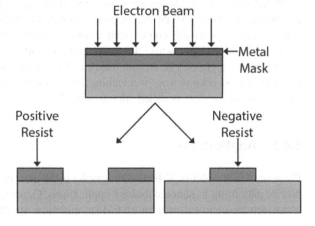

Fig. 5.25 Diagram showing the differences in the final resist profile for positive and negative resist exposed to the same electron beam pattern [42]

The degree of change caused by the electrons impacting on the resist is affected by the number of electrons or the electron dose, and on the accelerating voltage, since this controls the production of secondary electrons which actually expose the resist. Similarly, the rate of dissolution and the difference in final resist thickness can be controlled by changing the concentration of the developer and the amount of time used for development. In this way, a set of process parameters sometimes referred to simply as a "process," can be defined to give well-defined features. In general, there exists two commonly used E-beam resists; polymethyl methacrylate (PMMA) and hydrogen silsequioxane (HSQ). PMMA has been used as a resist for electron beam lithography for nearly 40 years [43]. It is a positive resist at moderate electron dose, but it can also be used as a negative resist at much higher electron dose [44].

5.6.4.3 Pattern Processing

There are two main fabrication processes used frequently in small-scale semiconductor fabrication. The processes can be thought of as roughly inverse of each other as one of the processes is an additive process, "lift-off," with the other is a subtractive process, "etching." Lift-off allows the patterned metal to be added to a substrate. It relies on the resist having an undercut profile after development such that the top of the features in the resist are slightly narrower than the bottom. The undercut resist profile means that when a thin layer of metal is deposited on top of the whole substrate, the metal becomes discontinuous at all the edges of the pattern. Submerging the substrate in a solution which dissolves the resist removes the remaining resist along with any metal on top of it, resulting in a metal layer attached to the substrate that matches the pattern of the electron beam exposure. This process is illustrated in Fig. 5.26a. A pattern produced by E-beam lithography is shown in Fig. 5.26b.

The metal can be deposited in a number of ways: filament evaporation, electron beam evaporation, effusion cell evaporation, and laser ablation. Sputter-coated metal generally cannot be used for a lift-off process.

Etching is the second method of pattern transfer and it is a subtractive method. Starting with a blank substrate, a blanket deposition of metal is performed. The resist is coated onto the metalized substrate, and lithography is then performed. After development, the substrate has regions where the underlying metal surface is exposed and regions still coated with a resist. The exposed metal regions are then removed in an etchant, while the resist protects and prevents the removal of the covered metal regions. The etchant used can be either a chemical solution, in which case the process is known as "wet etching," or it could be a gas or plasma etchant, in which case the process is called "dry etching."

5.6.5 Applications

Because of its flexibility, E-beam lithography is the most commonly used method for precise patterning in nanotechnology applications. Generally, it could be used in the nanoelectromechanical system (NEMS), quantum structures, magnetic devices,

5.6 Electron Beam Lithography

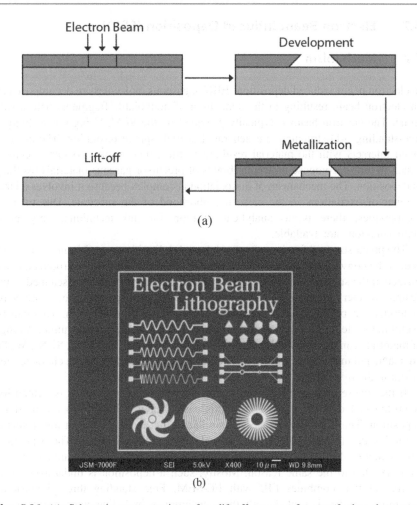

Fig. 5.26 (a) Schematic representation of a lift-off process for transferring the pattern [36]. (b) SEM image of a pattern produced using E-beam lithography. (Image courtesy of T. Siong, JEOL Ltd.)

solid-state physics, biotechnology, and transport mechanisms. It is used in the fabrication of many functional devices and products such as IC fabrication mask, nano-transistors, nano-sensors, and biological applications such as biomolecular motor-powered devices.

5.7 Electron Beam-Induced Deposition (EBID)

5.7.1 Mechanism

In electron beam-induced deposition (EBID), gaseous molecules are decomposed by an electron beam resulting in the deposition of nonvolatile fragments onto a substrate. The electron beam is typically provided by the SEM. Using this technique, free-standing 3-D structures are generated at high spatial resolution. The electron beam interacts with the material resulting in the emission of secondary electrons which in turn decompose molecular bonds of precursor gaseous materials resulting in deposition. The mechanism of dissociation is complex because it involves a large number of excitations in the close neighborhood of the substrate. Due to these complexities, there is no analytical solution for this technique; only gross approximations are available.

The precursor used for deposition can be solid, liquid, or gas. Liquids and solids need to be converted into gas form prior to deposition. The gas is introduced at the surface of the substrate in a controlled fashion, and electron beam is scanned in the desired manner to deposit the material in the required shape and size. In this way, materials can be deposited with a high spatial resolution of 1 nm. This can be useful in the electronics industry where nanoscaled structures are required. A range of materials can be deposited including Au, amorphous C, diamond, Si_3N_4, W, Pt, Pd, GaN, and many other elements. The deposits are clean and can be characterized in situ in the microscope.

If the focused ion beam is used to deposit materials, the process is termed ion beam-induced deposition (IBID). In this case, heavy gallium ions are used for deposition. The deposition process is similar to that in EBID with a lower spatial resolution due to the wider angular spread of secondary electrons. The deposition rate is higher due to heavier ions, but at the same time, the contamination rate increases due to the same reason. Mostly, such a deposition is undertaken in an instrument that combines FIB with FE-SEM. Free standing three-dimensional structures can be deposited with these techniques including nano-wires, nano-loops, nano-trees, etc.

A specially designed chamber is used because temperature rises during deposition. The chamber is isolated from the column, and beam comes in through a small hole in it. A general schematic of EBID technique is as shown in the Fig. 5.27.

It can be seen in Fig. 5.27 that the precursor gas comes into the chamber from the left-hand side of the substrate. A high-energy electron beam comes in and hits the precursor material placed on top of the substrate. During this process, multiple excitations take place and volatile part of the gas moves away. Only the nonvolatile part is left behind. In this way, once the precursor gas is led into the chamber, the electron beam scans over the substrate and deposits a layer of the precursor on the substrate. The scanning is controlled by a computer system. The rate and quality of deposition depend on the various factors such as pressure, the temperature of the material, and characteristics of the electron beam.

5.7 Electron Beam-Induced Deposition (EBID)

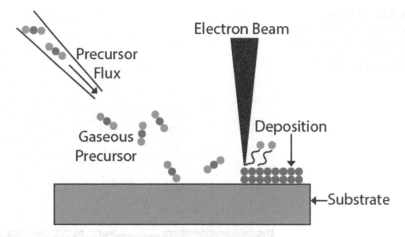

Fig. 5.27 Schematic diagram of electron beam-induced deposition

5.7.2 Advantages/Disadvantages of EBID

Advantages

1. This technique is very flexible with regard to the composition of the deposited material and its shape. Both of these factors are computer controlled.
2. The size control of the product after deposition and the accuracy of the process are high.
3. The characterization and deposition can be done simultaneously.

Disadvantages

1. It is a challenge to accurately control the chemical composition of the deposited material.
2. The structure broadening may happen due to proximity factors.
3. In the case of serial deposition, the rate of deposition is slow.

5.7.3 Applications

EBID is used to characterize, analyze, and fabricate nanomaterials and devices. The EBID structure can be developed using two separate gas injection systems (GIS) that introduces carbon and platinum precursor into the system. This is used as a hard mask in the production of quantum cellular automata (QCA) device structures [45]. Another application is the production of atomic force microscopy (AFM) cantilevers using EBID. If such a tip is produced in a cylindrical shape, it will lessen the effect of forces of attraction between specimen and probe during AFM. Carbon nanotubes have been attached to produce an ultra-sharp tip with radius from 6 to 25 nm using EBID on damaged AFM tips. A structure produced by EBID technique is shown in Fig. 5.28.

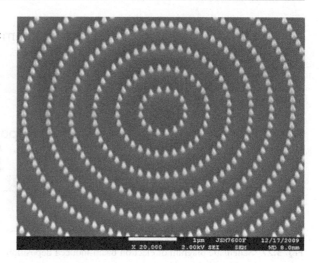

Fig. 5.28 SEM image of a structure produced by EBID: Pt nanodots deposited on Si wafer. (Image courtesy of T. Siong, JEOL Ltd.)

5.8 Cathodoluminescence

5.8.1 Introduction

A particular class of materials can emit light (photons of characteristic wavelengths in ultraviolet, visible, and infrared ranges) when bombarded with an electron beam in the SEM. This phenomenon is known as cathodoluminescence (CL) which occurs when atoms in a material excited by high-energy electrons in the beam return to their ground state, thus emitting light. Examples of cathodoluminescence from everyday life are light emitted from the inner surfaces of the cathode ray tubes in television set or computer monitors.

Cathodoluminescence can be observed with a special detector mounted in the SEM column. The detector collects the light or the wavelength emitted by the specimen. It can display the real color of visible light or an emission spectrum. Examples of materials that exhibit cathodoluminescence include zinc sulfide, anthracene, sedimentary rocks, semiconductors, Si wafers, synthetic crystals, fluorescent dyes, etc. Cathodoluminescence provides information about the distribution of trace elements in minerals, impurities in ceramics, and defects in crystals, etc. Some materials like plastics and glasses also show weak cathodoluminescence emission. This technique has found an important place in the microelectronics industry. It is used to study the optical and electronic properties of semiconductor materials. Semiconductors are bombarded with high-energy electron beam which transfers its energy into electrons that jump from valence into conduction band leaving behind holes. Recombination of electron holes at p-n junctions results in cathodoluminescence. In this manner, nanoscale features and defects of semiconductors can be studied.

Cathodoluminescence microscopy has become an essential tool in the petrographic description of sedimentary rocks. CL also has important applications in igneous-metamorphic petrography, ore deposits, and mineralogy [46]. When electron beam hits the sample, it absorbs most of the incoming energy, and atoms of the specimen get excited. Normally, the excited atoms (also termed cathodoluminescent centers) return to the ground state by transfer of the excess energy to adjacent atoms by inelastic collisions. Under certain circumstances, the absorbed energy is re-emitted as light energy in the visible range before these collisions can take place. The intensity of the light emitted from any particular point will be proportional primarily to the surface density of luminescent centers. The electron energy is readily absorbed in the sample, and little luminescence is emitted from below the surface. Transition metals and the rare earth elements are particularly susceptible to electron beam excitation. For instance, in transition metals, the 3-D electron shells are available for excited electrons to enter these levels. Thin sections, rock slabs, and loose grains can all be examined in the CL stage. Fine grains should be cemented to a glass slide so they will not enter the vacuum system. Thick samples are restricted to $50 \times 70 \times 17$ mm. The view area in both cases is 50×70 mm.

5.8.2 Instrumentation

Most of the parameters in SEM-CL are same as that in normal SEM. The electrons are generated and then accelerated toward the anode by providing a potential difference of 1–30 kV. The current on the surface of the sample can vary from 1 pA to 10 nA. The working distance can be in the range of 4–40 mm. The electron beam is focused to (5 nm to 1 µm) probe that can produce a CL image of that small area. The CL detector is placed in the chamber at an angle to the specimen as shown in Fig. 5.29.

A large surface area of the sample is scanned by the electron beam and CL image is produced. The SEM-CL detector works under high vacuum 10^{-4} Pa and is capable of producing high-resolution images. This image is just like a digital image showing different colors. The magnification of the CL-images can range from 10 to 10,000×, but it is difficult to obtain the lowest magnification due to instrumental configuration factors. The imaging process varies depending on the type of information that is required. The imaging process includes the CL (gray level image) in the range of ~200–800 nm. CL is frequently used to study the texture and chemical zone of the specimen. Three separate gray-colored images are obtained by using red, blue, and green color filters. Later, the image is recreated by using true colors. The *live* color image of the sample can be obtained by using an array detector system. A CL-SEM image is shown in Fig. 5.30.

Figure 5.31 shows various examples of CL images.

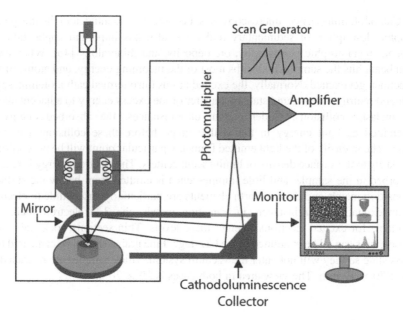

Fig. 5.29 Schematic showing the arrangement of cathodoluminescence detection in the SEM. The signal generated from the sample is reflected from the mirror into the CL detector and onto a photomultiplier tube for further processing

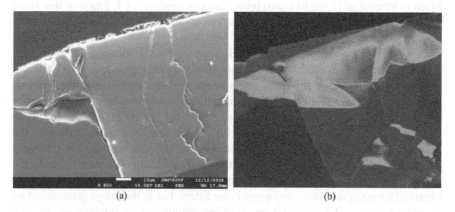

Fig. 5.30 SEM images showing cathodoluminescence of benitoite. (**a**) SEM image and (**b**) CL image. (Images courtesy of T. Siong, JEOL Ltd.)

5.8 Cathodoluminescence

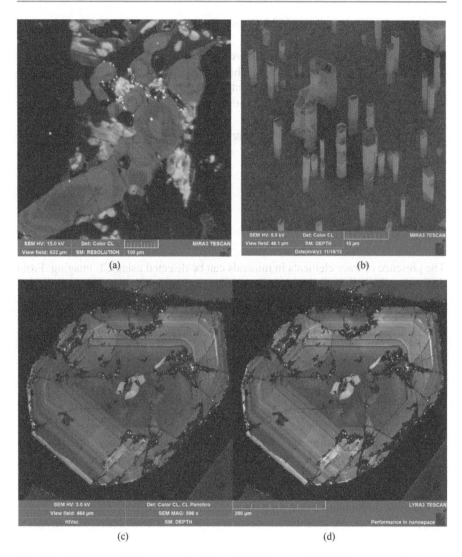

Fig. 5.31 CL images of (**a**) apatite and sodalite, (**b**) GaN wires with InGaN quantum wells and (**c**) zircon. (**d**) SEM image of zircon. (Images courtesy of TESCAN)

5.8.3 Strengths and Limitations of SEM-CL

The advantages of SEM-CL compared to optical-CL are the following:

- Spatial resolution is better.
- UV and IR response can also be obtained.
- Colorful image of the sample is created by using appropriate filters.

Some of the drawbacks of SEM-CL are the following:

- Complete setup with electron gun is required.
- The whole setup is more expensive than optical CL.
- The sample requires a conductive coating.
- Abnormalities in RGB color absorption filters and difficulties in proper color restoration.
- Difficult to obtain CL imaging of important CL-emitting minerals such as carbonate minerals and apatite.

5.8.4 Applications

The presence of trace elements in minerals can be detected using CL imaging. From different color patterns obtained from geological samples, the presence of trace elements can be confirmed. Some of the applications are as follows:

- CL imaging is very helpful in petrographic studies. For example, in clay cement, the bright blue luminescence indicates the presence of kaolinite [46].
- The zone analysis within the crystal can be done with CL imaging.
- Examination of cementation and diagenesis processes in sedimentary rocks can be undertaken [47].
- CL imaging is useful in studying the internal structure of fossils.
- Growth/dissolution analysis and deformation feature analysis in metamorphic minerals can be performed.
- With the help of this technique, various generations of the same minerals can be differentiated on the basis of trace amounts of activator elements. For example, sandstone can have many different types of quartz grains and compounds. Each different compound produces its own color pattern (different CL signal). These types of signals are invisible in secondary electron or backscattered imaging.

References

1. Boersch H (1954) Experimentale bestimmung der energieverteilung in thermisch ausgleoesten elektronenstrahlen. Z Phys 139:115
2. Erdman N, Bell DC (2013) SEM instrumentation developments for low kV imaging and microanalysis. In: Low voltage electron microscopy: principles and applications. Wiley, Chichester
3. Erdman N, Bell DC (2015) Scanning electron and ion microscopy of nanostructures. In: Kirkland AI, Haigh SJ (eds) Nanocharacterisation, RSC Nanoscience & Nanotechnology No. 37, 2nd edn. The Royal Society of Chemistry, Cambridge, p 311
4. Joens S (2001) Hitachi S-4700 ExB filter design and applications. Microsc Microanal 7:878–879

5. Sato M, Todokoro H, Kageyama K (1993) A snorkel type conical objective lens with E X B field for detecting secondary electrons. Proc. SPIE – Charged Particle Optics 2014:17–23
6. Kazumori H (2002) Development of JSM-7400 F: new secondary electron detection systems permit observation of non-conductive materials. JEOL News 37E(1):44–47
7. Steigerwald MDG, Arnold R, Bihr J, Drexel V, Jaksch H, Preikszas D, Vermeulen JP (2004) New detection system for Gemini. Microsc Microanal 10:1372–1373
8. Jaksch H (2008) Low Loss BSE imaging with the EsB Detection system on the Gemini Ultra FE-SEM. In: Luysberg M, Tillmann K, Weirich T (eds) Proceedings of EMC 2008, 14th European microscopy congress 1–5 September 2008, Aachen, Germany, vol 1. Springer-Verlag, Berlin Heidelberg, p 555 doi.org/10.1007/978-3-540-85226-1
9. Asahina S, Togashi T, Terasaki O, Takami S, Adschiri T, Shibata M, Erdman N (2012) High-resolution low-voltage scanning electron microscope study of nanostructured materials. Microsc Anal 26:S12–S14
10. Gestmann I, Kooijman K, Sakic A, Nanver L, van Veen G (2010) New solid state detector design for ultra-sensitive backscattered electron detection. In: Solorzano G, de Souza W (eds) Proceedings of the 17th international microscopy congress (IMC17), Rio de Janeiro, Brazil. International Federation of Societies for Microscopy (IFSM)
11. Sakic A, van Veen G, Kooijman K, Vogelsang P, Scholtes TLM, de Boer WB, Derakhshandeh J, Wien WHA, Milosavljevic S, Nanver LK (2012) High-efficiency silicon photodiode detector for sub-keV electron microscopy. IEEE Trans Electron Devices 59(10):2707–2714. https://doi.org/10.1109/TED.2012.2207960
12. Erdman N, Kikuchi N, Robertson V, Laudate T (2009) Multispectral imaging in a FEG-SEM. Adv Mat Proc 167(9):28–31
13. Schwandt CS (2010) Characterizing nanometer-scale materials using a low-angle backscattered electron detector. Amer Lab, Nov. 15, pp. 13–17
14. Mullerova I, Frank L (2003) Scanning low-energy electron microscopy. Adv Imag Elect Phys 128:309–443
15. Zach J (1989) Design of a high resolution low voltage scanning electron microscope. Optik 83 (1):30–40
16. Zach J, Haider M (1995) Aberration correction in a low voltage SEM by a multipole corrector. Nucl Instrum Methods Phys Res A 363:316–325
17. Honda K, Takashima S (2003) Chromatic and spherical aberration correction in the LSI inspection scanning electron microscope. JEOL News 38(1):36–40
18. Uno S, Honda K, Nakamura N et al (2005) Aberration correction and its automatic control in scanning electron microscopes. Optik 116:438–448
19. Kawasaki T, Tomonori N, Kotoko H (2009) Developing an aberration-corrected Schottky emission SEM and method for measuring aberration. Microelectr Eng 86:1017–1020
20. Kazumori H, Honda K, Matsuya M, Date M (2004) Field emission SEM with a spherical and chromatic aberration corrector. In: Proc. 8th Asian Pacific Conf. on Electr. Microsc. Council of Asia-Pacific Societies for Microscopy (CAPSM), pp 52–53
21. Thornley RFM (1960) Ph.D. thesis. University of Cambridge, Cambridge
22. Lane, WC (1970) Proceedings SEM Symposium (O. Johari, ed.), p. 43. IIT Research Institute, Chicago, IL
23. Danilatos GD (1988) Foundations of environmental scanning microscopy. Adv Electron Electron Phys 71:109–250
24. Johnson R (1996) Environmental scanning electron microscopy: an introduction to ESEM. Philips Electron Optics, Robert Johnson Associates, Eindhoven, The Netherlands
25. Melngailis J (2001) Ion sources for nanofabrication and high resolution lithography, proceedings of the 2001 particle accelerator conference, Chicago
26. Kikuchi S (1928) Diffraction of cathode rays by mica. Jpn J Phys 5:83–96
27. Boersch H (1937) Über Bänder bei Elektronenbeugung. Z Techn Phys 18:574–578
28. Alam MN, Blackman M, Pashley DW (1954) High-angle Kikuchi patterns. Proc R Soc Lond. Sect. A 221, pp. 224–242

29. Joy DC (1974) Electron channelling patterns in the scanning electron microscope. In: Holt DB, Muir MD, Boswarva IM, Grant PR (eds) Quantitative scanning electron microscopy. Academic Press, New York, pp 131–182
30. Biggin S, Dingley DJ (1977) A general method for locating the X-ray source point in Kossel diffraction. J Appl Crystallogr 10:376–385
31. Venables JA, Harland CJ (1973) Electron backscattering patterns. A new technique for obtaining crystallographic information in the scanning electron microscope. Philos Mag 27:1193–1200
32. Keller RR, Geiss RH (2012) Transmission EBSD from 10 nm domains in a scanning electron microscope. J Microsc 245(3):245–251
33. Trimby P (2012) Orientation mapping of nanostructured materials using transmission Kikuchi diffraction in the scanning electron microscope. Ultramicroscopy 120:16–24
34. Tallents G, Wagenaars E, Pert G (2010) Optical lithography: lithography at EUV wavelengths. Nat Photonics 4(12):809–811
35. Nabity J, Compbell L, Zhu M, Zhou W (2007) E-beam nanolithography integrated with scanning electron microscope. In: Zhou W, Wang ZL (eds) Scanning microscopy for nanotechnology, techniques and applications. Springer, New York, pp 120–151
36. Shwartz GC (2006) Handbook of semiconductor interconnection technology. CRC Press, Boca Raton
37. Pain L, Jurdit M, Todeschini J, Manakli S, Icard B, Minghetti B, Bervin G, Beverina A, Leverd F, Broekaart M, Gouraud P (2005) Electron beam direct write lithography flexibility for ASIC manufacturing: an opportunity for cost reduction (Keynote Paper). In: Emerging lithographic technologies IX, vol 5751. International Society for Optics and Photonics, pp 35–46
38. Pain L, Icard B, Manakli S, Todeschini J, Minghetti B, Wang V, Henry D (2006) Transitioning of direct e-beam write technology from research and development into production flow. Microelectron Eng 83(4):749–753
39. Todeschini J, Pain L, Manakli S, Icard B, Dejonghe V, Minghetti B, Jurdit M, Henry D, Wang V (2005) Electron beam direct write process development for sub 45nm CMOS manufacturing. In: Advances in resist technology and processing XXII, vol 5753. International Society for Optics and Photonics, pp 408–417
40. Zhou W, Wang ZL (eds) (2007) Scanning microscopy for nanotechnology: techniques and applications. Springer Science & Business Media, New York; London
41. Reichmanis E, Novembre AE (1993) Lithographic resist materials chemistry. Annu Rev Mater Sci 23(1):11–43
42. Chen YY, Chen CL, Lee PC, Ou MN (2011) Fabrication, characterization and thermal properties of nanowires. In: Nanowires-fundamental research. InTech, Rijeka
43. Hatzakis M (1969) Electron resists for microcircuit and mask production. J Electrochem Soc 116(7):1033–1037
44. Hoole ACF, Welland ME, Broers AN (1997) Negative PMMA as a high-resolution resist-the limits and possibilities. Semicond Sci Technol 12(9):1166
45. Bieber JA, Pulecio JF, Moreno WA (2008) Applications of electron beam induced deposition in nanofabrication. In: Proceedings of the 7th international Caribbean conference on devices, circuits and systems, ICCDCS http://ieeexplore.ieee.org/document/4542649/
46. Pagel M, Barbin V, Blanc P, Ohnenstetter D (2000) Cathodoluminescence in geosciences: an introduction. In: Cathodoluminescence Geosciences, vol 1995. Springer, Berlin, pp 1–21
47. Coenen T (2016) Cathodoluminescence imaging on sedimentary rocks: quartz sandstones, June, pp 1–14

Characteristics of X-Rays

6

In addition to the generation of backscattered and secondary electrons, the interaction of an electron beam with the specimen material releases x-rays which are used to undertake elemental analysis using energy−/wavelength-dispersive x-ray spectroscopy. Characteristics of x-rays are described in this chapter.

6.1 Atom Model

The nuclei of the atoms of specimen material examined in the SEM are composed of protons and neutrons. Since neutrons do not carry a charge, a nucleus is characterized by a concentrated positive charge. The negative charge is carried by electrons that are placed around the nucleus in orbits located at a specific distance. Orbits are grouped together into shells known as K, L, M, etc., each with specific energy defined by the principal quantum number, n. The shells close to the nucleus have the lowest potential energy. Thus, the energy level increases moving away from the nucleus from K to L and M shell. In normal state, the number of electrons in an atom equals the number of protons, and thus an atom does not carry any charge. Electrons occupy shells based on minimum energy. Electrons populate the low-energy shells close to the nucleus before they move onto the higher-energy shells. The negative charge or energy of the electrons is distributed according to their location in orbits. The K shell is closest to the nucleus and is the most tightly bound compared to those (e.g., L or M shells) away from the nucleus. The K shell ($n = 1$, where n is the shell number) is the first shell which is filled in by the electrons followed by L shell ($n = 2$) and so forth. Each shell can hold up to $2n^2$ electrons. Large atoms or heavy elements contain a larger number of electrons and electron orbits.

Each shell is made up of ≥ 1 subshells. The K shell has one subshell "1s"; the L shell has two subshells "2s" and "2p"; the M shell has three subshells "3s," "3p," and "3d"; and so on. Shells are populated according to Pauli exclusion principle which states that only one electron can possess a given set of quantum numbers and that the

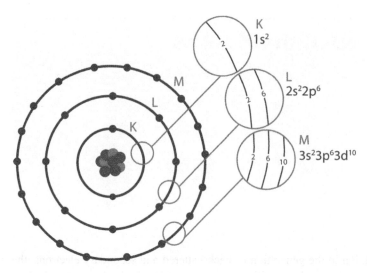

Fig. 6.1 Schematic showing atom model where the maximum numbers of electrons in K, L, and M shells are 2, 8, and 18, respectively

maximum number of electrons in a shell coincides with the number of states possessing the relevant principal quantum number. A subshell is the state defined by azimuthal quantum number "l" within a shell. The values $l = 0, 1, 2, 3$ correspond to s, p, d, and f subshells, respectively. The maximum number of electrons in a subshell is $2(2l + 1)$. This results in 2, 6, 10, and 14 electrons in s, p, d, and f subshells, respectively, as shown in Fig. 6.1.

In x-ray spectroscopy, shells are designated by the letters K, L, M, N, etc. along with their division into subshells. The K shell has no subshell, while L shell contains three subshells (L_I, L_{II}, L_{III}) and M shell contains five subshells (M_I, M_{II}, M_{III}, M_{IV}, M_V). Based on the Pauli exclusion principle, the maximum number of electrons in K shell is 2, in L shell 8, and in M shell 18 as shown in Fig. 6.1.

6.2 Production of X-Rays

6.2.1 Characteristic X-Rays

Primary electron beam penetrates the specimen material and interacts with the inner shells of atoms. As a result, inner-shell electrons of target atoms are ejected from their orbits and leave the bounds of the atom. The process of electron ejection results in a vacancy in the orbital and turns the atom into an ion of excited or energized state. This vacancy is immediately filled when an outer shell electron is transferred to the inner shell, which brings the atom to its ground (lowest energy) state with an accompanying release of energy equal to the difference in the binding energy of the two shells. This excessive energy is released in the form of an x-ray photon (see Fig. 6.2) which has

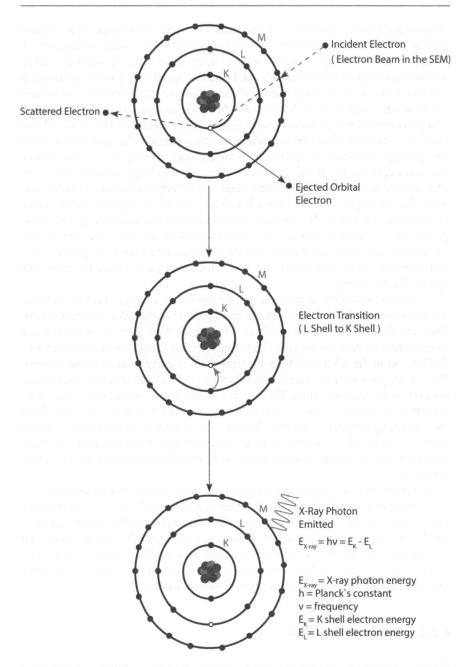

Fig. 6.2 Schematic shows that the electron beam incident upon the specimen knocks an orbital electron out of the K shell. The incident electron is scattered (changes direction and energy), and the knocked out electron is ejected out of the atom into the material. The vacancy created in the k shell is filled up by an electron from the L shell since it has higher potential energy. This is accompanied by the release of a photon whose energy is equal to the difference in the binding energies of the two shells, i.e., E_K–E_L

energy equal to the binding energy between the two shells. For example, if an electron from inner K shell is removed and the vacancy is filled in by an electron from outer L shell, the released x-ray photon will have an energy equal to E_K–E_L where E_K and E_L are binding energies of electrons of K and L shells, respectively. Vacancy produced in L shell is filled in by an electron from M shell giving rise to the emission of another x-ray photon. Large atoms with a large number of electrons and shells can give rise to a large number of x-ray photons resulting in an x-ray spectrum. The incident beam electron is scattered upon collision with the orbital electron changing direction and losing energy that is at least equal to the binding energy of the ejected orbital electron. For instance, in Fig. 6.2, the incident beam electron loses energy corresponding to the energy of the K shell (E_K). On the other hand, ejected orbital electron leaves the atom shell with an energy that varies from a few eV up to few keV depending on the nature of scattering interaction. The excited atom can release its excess energy and attain ground state by another process, i.e., through emission of Auger electron. In this process, the difference in the shell energies is not released as an x-ray photon but is transmitted to another outer shell electron, which then ejects out of the atom with specific kinetic energy.

Each shell around the atom has a specific amount of energy, which is known as the atomic energy level. It represents a characteristic energy of a specific element. Thus, the difference in the energy levels of these electron shells is considered as a characteristic value of an element. Therefore, electron transitions between any two shells result in the release of x-ray photons with an energy unique to an element. These x-ray photons have sharply defined energy values that occupy distinct energy positions in the x-ray spectrum. These x-rays are termed as *characteristic x-ray* lines since they are unique to the element they emanate from. Characteristic x-rays have specific energy depending on the elements that constitute the specimen. Distinct energy positions of x-ray lines form the basis for microchemical analysis where different elements in a specimen material are identified based on their unique orbital transitions.

The production of x-ray photons due to ionization process is known as *fluorescence yield*, which for K shells is higher than that for L shells. In addition, elements with a higher atomic number (e.g., heavy elements) have a higher fluorescent yield (see Fig. 6.3). Low yield for low atomic number (e.g., light) elements is responsible for their low detectability during microchemical analysis. Fluorescent yield for C is as low as 0.001, while for heavy elements it can be close to the value of 1. Fluorescence yield of some common elements is shown in Table 6.1.

6.2.2 Continuous X-Rays

Primary electron beam penetrates the specimen producing not only characteristic x-ray lines as stated above, but it also decelerates (brakes) due to interaction with atomic nuclei which have a positive field of a nucleus surrounded by bonded negative electrons. The energy loss due to deceleration is emitted as a photon of energy:

6.2 Production of X-Rays

Fig. 6.3 Plot showing an increase in fluorescence yield with atomic number

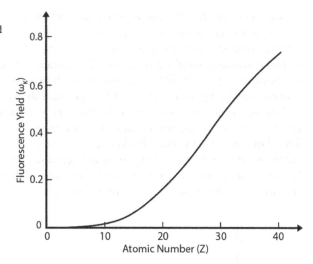

Table 6.1 Fluorescence yield for K_α transition lines of some common elements

Elements	Fluorescence yield for K_α lines
Carbon	0.001
Nitrogen	0.002
Oxygen	0.002
Sodium	0.020
Magnesium	0.030
Aluminum	0.040
Silicon	0.055
Phosphorus	0.070
Sulfur	0.090
Chlorine	0.105
Potassium	0.140
Calcium	0.190
Titanium	0.220
Chromium	0.260
Iron	0.32
Nickel	0.375
Copper	0.410
Zinc	0.435
Tin	0.860

$$\Delta E = h\nu \qquad (6.1)$$

where ΔE is the energy of the emitted photon, h = Planck's constant, and ν = frequency of electromagnetic radiation.

The interaction of the incident electron beam with the target atoms is occurring in a random manner; therefore, any electron deceleration may have different energy losses. Consequently, the continuum x-rays are produced at any value of energy ranging from zero to the maximum energy supplied by incident electrons, thereby forming a continuous electromagnetic spectrum called *continuum* or *white radiation or bremsstrahlung (braking radiation)*. For instance, if the incident electron beam reaches the sample with 20 keV of energy, it will generate continuum x-ray radiation extended from 0 up to 20 keV (see EDS spectrum shown in Fig. 6.4).

Continuum is generated due to all atoms in a specimen and appears as background in an x-ray spectrum. Since it is not unique to a particular element, it is devoid of any unique feature. Background intensity (i.e., continuum x-ray) is larger at the low-energy beam (see Fig. 6.5) and decreases with increasing x-ray energy. Not all of the generated x-rays are detected; some of them are absorbed inside the specimen material or in EDS detector window [1].

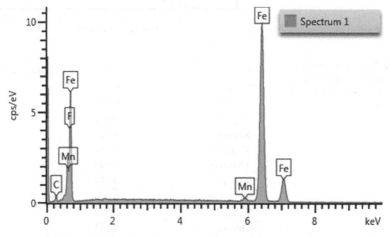

Fig. 6.4 Characteristic x-rays (peaks) and continuum (background) that together make up an energy-dispersive x-ray spectroscopy (EDS) spectrum obtained from a steel sample

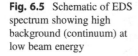

Fig. 6.5 Schematic of EDS spectrum showing high background (continuum) at low beam energy

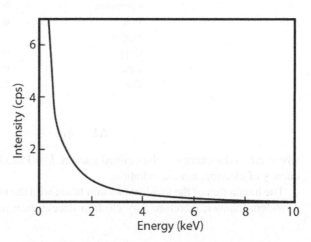

6.2 Production of X-Rays

6.2.3 Duane-Hunt Limit

Electron beam energy (E, keV) and wavelength (λ) of x-rays generated from a specimen are bound by the following relationship:

$$\lambda = \frac{1.2398}{E} \qquad (6.2)$$

where λ is the wavelength of the x-ray in (nm) and E is the electron beam energy measured in keV. The above equation shows that x-rays with higher energy will have a shorter wavelength. It can be seen in Fig. 6.6 (i) that the highest x-ray photon energy emanates when electron beam loses all of its energy in a single deceleration event. This introduces the concept of x-ray with short-wavelength limit (λ_{SWL}) or minimum wavelength ($\lambda_{\min}$). It states that x-ray photon of highest energy (or shortest wavelength) is generated when all of the incident beam (E_0) loses its energy instantly and is converted to photon energy. Since the wavelength varies inversely with the energy of the photons, this limit is referred to as short-wavelength limit. It is also called *Duane-Hunt limit* [2].

It can be seen in Fig. 6.7 that for spectra of carbon sample simulated at various beam energies (10, 15, and 20 keV), the Duane-Hunt limit increases with primary beam energy. The point where extrapolation goes to zero is considered the Duane-Hunt limit.

6.2.4 Kramer's Law

The intensity of the continuum x-rays also depends on the atomic number of specimen material as demonstrated by Kramer [3].

$$I_{\text{cm}} \approx i_p Z \frac{E_0 - E_v}{E_v} \qquad (6.3)$$

where I_{cm} represents x-ray intensity of the continuum background, i_p represents current of the probe, and the average atomic number value is represented by Z, while E_0 is the energy of the incident electron, and E_v is the energy of the continuum photon at a point in the spectrum. It is clear from Eq. 6.3 that the continuum intensity changes proportionally to the average atomic number of the specimen target and that could be explained by high Z targets having more charge. In addition, the intensity of the continuum is increased proportionally with the probe current and the amount of the beam energy. In addition, a significant increase of continuum intensity at the low-energy end of photons (E_v) is observed. This rapid increase in the continuum intensity is due to the higher probability for slight deviations in the trajectory resulting from the Coulombic field of the atoms at low E_v.

(i)

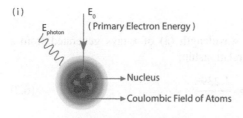

$E_{photon} = h\nu = E_0$ (Primary electron loses all energy in a single event, Photon energy equals the primary electron energy)

(ii)

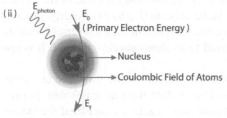

$E_f = E_0 - E_i > 0$ (Primary electron loses energy in multiple events)

$E_{photon} = h\nu = E_0 - E_f$ (Photon energy equals the primary electron energy less E_f, where $0 < E_{photon} < E_0$)

(iii)

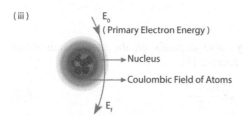

$E_f = E_0$ (Primary electron loses zero energy, no photon emitted)

Fig. 6.6 Generation of continuous x-rays due to incident beam deceleration within the specimen. The energy of the emitted x-rays depends on the nature of the interaction between the electron beam and specimen atoms. (i) Emitted x-rays will have the highest energy when the incident electrons lose energy instantly in a single scattering event. (ii) When primary beam loses energy in multiple scattering events, the x-ray energy is equal to the energy lost by the incident electron due to scattering. (iii) If the primary beam fails to lose any energy while passing through the sample, no photons will be generated as a continuum is produced due to deceleration of the primary beam in the sample matrix

6.2.5 Implication of Continuous X-Rays

The continuum x-rays represent the background of the spectrum, and the photons of this type of x-rays have no relationship to the sample component. Thus, it is considered a kind of noise. The photons that emanate due to the background

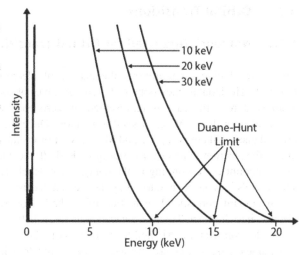

Fig. 6.7 Duane-Hunt limits for simulated spectra of carbon sample at a beam energy of 10, 15, and 20 keV. Duane-Hunt limit increases with beam energy. Duane-Hunt limit marks the end of continuum background, beyond which the continuum does not show any intensity [2]

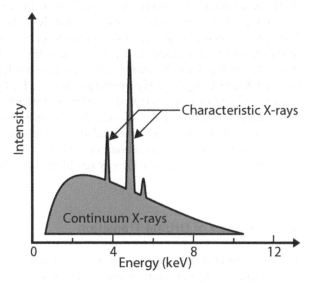

Fig. 6.8 Schematic representation of an EDS spectrum showing characteristic peaks superimposed on background that is formed due to continuous x-rays

(continuous x-rays) cannot be distinguished from those that originate due to orbital transitions (characteristic x-rays). They both contribute to the x-ray spectrum obtained from a specimen (see Fig. 6.8). Thus, the background present at an energy equal to that of a characteristic x-ray sets a limit to the minimum amount of an element that can be detected using x-ray spectroscopy. Therefore, in order to calculate the concentration of an element, the contribution of background needs to be removed from underneath a characteristic peak.

6.3 Orbital Transitions

6.3.1 Nomenclature Used for Orbital Transition

Energy level diagram represents the energy of electrons in specific shells (see Fig. 6.9). Horizontal lines denote the energy level of an electron state. When the atom is at rest, the case is represented by zero energy. Thus, the atom energy increases upon K, L, M, or N shell ionization. Once the atom returns to the usual level of energy, it emits K_α, L_α, and M_α x-rays. For instance, when an atom is ionized due to the ejection of an electron out of its K shell, the energy of the atom increases by an amount corresponding to the energy of K level. Then, if an electron from L shell travels into the K shell to fill up the created vacancy, the atom energy decreases by an amount corresponding to that of L shell. However, the movement of the electron into the K shell creates a vacancy in the L shell that is required to be filled up by an electron from the M shell level, and so on [1].

Characteristic x-rays are produced due to the transition of electrons between shells. X-ray lines are denoted by the shell from where the electron is originally ejected (i.e., shell of innermost vacancy) such as K, L, M, etc. This is followed by a line group written as α, β, etc. If the transition of electrons is from L to K shell, transition line is designated as K_α. If the transition is from M to K shell, it is designated as K_β. Since the energy difference between K and M is larger than that between K and L, K_β is of higher energy than K_α. Lastly, a number is written to signify the intensity of the line in descending order such as 1,2, etc. Therefore, the most intense K line is written as $K_{\alpha 1}$ and the most intense L line is denoted as $L_{\alpha 1}$. A schematic showing line transitions and their nomenclature is shown in Fig. 6.9. Typical line transitions are also listed in Table 6.2. Not all transitions of electrons between subshells are allowed, thereby resulting in the absence of several lines.

6.3.2 Energy of Orbital Transition

The energy of characteristic x-ray lines varies depending on the type of transitions. For instance, E_K–E_L transitions in a particular element give rise to K_α lines, and E_K–E_M transitions produce K_β lines, which have higher energy. This is followed by E_L–E_M transitions in that element giving rise to L lines. Similarly, E_M–E_N transitions produce M lines with lower energy compared to K and L lines in that particular element. For a given element K, L, M, N, etc., lines will always have different energies and therefore distinct positions in an x-ray spectrum. However, in a multiphase material, different lines from two elements can fall at the same energy position. For instance, M line of a heavy element might overlap with K line of a light element. This means that the energy of x-ray photons emitted due to M transition in a heavy element equals that emitted due to K transition in a light element. Due to the presence of subshells within K, L, M, N, etc., electron jumps (i.e., transitions) also take place within a particular shell.

6.3 Orbital Transitions

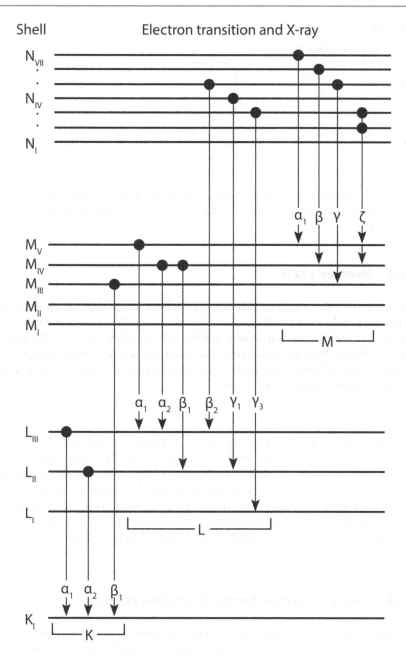

Fig. 6.9 Energy level diagram of an atom showing electron transitions

Transition energy is measured in electron volts (eV); 1 eV of energy corresponds to a change of 1 V in the potential of an electron and equals 1.602×10^{-19} J. Most transition x-ray lines of interest in EDS spectrum fall in the range 1–10 keV. The

Table 6.2 Typical line transitions and nomenclature used

Initial vacancy produced at		
Electron movement from	Electron movement to	X-ray line notation
L_{III}	K	$K_{\alpha 1}$
L_{II}	K	$K_{\alpha 2}$
M_{III}	K	$K_{\beta 1}$
M_V	L_{III}	$L_{\alpha 1}$
M_{IV}	L_{III}	$L_{\alpha 2}$
M_{IV}	L_{II}	$L_{\beta 1}$

total x-ray intensity emanating from a particular shell originates from several transition lines. For instance, for a K shell, >80% of the intensity comes from the $K_{\alpha 1,2}$ line.

6.3.3 Moseley's Law

The energy of a specific atom shell has a unique value that depends on the atomic number (Z) of the material. When x-ray photons are emitted from a material, it has energy characteristic of that atomic number (or material). In 1913, the English physicist Henry Moseley discovered that when the atomic number changes, the energy difference between the shells varies in a regular step. The energy of a photon can be given by Moseley's law below:

$$E = A(Z - C)^2 \tag{6.4}$$

where E is the energy of the x-ray line, Z is atomic number, and A and C are constants with specific values for K, L, M, etc., shells. This forms the basis for identification of elements in materials using x-rays. The above relationship describes energy required to excite any series of transition lines. For instance, x-ray photons of the highest energy in an atom are emitted from K_α shells. This energy equals the binding energy of 1 s electron which in turn is proportional to Z^2 as described above. This energy will be different for each element (depending on its atomic number) and can be used to identify it.

6.3.4 Critical Excitation Energy (Excitation Potential)

The minimum energy required to eject an electron from an atomic shell is known as critical excitation energy or excitation potential (E_c). Since the energy of electrons in a specific shell or subshell is a fixed value, so is the energy required to eject it. As the size of the atom increases (e.g., from light to heavy elements), the energy required to excite any particular transition line also increases. For instance, E_c of Ni K_α is much higher than that of Al K_α. Critical excitation energy of characteristic x-rays for common elements is shown in Table 6.3. The critical excitation energy is also known

6.3 Orbital Transitions

Table 6.3 Critical excitation energy of characteristic x-rays for common elements

Critical excitation energy (keV)				
Element	Atomic number	$K_{\alpha 1}$	$L_{\alpha 1}$	$M_{\alpha 1}$ (M_V)
C	6	0.283	–	–
O	8	0.531	–	–
Al	13	1.559	–	–
Si	14	1.838	–	–
Fe	26	7.112	0.709	–
Cu	29	8.979	0.932	–
Mo	42	20.002	2.520	–
Sn	50	29.200	3.929	–
W	74	69.524	10.204	1.809
Au	79	80.723	11.918	2.206
U	92	115.603	17.167	3.552

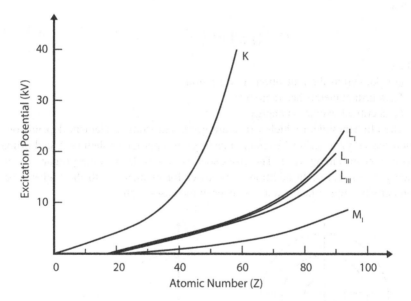

Fig. 6.10 The minimum energy required to eject an electron from an atomic shell (known as excitation potential, E_c) increases with atomic number

as excitation potential, critical ionization energy, and x-ray absorption edge energy. The excitation potential is used to calculate the intensities of characteristic x-rays.

The relationship between excitation potential E_c and atomic number Z for principal shells is shown in Fig. 6.10. It can be seen that E_c increases with Z.

It can be seen in Fig. 6.10 that the excitation potential of K shell is higher than other shells. In addition, the excitation potential of K shell increases extensively with a small increase in the atomic number. While for M_I shell that is located at some distance from the nucleus, the excitation potential is very small compared to the value of K shell for the same atomic number. Its increase with Z is also not significant. Therefore, the energy required to remove an electron from a specific shell depends on the atomic number of the specimen material.

6.3.5 Cross Section of Inner-Shell Ionization

Cross section of inner-shell ionization (σ or Q) is defined as the probability for an incident beam electron to be inelastically scattered by an atom per unit solid angle Ω. This is represented by the differential scattering cross section as a function of the scattering angle (θ) in Fig. 6.11a. Generally, the inelastic scattering cross section is higher than elastic scattering at low θ but decreases with increasing θ as shown in Fig. 6.11b [4].

The probability of an atom to get excited by the primary electron beam is shown below:

$$Q = \left(\frac{1}{E_0 E_c}\right) \log_e \left(\frac{E_0}{E_c}\right) \qquad (6.5) \quad [5]$$

where
Q is known as the ionization cross section
E_0 is instantaneous beam energy
E_c is critical excitation energy

The efficiency with which x-rays are generated from an element depends on its fluorescence yield, critical excitation energy for a particular shell (E_c), and primary electron beam energy (E_0). The cross-section values drop as the primary electron energy E_0 increases. In addition, it is lower for elements with the higher atomic number since the critical excitation energy increases with Z.

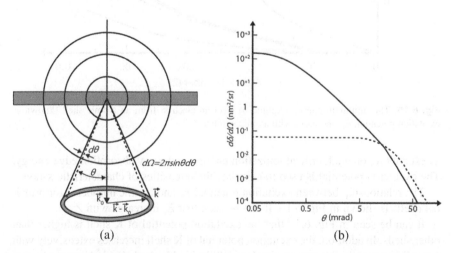

Fig. 6.11 (a) Schematic showing the dependence of electron scattering on scattering angle θ. (b) Plot showing the angular dependence of the elastic (dashed line) and inelastic (solid line) cross sections for C at 100 kV [4]

6.3 Orbital Transitions

Table 6.4 Energy loss due to inelastic scattering [1]

Element	Z	A	ρ (g/cm^3)	J (keV)	dE/ds (keV/cm)	dE/ds (eV/nm)
Carbon	6	12.01	2.1	0.1	2.24×10^4	2.24
Iron	26	55.85	7.87	0.285	6.32×10^4	6.32
Silver	47	107.9	10.5	0.487	6.94×10^4	6.94
Uranium	92	238.03	18.95	0.923	9.27×10^4	9.27

The cross-section of inner-shell ionization depends on the beam energy, which decreases as the beam penetrates further into the sample. The rate of beam electron energy loss (or continuous energy loss approximation) was determined by Bethe in 1930 [6] with the following expression:

$$\frac{dE}{ds}\left(\frac{\text{keV}}{\text{cm}}\right) = -2\pi e^4 N_0 \frac{Z\rho}{AE_i} \ln\left(\frac{1.166 E_i}{J}\right) \quad (6.6)$$

$$J(\text{keV}) = \left(9.76Z + 58.5Z^{-0.19}\right) \times 10^{-3} \quad (6.7)$$

where

e is the electron charge ($2\pi e^4 = 1.304 \times 10^{-19}$ for E in keV)
N_0 is Avogadro's number
ρ is the density (g/cm^3)
Z is the atomic number
E_i is the electron energy (keV) at any point in the specimen
A is the atomic weight (g/mole)
J is the average loss in energy per event

The loss of energy is indicated by the negative sign. Bethe's equation had limitations regarding low beam electron energies. These limitations were overcome later by Joy and Luo [7]. Typical values of the rate of energy loss (dE/ds) at 20 keV for various pure elements are shown in Table 6.4 [1].

It is important to determine the values of inner-shell ionization cross section since they have many applications in various fields like materials analysis using electron probe microanalyzer (EPMA) or EDS, thin-film analysis using electron energy loss spectroscopy (EELS), and surface analysis using Auger electron spectroscopy (AES) and in various fields of physics. These values can be determined by theoretical calculations or by experimental work [8].

6.3.6 Overvoltage

Critical excitation energy for a particular shell is approximately equal to the total sum of transition line energies for the outer surrounding shells. For instance, critical excitation energy for U K_α is equal to total line energies of $K_\alpha + L_\alpha + M_\alpha$ (i.e., 98.4 + 13.6 + 3.2 = 115.6 keV) for uranium. The primary electron beam energy must exceed the critical excitation energy (at least twice as much) to enable efficient excitation. Maximum efficiency is achieved when the primary beam is 2.7 times the critical excitation energy. The relationship of critical excitation energy with primary electron energy is given as follows:

$$U = \frac{E_0}{E_c} \tag{6.8}$$

where
E_0 is the primary energy
E_c is the critical excitation energy
U = Overvoltage

As an example, Fig. 6.12 shows the inner-shell ionization of Si K shell plotted as a function of overvoltage. It can be seen that the inner-shell ionization increases as the overvoltage increases from 1 to 3 followed by a decrease. The latter is attributed to a decrease in the energy of the primary beam due to inelastic scattering within the sample.

As the primary electron beam energy increases, the intensity of a particular x-ray line also increases as shown below [9, 10]:

$$I_c = i_p a \left(\frac{E_0}{E_c} - \frac{E_c}{E_c} \right)^n = i_p a (U - 1)^n \tag{6.9}$$

where
I_c = X-ray line intensity
i_p = Electron probe current
U = Overvoltage
a and n are constants specific for a particular element and shell

The generation of x-rays depends on $(U - 1)^n$ where n varies from 1.3 to 1.6. At small values of U, x-ray generation is minimal as shown in Fig. 6.13.

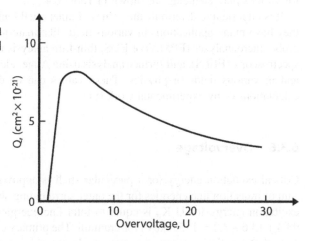

Fig. 6.12 Inner-shell ionization of Si K shell plotted as a function of overvoltage [1]

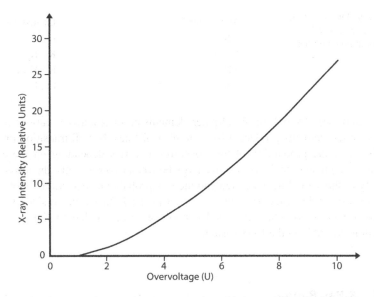

Fig. 6.13 Plot showing a decrease in x-ray generation with decreasing overvoltage, U

6.4 Properties of Emitted X-Rays

6.4.1 Excited X-Ray Lines

Characteristic x-ray peaks are usually very sharp and narrow. For example, the line width of the characteristic peak of calcium is 2 eV only. Because of the thin nature of these peaks, they are referred to in the literature as lines. The width of these lines depends on the resolution of the EDS spectrometer used to plot the spectrum. The width of a peak is typically 70 times wider than the natural width of the line [1].

The number of x-rays emanating from a specimen also depends on the beam current, specimen atomic number, interaction volume, etc. A primary electron beam of 15–20 keV is generally used in the SEM during energy-dispersive x-ray spectroscopy (EDS) to be able to excite characteristic x-ray lines that make detection of most elements possible. For elements with Z >35, an excessively high primary beam energy is required to excite K lines. This is avoided by detecting L and/or M lines (instead of K lines) for heavy elements, which require lesser primary beam energies. Excessively high primary beam energy results in deeper penetration of electron beam and larger interaction volume, which is undesirable in most cases. For light elements, only x-rays of K series are excited, for intermediate elements both K and L series are excited, while for heavy elements L and M series are excited. For a common element such as Zn, most intensely excited x-ray lines are $K_{\alpha 1}$ and $K_{\alpha 2}$ followed by $K_{\beta 1}$, $K_{\beta 2}$, and $K_{\beta 3}$ and then by $L_{\alpha 1}$, $L_{\alpha 2}$, $L_{\beta 1}$, and $L_{\beta 2}$. Series of lines excited for a range of atomic numbers are summarized in Table 6.5.

Table 6.5 Different types of line series excited for a range of atomic numbers

Atomic number	Excited line series
≤10	K_α
10–21	K_α, K_β
>21	K_α, K_β, L
≥50	L, M

As can be seen from Table 6.5, lighter elements produce a lesser number of lines, while heavier elements produce a large number of lines. Not all transition lines are possible, and the probability of their occurrence varies depending on the atomic number. The greater the difference in energy between two subshells, the lower is the intensity of the x-ray line generated and the less probable its detection. A lot of line transitions possible in theory cannot be seen in the EDS spectrum since they are located too close to other lines and cannot be resolved due to limited energy resolution available in the EDS system.

6.4.2 X-Ray Range

As stated in earlier sections, elastic and inelastic scattering events result in the penetration of electrons into the depth and distribute laterally across the specimen forming a relatively large *interaction volume*. Therefore, the information obtained from the specimen is not restricted to the size of the incident beam but is gathered from a much larger volume. The size of the interaction volume created depends on the specimen density, accelerating voltage of the beam and probe current density. The higher the accelerating voltage, the greater is the depth and the width to which the electrons can travel within the specimen. For specimens with a high atomic number, the elastic scattering is greater which deviates the electrons from their original path more quickly and reduces the distance that they travel into the specimen. Electron range is defined as the mean straight-line distance of the electron from the point of entry to the point of final rest in the specimen. The path length of an electron trajectory is primarily influenced by and inversely proportional to the atomic number and density of the specimen material for a given beam energy.

X-ray range is the depth of x-ray production within the interaction volume. It mainly depends on the beam energy, the critical excitation energy, and the specimen density. A significant part of the electron range (interaction volume) may produce x-rays depending on the critical excitation energy E_c. Characteristic x-rays are produced within electron range (see Sect. 3.2.6 and Eq. 3.8) for which E_c is exceeded for a particular x-ray line. The range of primary x-ray emission is smaller than electron range. Continuous x-rays are produced due to deceleration of electron beam within the specimen and do not require to surpass E_c for any particular x-ray line. Therefore, white radiation is produced until the electron energy becomes zero. This is shown as a schematic diagram in Fig. 6.14.

6.4 Properties of Emitted X-Rays

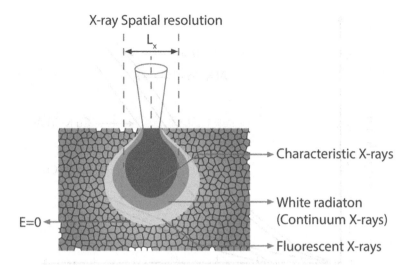

Fig. 6.14 Schematic showing x-ray range which makes up a considerable portion of electron range. Primary beam loses energy at a greater rate as it travels deeper into the specimen material. Characteristic x-rays are produced at depths where critical excitation energy for specimen element(s) is exceeded. Continuous x-rays are produced from greater depths until the primary beam completely loses its energy

X-ray range is given as:

$$R_x = \frac{0.064}{\rho}\left(E_0^{1.68} - E_c^{1.68}\right) \quad (6.10) \quad [11]$$

where
R_x = X-ray range, microns
ρ = Specimen density, gm/cm^3
E_0 = Incident electron beam energy, keV
E_c = Critical excitation energy, keV

Figure 6.15 shows Cu L_α and Cu K_α in a Cu sample and Al K_α and Cu K_α ranges in an Al specimen as a function of beam energy. It can be seen that electron range is larger than x-ray ranges in Al [11].

6.4.3 X-Ray Spatial Resolution

The depth and width from which x-ray lines are produced depend on the beam accelerating voltage, and atomic number and density of the specimen. X-ray spatial resolution is defined as the maximum width of the interaction volume generated by electrons or x-rays projected up to the specimen surface. Specimens with low atomic number and density allow deeper electron beam penetration and generation of x-ray lines from greater depths. As the depth of penetration increases, so does the lateral diffusion, which degrades the x-ray spatial resolution achieved. Figure 6.16 depicts the electron range and x-ray spatial resolution. It can be seen that materials with low

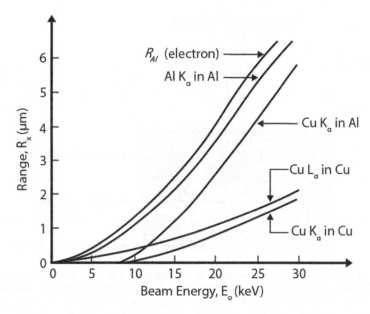

Fig. 6.15 Anderson-Hasler x-ray generation range for Cu L_α and Cu K_α in a Cu sample and Al K_α and Cu K_α ranges in an Al specimen as a function of beam energy. It can be seen that electron range is larger than x-ray ranges in Al [11]

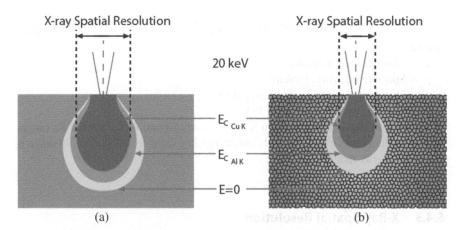

Fig. 6.16 Schematic showing electron range and x-ray spatial resolution for Al-Cu alloy of different compositions at a beam energy of 20 keV. (**a**) Low density 3 g/cm^3 and (**b**) high density 10 g/cm^3. For both Al K_α and Cu K_α x-ray lines, the range of x-ray production is deeper, and the x-ray spatial resolution is wider in the low-density sample than the corresponding range and spatial resolution in the high-density sample. X-rays are generated from a larger volume in low-density material, thus degrading the resolution of x-ray signal (Adapted from [1])

density (low Z) will produce x-ray signals with low spatial resolution as x-rays will be generated from a larger depth and width of the sample.

Higher accelerating voltage reduces the x-ray spatial resolution achieved in a specimen with a thickness typically used in SEM. The shape of interaction volume for the low-density specimen is pear-shaped, while that of a high-density specimen is spherical. X-ray interaction volume also depends on the critical excitation energy of the x-ray line. For example, for the same primary beam energy, the interaction volume for Ni K_α will be different from that for Ni L_α. In addition, x-ray generation within the interaction volume is not uniform and varies along its depth and width. It is higher at the point of penetration of electron beam into the specimen and decreases with distance from that point. It follows that, in order to increase the accuracy and precision of microchemical analysis, the specimen needs to be homogeneous over the entire interaction volume.

6.4.4 Depth Distribution Profile

It is clear from the shape of the sampling volume and Monte Carlo electron trajectories that the distribution of x-ray generation is not uniform both laterally and in depth. The lateral distribution is important for defining the spatial resolution of x-rays. Depth distribution is important because the deeper the generation point of an x-ray, the more distance it has to travel to escape the surface and reach the detector; thus the higher is the probability of the x-ray being absorbed by specimen atoms. Figure 6.17 shows the distribution of x-ray generation points for Al K_α x-rays in pure Al and Cu K_α x-rays in pure Cu, at 10 and 30 keV. On the left-hand side of the figure, there is a histogram of the distribution of generation points with respect to depth. This relation between generation point intensity and depth is called *depth distribution function* $\varphi(\rho z)$. In practice, it is difficult to measure the exact amount of x-ray generated at a certain depth. Therefore, an approximate approach is followed using mass depth, (ρz), method [1]. In general, as the beam energy increases and the atomic number decreases, the sampling volume or the total volume of x-ray generation increases; thus the depth distribution function $\varphi(\rho z)$ is affected.

It can be seen in the histograms in Fig. 6.17 that the depth distribution function is higher directly beneath the surface and decreases to zero as the electron energy becomes less than the critical excitation energy. Nevertheless, we can also see that $\varphi(\rho z)$ becomes higher close to the surface as the atomic number increases with the points of x-ray generation becoming more dense in that area. This is mainly due to two reasons: firstly, as the atomic number increases, the elastic scattering becomes more dominant, and beam electrons get scattered at early stages of sample penetration at a higher probability than for lower atomic number elements. Thus, fewer electrons penetrate to greater depths, and fewer x-rays are generated at those depths. Secondly, as the atomic number increases, critical excitation energy of a particular x-ray line in a sample also increases. The maximum excitation occurs when the electron beam has an overvoltage, i.e., two to three times more energy than the critical excitation energy. So as the critical excitation energy increases, fewer x-rays

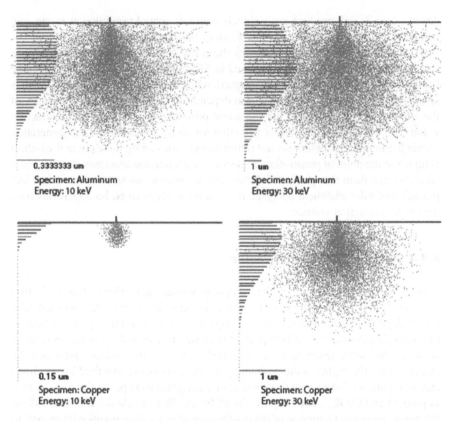

Fig. 6.17 Depth distribution profiles of x-ray generation points for Al and Cu Kα x-rays in pure metals at 10 and 30 keV. Histogram shows the x-ray generation function with mass depth, (ρz). It can be seen that the x-rays are generated from larger vertical and lateral dimensions with an increase in beam energy. The increase is more dramatic in materials with high mass depth (ρz), i.e., Cu. Similarly, the sampling volume is smaller in Cu (high ρz) compared to Al at both beam energies. Moreover, Al has lower critical ionization energy than that of Cu. Therefore, the same primary beam energy generates more x-rays of Al than Cu affecting depth distribution

are being generated, especially at greater depth where beam electrons would have lost much of their energy.

6.4.5 Relationship Between Depth Distribution $\varphi(\rho z)$ and Mass Depth (ρz)

The mass depth (ρz) is the product of the density ρ (g/cm^3) and the linear depth z (cm). The use of the mass depth term ρz is more common than the use of linear depth term z because the mass depth eliminates the need for distinguishing different materials because of their different densities when illustrating the relation with the

6.4 Properties of Emitted X-Rays

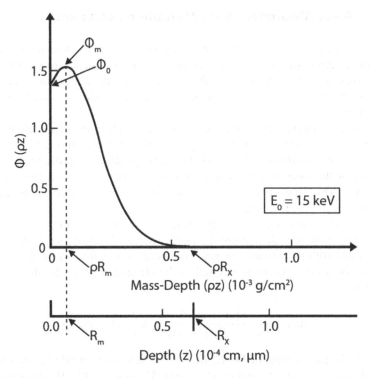

Fig. 6.18 Schematic for the measurement of $\varphi(\rho z)$ curve as a function of ρz and z [1]

depth distribution $\varphi(\rho z)$. Schematic in Fig. 6.18 shows the relationship between depth distribution $\varphi(\rho z)$ and mass depth (ρz).

As an electron penetrates the specimen, it gets scattered or strikes an orbital electron with enough energy to eject it. In the ejection case, the atom will be excited and it will release characteristic x-ray as it de-excites. In the scattering case, either the electron will travel deeper into the specimen, or it will be backscattered. If the electron is backscattered, it will generate x-rays as it leaves the specimen surface. Deeper traveled electron will repeat the scattering process until it is backscattered or its energy becomes lower than the excitation energy.

From the above explanation, we can see that there is a higher probability to generate x-rays near the surface. At greater depths of the specimen, there are fewer electrons because some of them are backscattered, and a lesser number of backscattered electrons is available to generate x-rays compared to the area directly beneath the surface. The generation of x-rays gets a maximum peak at ρR_m. At greater depths, the production of x-ray radiation begins to decrease as the depth increases. This is because the backscattered electrons of the incident beam reduce the number of electrons available at further depths. The electrons that succeed to penetrate deeper lose energy, and therefore they possess less excitation power as they scatter. Finally, x-ray generation points go to zero at $\rho z = \rho R_x$, where the electrons no longer possess an energy that exceeds E_c.

6.4.6 X-Ray Absorption (Mass Absorption Coefficient)

X-rays generated within the specimen target by incident electron beam can—as photons of electromagnetic radiation—undergo absorption by specimen atoms. Three types of x-ray absorption can take place as x-rays travel from their generation point to the detector, namely, elastic scattering, inelastic scattering, and photoelectric absorption.

In elastic scattering, the x-rays are absorbed by electrons of the atom. If the atomic forces—or the ionization energy—are high, the electrons are not ejected; rather, they are forced to oscillate about their mean positions. This oscillation emits a radiation of the same frequency and with no loss of energy in a new direction (Fig. 6.19). This type of scattering is dominant in atoms of high atomic number like gold ($Z_{Au} = 79$) [12, 13].

In inelastic scattering, the x-ray incidents on orbital electrons do not get completely absorbed; rather, part of the x-ray energy is absorbed causing the electron to be ejected with some kinetic energy. The energy loss of the x-ray radiation ΔE is equal to the kinetic energy transferred to the electron as given by the following equation:

$$\Delta E_x = \frac{E_x^2(1 - \cos\theta)}{m_0 c^2 + E_x(1 - \cos\theta)} \cong \frac{E_x^2}{E_0}(1 - \cos\theta) \tag{6.11}$$

If we take the example of Mo K_α ($E_x = 17.5$ keV) and considering the electron rest energy $E_0 = 511$ keV, we find $\Delta E_x = 600$ eV when $\theta = 90°$. This energy can be detected by an energy-dispersive x-ray detector. This type of scattering dominates in materials with a low atomic number [13].

In photoelectric absorption, photons interact with specimen material in a way that their energy is completely transferred to specimen atoms. In this way, an x-ray photon loses all its energy to an orbital electron, which is ejected with a kinetic energy equal to the difference in photon energy and critical ionization energy required to eject the electron. In photoelectric absorption, either the photon is completely absorbed in a single event, or it continues to propagate without any change in its energy. The intensity of the x-ray radiation—not the energy—will

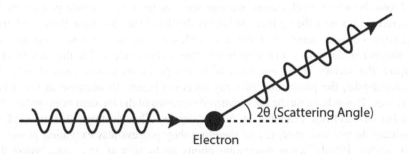

Fig. 6.19 Schematic showing elastic scattering of x-ray

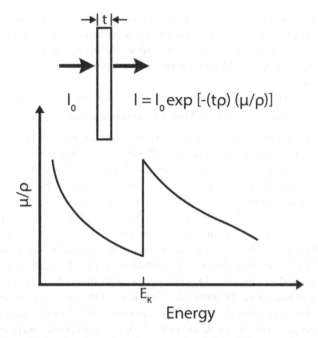

Fig. 6.20 Schematic showing absorption of x-rays and variation of the absorption coefficient with energy. The higher the energy of the x-rays generated within the specimen material, the lower is the mass absorption coefficient, i.e., x-rays will pass through the specimen easily. However, if the x-ray is energetic enough to overcome critical ionization energy and knock out an orbital electron of the constituent element, the mass absorption coefficient increases dramatically, i.e., absorption of x-ray occurs. This is indicated as a sudden increase in the coefficient at E_k in the schematic above. Further increase in x-ray energy will not increase absorption, as the x-ray energy will become too high for proper coupling (between the x-ray and atom) to initiate ejection of the orbital electron [12]

decrease due to photoelectric effect with an exponential decay (see Fig. 6.20) while it travels through a sample, according to the following equation:

$$I = I_0 \exp\left[-\left(\frac{\mu}{\rho}\right)(\rho t)\right] \tag{6.12}$$

where
  I is the intensity of x-ray photons when leaving the specimen surface
  I_0 is the original intensity of x-ray photons
  μ is the absorption coefficient
  ρ is the density of the specimen
  t is the thickness of specimen traveled
  ρt is the area density known also as the mass thickness
  $\left(\frac{\mu}{\rho}\right)$ is the mass absorption coefficient of the absorber for the specific x-ray energy, cm²/g

Absorption takes place by way of ejection of an orbital electron from its respective shell with a transfer of energy from the x-ray photons resulting in their complete absorption. A sharp increase in mass absorption coefficients is observed at energies corresponding to K, L, and M shell energies. These points of strong x-ray absorption are known as *x-ray absorption edges*.

Different materials absorb x-rays to different degrees and are defined by their mass absorption coefficients $\left(\frac{\mu}{\rho}\right)$. Mass absorption coefficient is a measure of how quickly x-ray intensity is lost within a specimen due to absorption. X-rays leaving the specimen at a high takeoff angle (θ) will travel less within the specimen and will be absorbed to a lesser extent compared to those that leave at a small θ after travelling a longer distance within the specimen. The mass absorption coefficient is equal to the absorption coefficient divided by the density. The absorption coefficient has units of inverse length and density has units of mass per volume. Unit of mass absorption coefficient is (length)^2/mass. The SI unit is cm^2/g or m^2/kg.

The probability of absorption is the highest when generated x-ray photons have energy slightly higher than the critical excitation energy of a particular shell of the specimen material or absorber. In other words, high mass absorption occurs when x-ray energy is just above the absorption threshold of the absorbing element. Unlike electron excitation of inner shells, where the maximum excitation occurs when beam electrons have an overvoltage of the order 2–3, the absorption coefficient of x-ray shows a steep increase when the x-ray energy exceeds only slightly above the critical ionization energy. This effect is stronger for x-rays with low energies. At the other extreme, low absorption occurs for high-energy x-rays that are farther away from the absorption threshold [12, 13]. Values of $\left(\frac{\mu}{\rho}\right)$ are widely variable, ranging from <100, for x-rays of high energy and absorbers of low atomic number, to >10,000 for x-rays of low energy and absorbers of high atomic number. In the latter case, severe absorption occurs even for thickness t less than 1 μm [14].

Characteristic x-rays belonging to a particular element (say Ni K$_\alpha$) will always have less energy than critical excitation energy E_c for that element (e.g., for Ni K$_\alpha$). Therefore, a matrix of Ni does not absorb too many of its K$_\alpha$ x-rays. In other words, the mass absorption coefficient of a Ni specimen for Ni K$_\alpha$ will be low. The absorption coefficient of an element for its own radiation is always low because the energy of an element's characteristic radiation is less than the excitation energy of the element. Thus, characteristic radiation of an element passes through it with little absorption.

In another example, the mass absorption coefficient of Cu K$_\alpha$ radiation is highest for cobalt (Co) (see Table 6.6). This is because the energy of Cu K$_\alpha$ is slightly higher than excitation energy E_c for Co. On the other hand, the absorption coefficient is smallest for Cu as an absorber since the energy of Cu K$_\alpha$ is lower than E_c for Cu.

In addition to absorption, some of the x-ray intensity is also lost within the specimen due to inelastic scattering. However, this can be ignored since interaction volume used for microchemical analysis is small. Some of the absorption occurs after x-rays leave the specimen. This absorption takes place in the environment or

6.4 Properties of Emitted X-Rays

Table 6.6 Mass absorption coefficients of Cu K_α for various elements [1]

Element (Z)	X-ray energy, keV		$\left(\frac{\mu}{\rho}\right)$ of Cu K_α in a given element (cm²/g)
	K_α	$E_c = E_K$	
Mn (25)	5.895	6.537	272
Fe (26)	6.4	7.111	306
Co (28)	6.925	7.709	329
Ni (28)	7.472	8.331	49
Cu (29)	8.041	8.980	52

while passing through the x-ray detector window that is usually made of beryllium. Lighter elements with low x-ray energies are absorbed in this manner more readily than heavy elements. For instance, x-rays from light elements such as Li cannot pass through Be window used in EDS detector and therefore cannot be identified or measured. Similarly, if the thickness of Be window is increased, more and more x-rays of elements even heavier than Li will be absorbed.

6.4.6.1 Mass Absorption Coefficient in a Single Element

In order for the photoelectric absorption phenomena to occur, the energy of emitted x-ray has to exceed the critical ionization energy of the electron orbiting in the specific shell. Different shells require different x-rays energies for absorption. The maximum effect of photoelectric absorption occurs when the energy of the emitted x-ray slightly exceeds the critical ionization energy of the electron. This is the energy where it is most probable for the absorption of x-ray to occur. Different sample materials also possess different ionization energies. These factors can be correlated by the expression:

$$\frac{\mu}{\rho} = KZ^4 \left(\frac{1}{E}\right)^3 \quad (6.13)$$

The mass absorption coefficient (μ/ρ) can be used to represent how probable it is for photoelectric absorption phenomena to occur. The higher is the absorption coefficient, the more probable is for absorption to occur. It can be seen from Eq. 6.13, as the energy of the x-ray increases, the mass absorption coefficient decreases. However, a sharp jump in absorption coefficient occurs in the energy region slightly exceeding the critical ionization energy for each shell. We can see that as an example in Fig. 6.21, the energy of incident x-ray versus the absorption coefficient for lanthanum ($Z = 57$) as the absorber material. It can be observed that, generally, there is a smooth decrease in absorption coefficient with some sharp jumps at certain energies. These jumps are the x-ray absorption edges for lanthanum, namely, the K edge at $\approx$38.9 keV, the L edges at $\approx$5.9 keV, and the M edges at $\approx$1.1 keV.

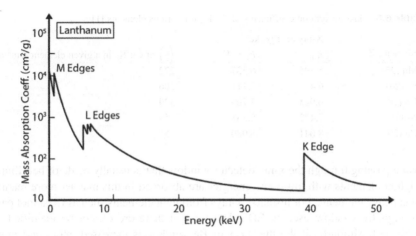

Fig. 6.21 Incident x-ray energy versus mass absorption coefficient in an absorber of lanthanum ($Z = 57$) [1]

6.4.6.2 Mass Absorption Coefficient in a Mixer of Elements

The mass absorption coefficient of a specimen containing more than one element is the sum of mass absorption coefficients for each element multiplied by its respective weight fraction. The absorption coefficient can be calculated using the expression:

$$\left(\frac{\mu}{\rho}\right)^A_{spec} = \sum_i C_i \left(\frac{\mu}{\rho}\right)^A_i \tag{6.14}$$

where $\left(\frac{\mu}{\rho}\right)^A_i$ is the mass absorption coefficient for the x-ray energy line from element A passing through element i and C_i is the concentration for each element used in the sample. All elements where absorption is possible to a significant extent should be considered. This consideration is critical, especially for low-energy peaks in the presence of light elements where absorption is important.

For example, in the case of Cu K_α x-ray line passing through a sample of SiO_2, the mass absorption coefficient can be calculated using the following equation:

$$\left(\frac{\mu}{\rho}\right)^{CuK_\alpha}_{SiO_2} = (wt.\text{fraction Si}) \left(\frac{\mu}{\rho}\right)^{CuK_\alpha}_{Si} + (wt.\text{fraction O}) \left(\frac{\mu}{\rho}\right)^{CuK_\alpha}_{O} \tag{6.15}$$

Inserting appropriate values into the equation:

$$\left(\frac{\mu}{\rho}\right)^{CuK_\alpha}_{SiO_2} = (0.468)\left(63.7\frac{cm^2}{g}\right) + (0.533)\left(11.2\frac{cm^2}{g}\right) = 35.8\frac{cm^2}{g}$$

It can be seen that the resultant mass absorption coefficient is affected by both elements' absorption coefficient values and elemental concentration.

As the primary beam energy increases with respect to the critical excitation energy (i.e., E_0–E_c increases), the peak-to-background (P/B) ratio obtained for an x-ray spectrum also increases. This is the ratio of intensities of the characteristic line over the continuum (background). It increases with increasing difference between the beam and critical energies and decreases with increasing atomic number. It is important because it determines the detectability limits of x-ray spectrometer. High P/B ratio has a positive influence on the ability to distinguish or remove continuum background from characteristic x-rays, in order to accurately determine the concentration of a particular element. However, as stated earlier, increasing primary beam energy will also result in its deeper penetration into the material. This will adversely affect the x-ray spatial resolution and increase absorption of x-rays within the specimen material. Absorption is one of the most crucial limiting factors to undertake the accurate microchemical analysis. It reduces the measured x-ray intensity, affects the detectability limits of elements, and necessitates absorption factor corrections during quantitative analysis. Therefore, the optimum value of primary beam energy is not more than two to three times the E_c for a given element. A sample is considered thin if its thickness is small in comparison with the elastic mean free path. It can be approximated using the cross-section of inner-shell ionization. Samples that have a thickness of 100 nm or more are considered thick. The thickness of 10 μm is considered infinite thickness when using SEM electron beam.

6.4.7 Secondary X-Ray Fluorescence

When primary electron beam penetrates a specimen, it ionizes atoms to generate characteristic x-ray photons. These photons, while on their way out of the specimen, may interact with other specimen atoms to cause secondary ionization resulting in the generation of additional characteristic x-rays or Auger electrons. The process by which x-rays are emitted because of interaction with other x-rays is called *secondary x-ray fluorescence* (see Fig. 6.22). Secondary x-rays will have a lower energy than the primary x-ray photons that induce x-ray fluorescence. Both characteristic and continuum x-rays can produce secondary x-ray fluorescence. The energy of the primary x-rays needs to exceed the critical excitation energy of secondary x-ray lines emitted from a particular element in the specimen. Fluorescence is significant only if the primary x-ray energy produced is within 3 keV range of the critical excitation energy of the element producing secondary radiation. The degree of x-ray fluorescence depends on the accelerating voltage, the concentration of the exciting element in the specimen, and the atomic number of the exciting and excited elements [15].

Fluorescence is a consequence of photoelectric absorption effect, and thus, as the mass absorption coefficient of the absorber increases, the fluorescence effect becomes stronger. This can be shown in Fig. 6.23 where the primary radiation is Zn K_α and the absorber is Ni. Nickel K_α fluorescent radiation is produced, as Zn K_α is strongly absorbed by Ni.

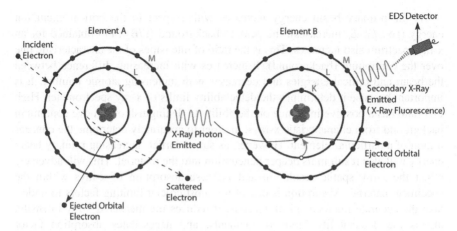

Fig. 6.22 X-ray photons generated by the primary electron beam may interact with other specimen atoms to cause secondary ionization resulting in the emission of additional x-rays. This phenomenon is called *secondary x-ray fluorescence*

Fig. 6.23 Mass absorption coefficient of Ni absorber as a function of x-ray energy. The position of Zn K_α energy line can be seen higher and very close to the critical ionization energy; thus it is strongly absorbed [1]

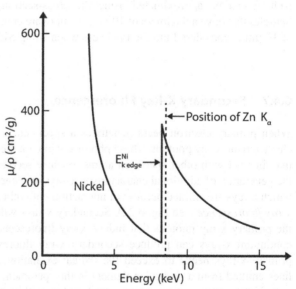

The fluorescence effect of different energy lines on a certain element can be estimated by comparing the mass absorption coefficient for each energy line; vice versa, the fluorescence of a certain energy line on different elements can be estimated by comparing different absorption coefficients. For example, Cu K_α line ($E = 8.04$ keV) is strongly absorbed in cobalt ($Z = 27$, $E_c = 7.709$ keV, $\mu/\rho = 326$ cm^2/g), while for nickel, which is only one atomic number higher than cobalt ($Z = 28$, $E_c = 8.331$ keV, $\mu/\rho = 49$ cm^2/g), the absorption coefficient is less by almost a factor of seven. This is because the energy of Cu K_α line is less than the critical ionization energy of nickel. We can say that the fluorescence effect of Cu K_α in cobalt is much higher than in nickel [1, 12, 14].

Characteristic fluorescence effect occurs in element B purely by characteristic x-ray of element A and not continuum x-ray. For this type of fluorescence to occur, it is necessary that the energy of the characteristic x-ray of element A exceeds the critical ionization energy of element B. For example, the Fe K_α (6.4 keV) can generate characteristic fluorescence of Cr (5.9 keV) x-rays, but not Mn K_α (6.5 keV) lines. However, the Fe K_β (7.0 keV) line can generate Mn K_α x-rays.

The fluorescence effect due to continuum x-rays is referred to as continuum fluorescence, which can take any energy up to the incident beam energy. Therefore, continuum x-rays will always contribute to fluorescence effect if the electron beam has higher energy than the critical ionization energy of the element of interest. However, since the continuum x-rays have a wide range of energies with low intensities, only a small portion of these intensities can cause fluorescence to occur. In practice, the extra intensity of induced characteristic x-ray caused by continuum fluorescence ranges from 1% to 7% for $Z = 20$ to 30 at a beam energy of 20 keV [1].

X-ray fluorescence can complicate quantification of elemental concentrations present within specimen material. For example, in the example cited above, the Zn K_α is strongly absorbed by the Ni specimen to produce Ni K_α fluorescent radiation. This can suppress the primary Zn K_α x-ray line and enhance the Ni K_α line creating a challenge to the accurate measurement of elemental concentration. In another example, the K_α x-ray of Cu element has an energy value of 8.05 keV, and it can be generated by K_α x-ray of Zn that exists in a brass sample. In 70Cu-30Zn alloy, more than expected Cu K_α and less than anticipated Zn K_α x-rays will be generated due to the fluorescence effect. In this way, Cu will be overrepresented, and Zn will be underreported unless corrections are made to the calculations. X-ray fluorescence acquires importance in alloys that have elements with similar Z because it affects the relative amount of characteristic x-rays emanating from compounds. Since x-rays travel farther into the material compared to electrons, the range of x-ray induced fluorescence within the specimen is larger compared to electron-induced range.

References

1. Goldstein J, Lyman CE, Newbury DE, Lifshin E, Echlin P, Sawyer L, Joy DC, Michael JR (2003) Scanning electron microscopy and X-Ray microanalysis, 3rd edn. Springer Science + Business Media, Inc., New York, USA
2. Duane W, Hunt FL (1915) On x-ray wavelengths. Phys Rev 6:166
3. Kramers HA (1923) On the theory of x-ray absorption and of the continuous X-ray spectrum. Phil Mag 46:836. https://doi.org/10.1080/14786442308565244
4. Bell DC, Erdman N (2013) Introduction to the theory and advantages of low voltage Electron microscopy. In: Bell DC, Erdman N (eds) Low voltage electron microscopy: principles and applications. Wiley, UK
5. Green M, Cosslet VE (1961) The efficiency of production of characteristics x-radiation in thick targets by a pure element. Proc Phys Soc 78:1206

6. Bethe H (1930) Zur Theorie des Durchgangs schneller Korpuskularstrahlen durch Materie. Annalen der Physik, Leipzig 397(3):325–400
7. Joy DC, Luo S (1989) An empirical stopping power relationship for low-energy electrons. Scanning 11(4):176–180
8. Llovet X, Powell CJ, Salvat F, Jablonski A (2014) Cross sections for inner-shell ionization by electron impact. J Phys Chem Ref Data 43:013102. https://doi.org/10.1063/1.4832851
9. Green M (1963) In: Pattee HH, Coslett VE, Engstrom A (eds) In proc. 3rd International symposium on x-ray optics and x-ray microanalysis. Academic Press, New York, p 361
10. Lifshin E, Ciccarelli MF, Bolon RB (1980) In: Beaman DR, Ogilvie RE, Wittry DB (eds) In Porc. 8th international conference on x-ray optics and microanalysis. Pendell, Midland, Michigan, p 141
11. Anderson CA, hasler MF (1966) In: Castaing R, Deschamps P, Philibert J (eds) Proc. 4th international conference on X-ray optics and microanalysis. Hermann, Paris, p 310
12. Hawkes P, Spence J (2008) Science of microscopy. Springer, New York
13. Reimer L (1998) Scanning electron microscopy: physics of image formation and microanalysis, 2nd edn. Springer, Berlin
14. Reed S (1993) Electron microprobe analysis and scanning electron microscopy in geology, 2nd edn. Cambridge University Press, Cambridge
15. Wittry DB (1962) Fluorescence by characteristic radiation in electron PRO micro-analyzer, USCEC Report. 84–204. University of Southern California, Los Angeles

Microchemical Analysis in the SEM

7

In most cases, it is desirable to obtain chemical information from specimens that are examined in the SEM. This is usually accomplished using energy dispersive x-ray spectrometry (EDS) or wavelength dispersive x-ray spectroscopy (WDS) technique. The microchemical analysis is accomplished by EDS detector or WDS spectrometer fitted in the column of the SEM. Integration of this detector or spectrometer with the SEM enables a user to determine the localized chemistry of a region. For example, the microchemical make-up of features that are only a few microns in size can be determined with a high degree of sensitivity. Not only the elements that make up a phase are detected (qualitative analysis) but also their concentrations are determined (quantitative analysis). The microchemical analysis is efficient and nondestructive and thus plays an important role in materials verification and phase identification. The EDS detector and WDS spectrometer are incorporated into the SEM in a way that does not disturb or affect the imaging capability of the instrument. The EDS and WDS identify the quantum characteristic x-ray energy and wavelength, respectively for elemental analysis. Their mode of operation is controlled by computers. This chapter describes the techniques used to undertake microchemical analysis in the SEM.

7.1 Energy Dispersive X-Ray Spectroscopy (EDS)

Interaction of primary electron beam with the specimen material results in the generation of characteristic x-rays and white radiation (background x-rays) which collectively form an x-ray signal. An x-ray detector is used to collect x-ray signal, measure its energy and intensity distribution, and analyze it in a manner that identifies elements and determines their respective concentrations in the analyzed region of the specimen material. Most commonly used x-ray detector in the SEM is the energy dispersive x-ray spectrometer (EDS).

Historically, Fitzgerald [1] successfully demonstrated the use of an EDS detector coupled to an electron microprobe analyzer. The resolution of EDS detectors at the time was no better than 500 eV which has now improved to 122 eV (using Mn k_α as reference peak) making most of the present-day microchemical analysis possible. In the past, energy dispersive x-ray spectrometers coupled with the SEM were primarily single-crystal Si(Li) (lithium-drifted Si) solid-state semiconductor devices. At present, however, Si drift detectors (SDD) have become commonplace. A photograph of a modern-day EDS detector used with the SEM is shown in Fig. 7.1a. The output of an EDS detector in the form of an EDS spectrum is shown as an example in Fig. 7.1b.

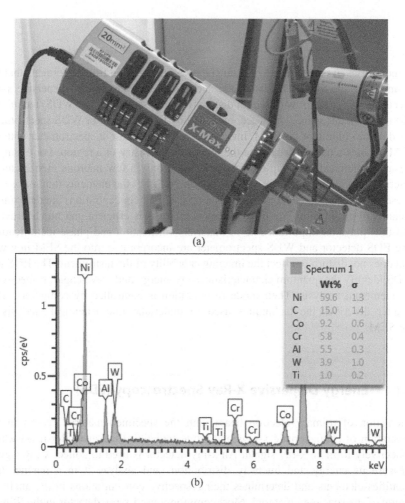

Fig. 7.1 (a) Photograph of Peltier-cooled EDS detector mounted on an SEM. (b) Energy dispersive x-ray spectrum obtained from a high-temperature Ni-based alloy showing the presence of various elements

7.1 Energy Dispersive X-Ray Spectroscopy (EDS)

7.1.1 Working Principle

A schematic diagram of the EDS detector setup commonly used in the SEM is shown in Fig. 7.2a. X-rays cannot be deflected into the detector thereby restricting collection to only those x-rays that are in line-of-sight of the detector. For this reason, the detector needs to be placed as close to the specimen as possible to increase the efficiency of x-ray collection. The distance between the specimen and the detector is normally 20 mm. X-rays emanating from a specimen are collected by a collimator tube which is located at the front end of the detector. The collimator acts as a limiting aperture and ensures that x-rays that originate only from the specimen are collected while stray x-rays from the specimen chamber or backscattered electrons do not find their way into the detector. Collimators can come in various shapes. One typical design is shown in Fig. 7.2b. A pair of permanent magnets is placed after the collimator to deflect any incoming electrons that can cause background artifacts in the x-ray spectrum.

Following the electron trap, there is a thin opaque window which serves to isolate the environment of the SEM chamber from the detector (see Fig. 7.2a). The window is followed by a semiconductor crystal that is sensitive to light. Thin window acts as a shield to protect the crystal from visible radiation. It also forms a barrier to maintain a vacuum within the detector assembly. Until 1982, the only available window was made of beryllium. Due to its high mechanical strength, Be window did not require a support structure. However, lack of support necessitated the use of a thick (around 8 μm) window which would absorb x-rays of energy less than 1 keV, thus preventing the detection of light elements such as boron, oxygen, nitrogen, carbon, etc. To enable light elements detection, an "ultrathin" window (UTW) made of thin (tens to a few hundred nanometers) organic film Formvar coated with gold was used instead of beryllium. This window is unable to withstand atmospheric pressure, and the detector assembly is kept under vacuum. The window can be removed altogether, and the detector can be used in a "windowless" mode. However, this leaves the detector exposed to contamination. In this situation, if the SEM chamber is vented, hydrocarbon condensation and ice formation will occur on the detector surface. The light will also be transmitted onto the semiconductor surface.

Presently, the ultrathin window of polymer covered with a thin layer of evaporated Al and supported with Si grid at the detector side is used as a standard. Due to grid support, the window is able to withstand the pressure of >100 Pascal in the SEM chamber. Support structure blocks part of the low-energy radiation thus reducing the detector efficiency to some extent. The grids are therefore designed to have up to 80% area available for x-ray transmission. This type of window can transmit low-energy ($\approx$100 eV) x-rays and is a preferred choice for light element analysis. Evaporated Al coating serves to restrict the passage of light through the polymeric material which otherwise exhibits high optical transparency. More recently, SiN [2, 3] and graphene [4, 5] have been investigated as potential window materials. Modern EDS detectors routinely detect elements from beryllium to uranium. First three elements of the periodic table H, He, and Li are not detected since they do not have enough electrons to produce characteristic x-rays.

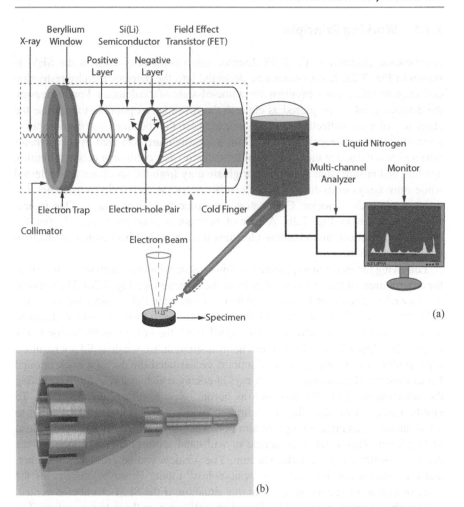

Fig. 7.2 (a) A schematic diagram of the EDS detector setup commonly used in the SEM. X-rays emanating from the specimen enter the EDS detector assembly through a tube called the collimator. (b) Photograph of a typical collimator assembly used to collect x-rays in the EDS detector

X-rays emanating from the specimen pass through thin window shield and reach the semiconductor diode detector made of single crystal of Si (or Ge). The energy gap between valence and conduction band is relatively small in semiconductors (1.1 eV in Si). X-ray photons striking the detector surface ionize Si atom through photoelectric effect creating electron-hole pairs (Fig. 7.3a). Upon application of a bias voltage between the thin gold contacts present at opposite ends of the semiconductor, these electrons and holes move in opposite directions toward the collection electrodes. The negative bias voltage applied to the front contact drives the electrons to the back contact and into the field-effect transistor (FET). This flow of current between the electrodes takes about 1 μs and is referred to as a charge pulse

7.1 Energy Dispersive X-Ray Spectroscopy (EDS)

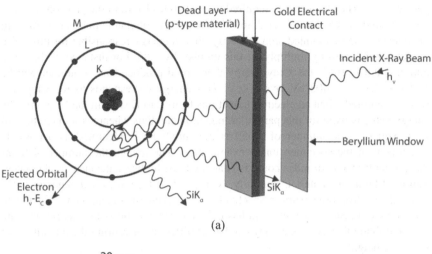

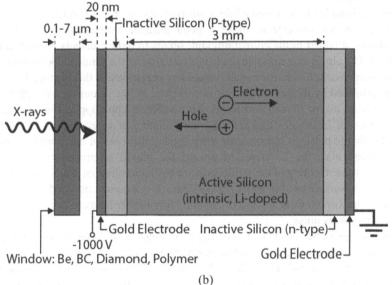

Fig. 7.3 (a) X-rays pass through thin window that protects the detector surface from visible radiation. Interaction of the x-rays with the Si(Li) detector results in the generation of electron-hole pairs whose number is proportional to the energy of the x-ray photons. (b) Schematic illustrating the working of p-i-n junction reversed biased semiconductor detector. Electrons and holes move in opposite directions and result in the generation of a pulse. The number of pulses is counted and correlated to the energy of the photons which created these pulses. Elements are identified since these generate photons with unique energy values

(Fig. 7.3b). The higher the energy of the x-ray photons that arrive from the specimen, the greater is the number of the charge pulses generated. Thus, the electrical charge that flows through the semiconductor is proportional to the number of electron-hole pairs created. The mean energy required to create one electron-hole pair (one electric pulse) in undoped Si is taken as 3.86 eV. The number of charge pulses generated in the detector can be counted, and the x-ray photon energy responsible for this pulse output is calculated by multiplying this number by 3.86. For instance, if the pulse output count is 1642, the x-ray energy that would produce such a number will be $1{,}659 \times 3.86 = 6{,}403$ eV or 6.4 keV. This energy corresponds to K_α x-ray line which is emitted when an electron transitions from L to K shell in the Fe atom. The energy value is fixed for this particular transition and thus whenever a magnitude of pulse equaling the number of 1,659 is measured; Fe is identified as a possible constituent of the specimen under examination. The greater the number of times this particular value of pulse count is generated, the higher is the elemental concentration of Fe in the material. Similarly, Ni K_α and Al K_α x-rays will generate 1,927 and 385 electron-hole pairs, respectively, as a result of ionization within the Si semiconductor. Since each element has unique characteristic x-ray energies that are different than other elements, they can be identified by measuring the magnitude of the pulse height.

Lithium is added in Si to make lithium-drifted silicon or Si(Li) detectors since, in practice, it is not possible to make a good intrinsic semiconductor from pure Si. Lithium, if added in the correct amount, serves to reduce defects present in the Si lattice. The aim is to create a large charge-free zone in the semiconductor using Li which compensates for charge carriers created by impurities. In this manner, the only charge exhibited by the semiconductor is generated by the incoming x-ray photons. Lithium is an n-type dopant and forms a p-n junction upon application onto pure Si. Addition of Li ensures that the maximum number of x-ray photons is used to generate charge pulses in the detector. One disadvantage is that upon application of high voltage bias, the Li is pulled toward the biased electrode giving rise to electronic noise. Cooling of the semiconductor by liquid nitrogen limits the mobility of Li ions. Si surface used in a Si(Li) EDS detector is around 3 mm in thickness and provides a maximum working area of 30 mm^2. Any further increase in the size of the detector increases noise.

The charge pulse created at this stage is small with large noise making it impossible to measure the energy of the x-ray photon. The Si(Li) crystal is connected to a FET at the rear end (Fig. 7.2a), which acts as the preamplifier to increase the signal strength and signal-to-noise ratio. The charge pulse created by the electron-hole pairs is converted into voltage steps (in mV) with the help of the preamplifier. The output of the FET is in the form of a *staircase waveform*. The size of a voltage step is proportional to the energy of the x-rays incident on the detector surface and the number of electron-hole pairs created. The voltage step is converted into a signal pulse by a pulse processor. The height of the signal pulse is proportional to the voltage step or to the energy of the photon striking the diode surface. The signal is averaged to reduce noise and improve pulse shape. An analog-to-digital converter (ADC) is used to convert pulses with various heights into pulses with constant heights. A number of pulses created is proportional to the heights of the input pulses. This process is known as *pulse height analysis* (PHA). The peak height of each signal pulse is converted into a digital value and assigned to the appropriate channel

7.1 Energy Dispersive X-Ray Spectroscopy (EDS)

in a computer *multichannel x-ray analyzer* (MCA) which displays the data in the form of a plot between voltage and intensity. The voltage range (displayed as units of energy, e.g., 10 keV, 20 keV, etc.) on the *x-axis* is divided into a number of channels (e.g., 1024, 2048, etc.). Each channel corresponds to a specific range of energy (e.g., Fe K_α x-ray line is from 6400–6410 eV). In this manner, one count is recorded at that particular energy level. Due to statistical variation in the energy of the electron-hole pairs created, a single peak with a Gaussian profile occupies several channels and can be roughly 150 eV wide. The number of times a particular voltage pulse is generated is plotted as intensity on the *y-axis* in the units of counts or counts per second (Fig. 7.1b).

In addition to the qualitative identification of an element, a quantitative measure of the concentration of that element can be undertaken. This is accomplished by counting the number of times a voltage pulse corresponding to a particular characteristic x-ray photon is generated and received in a channel of MCA reserved for that energy. The higher the count, the higher is the elemental concentration within the specimen volume analyzed. The higher the intensity of a peak, the greater is the concentration of element represented by that peak. The process of pulse generation, counting, identification of an element, and measurement of concentration is more or less automated in most cases. Output can be printed in the form of a labeled x-ray spectrum or transferred as data files onto another storage device such as a USB memory device.

The semiconductor crystal needs to be kept cool (at around −140 °C) to function properly; otherwise, electron-hole pairs are created at room temperature without any bombardment of x-ray photons. This will result in the addition of electronic noise to the x-ray spectrum. Cooling is usually achieved by mounting the detector and FET onto a cold finger (copper rod) that is connected to a Dewar of liquid nitrogen (LN_2) kept at −195.80 °C (Fig. 7.2a). Liquid nitrogen needs to be replenished every few days. Liquid nitrogen-free detectors such as Peltier-cooled (Fig. 7.1a) have become popular. The whole assembly is kept under vacuum at all times to avoid picking up contamination from the SEM chamber. The vacuum helps to maintain a low temperature as well. Water vapor and hydrocarbon molecules present within the SEM chamber are prevented from condensing on the surface of the semiconductor device by the thin window mounted before the diode. To avoid damage, a detector is not to be used in a "warmed-up" condition or if the vacuum is not present. A temperature sensor switches off the bias voltage if the detector is warm. The detector is constructed in a way that the semiconductor crystal and the cold finger are separated from the housing assembly. Retractable EDS detectors are usually employed whereby it is possible to move the detector close to and away from the specimen without breaking the vacuum in the SEM chamber.

Improvement in silicon detector technology has allowed the development of silicon drift detectors (SDD) where *n*-type large-area silicon wafer receives the incoming x-rays. The other side of the Si is decorated with concentric shallow rings of *p*-type drift material surrounding a small central anode contact. Upon application of bias, electrons *drift* through a field gradient that exists between the concentric rings and are collected at the central anode. These detectors are cooled using moderate thermoelectric cooling (e.g., Peltier technology) thus eliminating the need to use liquid nitrogen as a coolant. These detectors exhibit faster analysis times with higher count rates compared to conventional detectors. High count rates are possible due to the large surface area (up to 100 mm^2) of semiconductors used in SDD enabling fast data collection. Surface area in conventional Si(Li) detectors is limited to 30 mm^2 due to an increase in anode capacitance and noise with an increase in size. Larger sensors are possible with SDD detectors giving a superior resolution. The main disadvantage of these detectors is low detectability of light elements due to the presence of noise at low energies of the x-ray spectrum. The large size of the detector necessitates an equally large port opening in the SEM chamber which reduces flexibility in equipment design.

7.1.2 Advantages/Drawbacks of EDS Detector

It is common to find an EDS detector attached to an SEM due to its large number of advantages. An EDS detector is simple, robust, versatile, easy-to-ease and do not take up a large amount of space. Its functionality is seamlessly integrated into SEM operation. It undertakes a simultaneous analysis of all elements. The high efficiency of the detector combined with the large solid angle of collection (typically 0.5 steradian) results in small analysis time (e.g., less than 1 min). Due to this reason, low probe currents can be employed to extract elemental information from sensitive specimens. EDS technique is sensitive to light elements (can detect Be and higher) and can efficiently perform quantification of elemental data. The working distance setting when using EDS is not as critical as it is for WDS.

Disadvantages include low-energy resolution (122 eV) compared to wavelength dispersive x-ray spectrometer due to which closely spaced x-ray peaks cannot be distinguished, low detectability of elements (0.1–0.2 wt%) compared to WDS (0.001–0.002 wt%), low sensitivity to minor/trace elemental concentrations and lighter elements, and decreased resolution at high count rates.

7.2 Qualitative EDS Analysis

Qualitative EDS analysis is the identification of elements present within a specimen using energy dispersive x-ray spectroscopy. Qualitative EDS analysis in the SEM is a powerful tool that quickly determines the microchemical constituents of a specimen in a nondestructive manner. Since the x-ray signal resulting in an EDS spectrum is generated from a limited (in the order of microns) volume of material, it can be used to identify heterogeneity or segregation in specimens and also determine the chemistry of small objects or areas of interest.

7.2.1 Selection of Beam Voltage and Current

The SEM-EDS analysis is conducted by selecting a region of interest in the specimen. Usually, a high accelerating voltage (such as 20 kV) is selected for EDS analysis in order to provide adequate energy to the primary beam for it to excite characteristic x-rays of all elements of interest within 0–15 keV spectrum range. Occasionally, it is necessary to acquire spectrum up to 20 keV range for which higher beam energies (e.g., 30 keV) are required. Higher beam energies allow for higher peak intensities and a complete coverage of x-ray peaks from light to heavy elements. However, at the same time, it increases specimen interaction volume hence decreasing x-ray spatial resolution and increasing absorption. Generally, a primary energy of 2.7 times greater than critical excitation energy of a particular x-ray peak is optimum for analysis. Similarly, the probe current used during EDS analysis is usually higher than that recommended for imaging. These two parameters need to be considered in conjunction with the specimen's ability to resist beam damage since higher voltages and currents and longer analysis time increase the probability of contamination and beam-induced specimen damage.

7.2.2 Peak Acquisition

The whole EDS spectrum usually up to 20 keV can be acquired within approx. 100 s. The computer checks peak energies against values of characteristic x-ray energies for different elements saved in its database and can label these peaks with the names of the elements during the acquisition process itself. Most of the major (high-intensity) peaks are automatically identified quickly in this manner. Minor (low intensity) peaks may need some operator input. Thus, from a user's point of view, the process of acquisition and qualitative analysis of x-ray spectra is efficient and fairly straightforward. However, it is necessary to understand the process behind this identification in order to be able to verify results and also resolve any complications arising due to any overlapping or low-intensity peaks and artifacts in the spectrum.

7.2.3 Peak Identification

In an EDS spectrum, x-axis displays x-ray energy in keV, and the y-axis shows intensity in counts or counts per second as shown in a typical EDS spectrum shown in Fig. 7.4. Different peaks are positioned at different x-ray energies. Elements are identified on the basis of their peak positions or x-ray energies. For instance, a peak occurring at 7.47 keV is identified as Ni K_α as it is known that the latter falls at this specific position. In order for peak positions to be accurately identified, it is necessary that EDS is properly calibrated which is done by using a pure metal standard such as Ni. Once the EDS spectrum is acquired by the computer, the

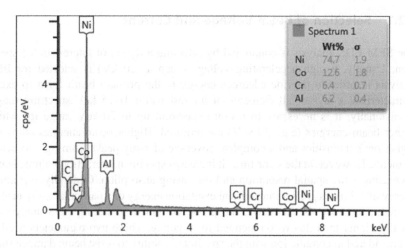

Fig. 7.4 Typical EDS spectrum a showing plot of intensity (counts or counts per second, cps) on y-axis and energy of emitted x-ray photons (keV) on the x-axis. Quantified elemental concentration is shown in the inset

specific energy of each characteristic x-ray peak is determined and compared with the values present in the database. The database has reference peak values stored for all elements. Most of the times, this database suffices for peak identification. Peaks are identified automatically by the computer, or it can be overridden by the user. With increasing atomic number, the number of peaks emanating from elements also increases. Heavy elements give rise to a large number of x-ray peaks. In order to identify a given element with confidence, most of this family of peaks need to be identified. Detection of a single peak for a particular element may lead to erroneous identification. All peaks belonging to a particular element need to be sought. Acquired spectra are stored in the computer and can be processed at any time later to reconfirm analysis or reassign peak identities, if necessary.

Peak intensity should be at least three times the intensity of the background in order to be identified properly. If necessary, analysis time is increased to acquire adequate peak height. Peaks falling at the high-energy end of the spectrum are identified first as they are more widely separated and easily determined. High-intensity peaks are identified as K_α, L_α or M_α depending on atomic numbers of elements present in the specimen. This is followed by the identification of corresponding K_β, L_β, or M_β whose intensity is many times lower. Next, low-energy peaks are identified. EDS detector's resolution at low-energy end is less, therefore restricting the number of peaks originating from light elements to one only. This end of the spectrum will also have L peaks for elements displaying corresponding K peaks at higher-energy end. If the low concentration of elements are present and need to be identified, then a spectrum with very high counts need to be acquired by increasing the analysis time.

7.2.4 Peak to Background Ratios

The high peak-to-background ratio in the EDS spectrum is desirable as it increases the detectability limit of elements. Small probe size generally results in high peak-to-background ratio, which also increases with increasing values of $E_0 - E_c$ where E_0 is the primary beam energy and E_c is the critical excitation energy of x-ray line. However, E_0 can only be increased to an optimum level beyond which the beam will travel to a greater depth within the specimen deteriorating the spatial resolution and increasing x-ray absorption. This will in turn decrease the number of x-ray photons emitted from the specimen and degrade the detectability limit. Therefore, an overvoltage of 2–3 times is optimum for most materials.

7.2.5 Background Correction

Both characteristic and continuum intensities make up the x-ray spectrum. Gaussian peak tail extends over a substantial range of energy and interferes with the adjacent background. Therefore, measurement of background becomes difficult due to the challenge of pinpointing its exact level adjacent to the peak under observation. This situation becomes more complex for a mixture of elements which also cause less accurate interpolation. Background correction is undertaken by the software as follows:

Background Modeling Continuum energy distribution function can be measured and also calculated. It is then combined with a mathematical description of the detector response function which is used to find the background. Finally, subtraction from detected spectral distribution is undertaken.

Background Filtering Mathematical filtering or modification of frequency distribution is also used for background removal. Digital filtering and Fourier analysis are examples of this method.

7.2.6 Duration of EDS Analysis

Length of analysis can be 10–100 s depending on the required strength of the x-ray signal. For major elements, shorter counting times can be used while longer counting times are required for minor or trace element detection. Major elements can display peaks of reasonable intensity in shorter times, while minor elements need longer analysis time to achieve equivalent or reasonable peak intensities. Solid-state detectors are placed close to the specimen thus increasing solid angle for x-ray collection raising count rate and sensitivity to detect the small concentration of elements or light elements in a given acquisition time.

7.2.7 Dead Time

X-rays emanating from the specimen enter into the EDS detector, processed and displayed on the computer in the form of a spectrum. EDS detector's capacity to receive and process x-ray photons is not unlimited. While one x-ray event is received and processed, other simultaneous incoming x-rays are not processed. The duration for which these x-ray signals are not processed is known as *dead time*. The stronger the x-ray signal the longer is the dead time of the detector, i.e., the longer it takes to process x-rays. Dead time appears as a percentage and is normally kept below 25% for efficient analysis. The analysis is usually done in *live time* mode which indicates the duration during which x-ray signals are actually processed. Modern SDD detectors tend to have shorter dead time intervals meaning they can process x-ray signals relatively quickly. The dead time is calculated as follows [6]:

$$\%\text{Dead time} = \left(1 - \frac{\text{Count rate of the output}}{\text{Count rate of the input}}\right) \times 100\% \quad (7.1)$$

Alternatively, it is defined as:

$$\%\text{Dead time} = \left(\frac{\text{Total clock time} - \text{Live time}}{\text{Live time}}\right) \times 100\% \quad (7.2)$$

Live time is the time required for signals collection, and the total clock time is the time required for the signals collection in addition to the signals processing time. The count rate of the spectrum will change depending on the rate of continuous x-rays arriving from the specimen, which varies with the sample's elemental composition resulting in changes in the dead time [7].

7.2.8 Resolution of EDS Detector

The energy resolution of an EDS detector is its ability to distinguish two adjacent peaks in the EDS spectrum. It is measured at full width at half maximum (FWHM) and quoted for a peak at 5.9 keV (Mn K_α) energy position. The energy resolution of present-day EDS detectors is quoted to be around 122 eV. The lower number (in eV units) indicates a higher resolution. Narrow peaks represent better resolution as the overlap between peaks decreases at increasing resolution. Peaks formed by low-energy x-rays show better resolution. For example, in silicon drift detectors, the resolution of the F K_α peak and C K_α peaks is between 60–75 eV and 56–72 eV, respectively. Energy resolution is also related to the live time used when collecting the EDS spectrum. Narrower peak and better resolution are obtained when the process time is long. However, this leads to longer dead time; thus increasing the total time required to acquire the spectrum. Good spectral resolution is desirable in order to identify and quantify the elements present in a specimen.

Electronics used in pulse processing plays an important role in achieving good energy resolution by way of eliminating peak shifts and peak distortions [6]. The

7.2 Qualitative EDS Analysis

relationship between the EDS detector's energy resolution, the quality of the electronics used, the width of the intrinsic line, and FWHM is expressed as follows [6]:

$$R^2 = I^2 + P^2 + X^2 \qquad (7.3)$$

where

R = the detector's energy resolution
I = width of the intrinsic line of the detector
P = indicator of the quality of the electronics used (FWHM of the electronics generating the pulse)
X = the equivalent FWHM related to incomplete charge collection (ICC) and leakage current of the detector

Modern EDS software can automatically measure the resolution. On the other hand, Mn or Cr peaks can be collected with the selection of a proper window that can contain the peaks on both sides. The peaks should include 50% of the maximum count in the channel at the center. Figure 7.5 shows the energy resolution measurement by specifying the number of channels which contain the FWHM of Mn K_α peak for a specific EDS detector.

X-ray lines acquire the shape of a peak since there is a statistical distribution of energies associated with x-ray photons emanating from a given element due to a particular type of transition. Natural width of an x-ray peak is small, but it gets broadened after passing through the EDS detector electronics. Peak broadening leads to a decrease in peak height and peak-to-background ratio which adversely affects

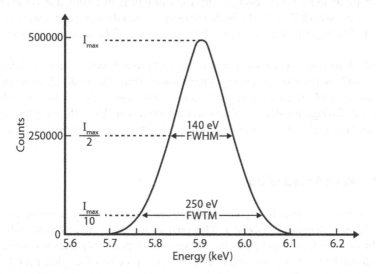

Fig. 7.5 Energy resolution is measured by identifying the channels that encompass the FWHM of Mn K_α peak. Full width at tenth maximum (FWTM) can also be measured which indicates the extent of distortions in the peak [6]

detection of elements in a specimen. It also contributes to peak overlaps since the wider the peaks the greater is the chance that some peaks overlap each other.

7.2.9 Overlapping Peaks

Characteristic x-ray peaks from different elements can have the same energy in which case they will overlap in the EDS spectrum. It is difficult to distinguish between peaks that fall within 100 eV of each other, especially if there is a substantial difference in their heights. Small peaks in the neighborhood of large peaks also present a similar challenge. Frequently encountered overlapping peak pairs include $SK_{\alpha,\beta}$-MoL_α, $SK_{\alpha,\beta}$-PbM_α, TiK_α-BaL_α, CrK_β-FeK_α, MnK_α-CrK_β, FeK_α-MnK_β, $WM_{\alpha,\beta}$-$SiK_{\alpha,\beta}$, $TaM_{\alpha,\beta}$-$SiK_{\alpha,\beta}$, YL_α-PK_α, etc. The user should be aware of these overlaps in order to avoid incorrect peak assignments. Peak overlaps appear as unusually broad peaks or shoulders in large peaks. Peak stripping feature is provided in EDS software which can help strip one peak based on stored peak positions to reveal hidden peaks underneath.

7.3 Artifacts in EDS Analysis

7.3.1 Peak Distortion

Thin "dead" layer present on the Si crystal can lead to self-annihilation of electron pairs resulting in the loss of charge. This can cause peak distortion due to incomplete charge collection (ICC) [7]. The peak will deviate from the Gaussian shape as seen in Fig. 7.6. Heating the detector will reduce the effect of the dead layer on the peak shape.

Peak distortion can also occur due to *background shelf* which is defined as background increments at energy range lower than the peak of concern. The background shelf occurs when continuous x-rays are scattered inelastically and spread out through the detector thus escaping detection. This effect is prominent in the spectra of radioactive materials [7] such as (^{55}Fe) [8] as seen in Fig. 7.7.

7.3.2 Peak Broadening

The number of electron-hole pairs produced for specific energy is not absolute but depends on statistical distribution. The final count shows the average only. This introduces uncertainty leading to the broadening of peaks [8]. Another uncertainty arises from the thermal noise originating from the process of amplification [7]. For photon energy, the Gaussian distribution is used to describe the distribution of the number of charge carriers and is shown in Fig. 7.8. The description is given by the following equation:

7.3 Artifacts in EDS Analysis

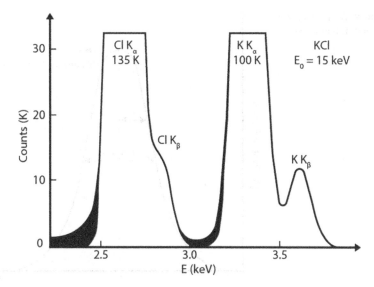

Fig. 7.6 EDS spectrum of potassium chloride with ICC artifacts. The dark region in the spectrum shows the effect of the incomplete charge collection resulting in the peak deviation from the perfect Gaussian shape [8]

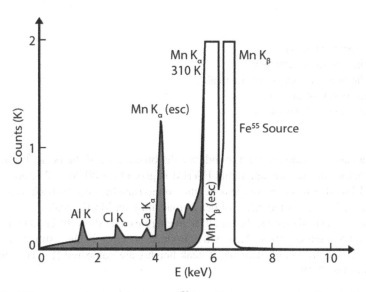

Fig. 7.7 EDS spectrum of radioactive ferrous (^{55}Fe) with background shelf effect visible for the energy range lower than Mn K$_\alpha$ peak [8]

Fig. 7.8 Peaks are defined by Gaussian distribution [8]

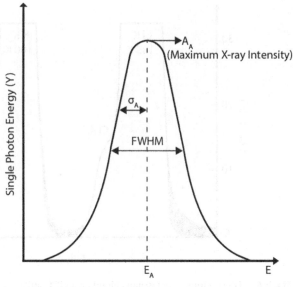

$$Y = A_A \exp\left[-\frac{1}{2}\left(\frac{E_A - E}{\sigma_A}\right)^2\right] \tag{7.4}$$

where

Y is the peak intensity
A_A is the maximum intensity
E_A is the average energy for the peak
E is the energy of the x-ray
σ_A is the standard deviation

The standard deviation is used to indicate the broadening of the peak. The relation between the standard deviation and FWHM is given by FWHM = $2.355\sigma_A$.

Peak broadening decreases peak height (counts) and the peak-to-background ratio [8]. Figure 7.9 shows the effect of peak broadening on Mn K_α peak.

Peaks with similar counts but at different energies may show different heights due to peak broadening effect. This will introduce an error in the estimation of the relative concentration of elements if peak heights are compared [8]. This effect is shown in Fig. 7.10.

7.3 Artifacts in EDS Analysis

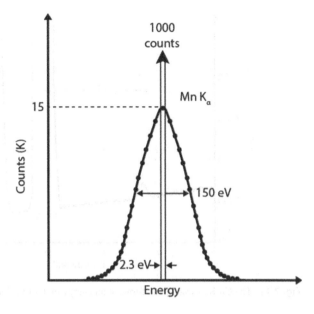

Fig. 7.9 Schematic illustrating the peak broadening effect on the Mn K_α peak. Width increases from 2.34 to 150 eV. Counts are reduced from 1000 to 15 [8]

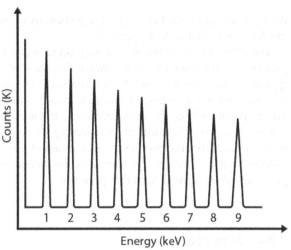

Fig. 7.10 Variation of height in the peaks with different energy but same counts [8]

7.3.3 Escape Peaks

It is statistically possible that x-ray photons emanating from specimen enter the detector and ionize Si releasing K-type x-ray photons. If this transition occurs close to the detector surface, the photons can escape the detector. This will decrease the energy of the x-rays emanating from the specimen by an amount equal to that required for the Si K transition event. Due to this event, an escape peak is generated in the x-ray spectrum at energy $E_{(\text{specimen})}\text{-}E_{\text{SiK}\alpha}$. For instance, if Cu is the specimen material tested, an escape peak at 8.04 ($E_{\text{CuK}\alpha}$) − 1.74 ($E_{\text{SiK}\alpha}$) = 6.3 keV can form in

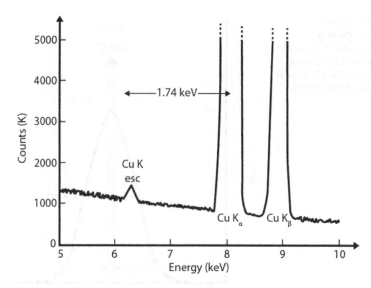

Fig. 7.11 The Cu K_α escape peak forms at an energy of 1.74 keV less than that of Cu K_α

the x-ray spectrum (see Fig. 7.11). The probability of the formation of Si K_β peak is much less than that of Si K_α peak [8].

The fluorescence of the Si can only occur when the incident x-ray energy is greater than that required for the critical excitation of Si. The escape peaks size is typically related to the parent peak (between 1% and 2% of the parent peak) [9]. Since the x-rays emitted from high atomic number atoms will lose energy in greater depths within the detector, hence, in case of Si fluorescence x-rays are produced, they will face difficulties in leaving the detector, thus reducing the escape peaks artifact [10]. The user needs to be aware of this escape peak phenomenon in order to be able to recognize it when it occurs and not to confuse it with some other genuine peak emanating from the specimen.

7.3.4 Sum Peaks

When two characteristic x-rays arrive at the detector simultaneously, the detector might consider them as one and display it at an energy equal to the sum of the energies of the two x-rays. Accumulation of such events might lead to a peak in the spectrum at sum energy position [6, 8, 9, 11]. This peak is known as *sum peak* (or double or coincidence peak) which is an artifact. The sum peaks are more probable to occur when the count rate for the input photons is large and the dead time exceeds 50–60%. Presence of major speaks in the EDS spectrum also

7.4 Display of EDS Information

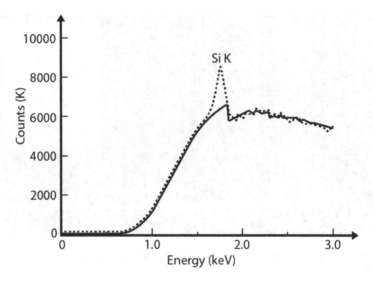

Fig. 7.12 Si internal fluorescence peak artifact in pure carbon spectrum [6]

contributes to the formation of sum peaks [6, 11]. Due to this artifact, the presence of certain elements in the test specimens was misreported in the past. With the development of more reliable electronics, these major errors are now eliminated except perhaps for very low-energy EDS analysis [6].

7.3.5 The Internal Fluorescence Peak

This artifact originates from the dead layer of the Si or Ge detector. It occurs when the x-rays entering the detector strike the detector atoms and cause them to fluoresce, resulting x-rays of Si K_α, Ge K, or Ge L appearing in the spectrum. This effect is known as *internal fluorescence peak* artifact. The advancements in detectors manufacturing resulted in the reduction of the thickness of dead layers which in consequence decreased the artifact of the internal fluorescence peak. Nonetheless, this artifact has not disappeared completely [6] especially when trace amounts of Si are analyzed [8, 11]. Figure 7.12 shows the internal fluorescence peak artifact.

7.4 Display of EDS Information

Information obtained from EDS analysis can be displayed in the formats summarized below.

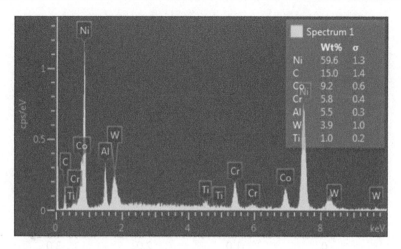

Fig. 7.13 EDS spectrum obtained from the superalloy sample showing the presence of various elements

7.4.1 EDS Spectra

The most common form of the visual format used to display microchemical information obtained from the analyzed area of a sample using EDS is the x-ray spectrum as shown in Fig. 7.13.

To obtain an EDS spectrum, the beam is usually placed over the feature of interest in the form of a focused circular spot or quadrilateral format (e.g., *spot analysis*). Irregular shapes can also be analyzed in modern microscopes. The x-rays emanating from the area of interest pass through the detector electronics, and the processed information is displayed in the form of a plot on the viewing monitor. The horizontal axis shows the energy of x-ray photons emitted from the sample, and the vertical axis shows the intensity of photons in the form of counts or counts per second. The characteristic x-rays peaks are superimposed on the background formed by continuous x-rays. The energy scale is usually displayed up to 10 keV, although it can be increased to coincide with the primary beam energy used during analysis. The EDS spectrum takes shape within a minute and serves as a quick qualitative visual indicator of the sample constitution. Further, the spectrum can be processed by the software seamlessly to display quantitative chemical information about the sample in the form of relative elemental concentrations.

7.4.2 X-Ray Maps

X-ray map displays the elemental distribution information visually in a two-dimensional plot. The process of acquiring x-ray maps is generally known as *x-ray mapping*. Area of interest is scanned by the electron beam, and from each

7.4 Display of EDS Information

discrete location (pixel), an EDS spectrum is obtained and stored. The number of discrete locations and beam dwell time for acquiring EDS data from within the area of interest can be selected by the user thus controlling the final resolution of the x-ray map. The EDS data from each location is stored in the computer memory thus making it possible to recall and analyze that particular data set offline later. Multitudes of frames (sometimes up to hundred) are taken from the same area to improve map resolution. High probe currents are employed for x-ray mapping to attain good contrast.

Maps are displayed in multiple windows. Data for all elements is captured in each pixel due to which different elements can be mapped simultaneously. Each window displays the total scanned area. Each window is reserved for one element of interest only that is predetermined using spot or area EDS analysis. The area(s) that appears bright within a window is the region where a particular element (for which that window is reserved) is concentrated (see Fig. 7.14).

Alternatively, color contrast can be used in a single window for clearer visualization and better understanding (see Fig. 7.15). Primary color superposition displays images with three elements where maps are obtained by assigning color red, green, and blue, respectively. Another way is pseudocolor scales which are based on either thermal scale or on logarithmic three-band scale. In this way, elements present in a sample are colored maps of estimates of elements; each color in a window represents one element.

X-ray mapping sorts out and visualizes the elemental dispersion in a multiphase sample. X-ray mapping can be considered to be an image of the scanned area of interest formed by x-ray spectra. Images are usually produced with a file extension of TIFF. Multiple scans at extremely high magnification can produce image drift during the scan to make the topographies appear smeared in x-ray maps. The drift could distort the image even during a single scan, whereas discrepancies become noticeable when multiple scans are performed. In such a case, the operator has the option to stop and continue as appropriate rather than wait until the end of the full scan. Electron beam stability becomes important during x-ray mapping which can take anything from tens of minutes to a few hours depending on the number of elements scanned, beam dwell time, and the number of frames employed.

Continuous x-rays form part of the x-ray maps. Background in the peak constitutes roughly 6% count in major peaks. It is difficult to attain high-resolution and good detection limits during x-ray mapping, thereby precluding detection of minor or trace elements in x-ray mapping. High-resolution x-ray maps can be obtained with WDS technique but at the expense of time.

7.4.3 Line Scans

Line scanning is another type of elemental mapping where only selected area through selected line is mapped. In a line scan, probe travels linearly along a line on the specimen, and the change in count rate is measured in relation to the probe

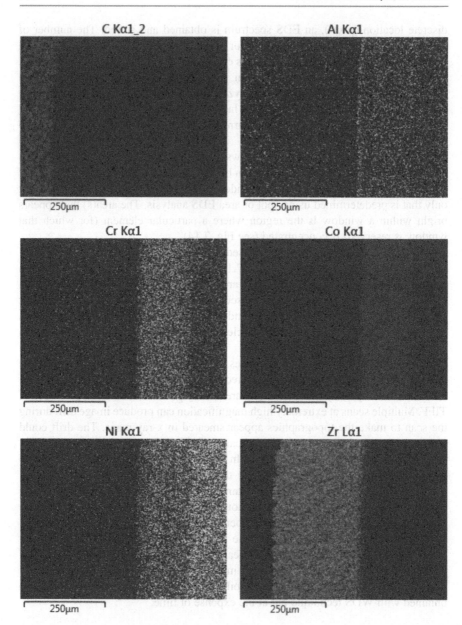

Fig. 7.14 X-ray map showing elemental distribution within the thermal barrier coating (TBC) sample. Each window represents one particular element. The bright colored region is the area of the sample where an element is concentrated

7.5 Quantitative EDS Analysis

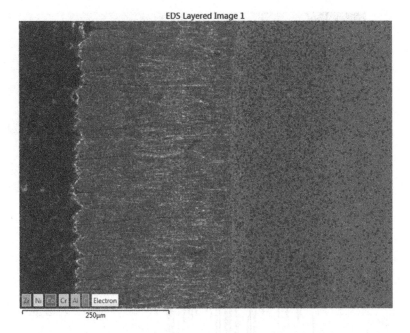

Fig. 7.15 X-ray combination map of thermal barrier coating (TBC) where window with six colors represents the extent of distribution and concentration of six different elements in one image

position. The essential variables are number of points, dwell time per location, and number of passes. Line scans are used to examine elemental variations in features such as impurities, precipitates, and grain boundaries. Similar to x-ray maps, an EDS spectrum is taken and recorded at each pixel on the selected line. An example of a line scan is given in Fig. 7.16.

7.5 Quantitative EDS Analysis

7.5.1 Introduction

Once the elements present within a specimen are identified using qualitative EDS analysis, it is generally required to determine their level of concentration. This is undertaken using quantitative EDS analysis. The concentration of an element in a specimen can be high in which case it is called a major element (generally taken as more than 10 wt%), lower concentration element is termed as minor element (1–10 wt%), and a concentration below 1 wt% is usually designated as trace element [8]. The lower limit of detection for EDS is around 0.2 wt%. The higher the concentration of an element in a specimen, the greater is the accuracy with which it can be quantified. It follows that trace elements are difficult to detect as well as hard to quantify with a high degree of accuracy. Likewise, quantification of

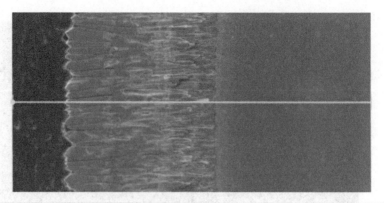
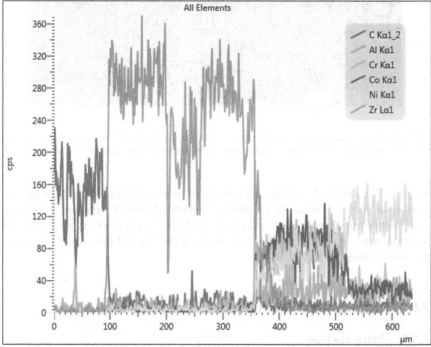

Fig. 7.16 Line scan through thermal barrier coating (TBC) sample showing variation in elemental composition of various elements along a line

concentrations of light elements presents bigger challenge compared to that of heavy elements. Generally, if correct specimen preparation and data acquisition procedures have been adopted, 1–2% accuracy of quantification is achieved using EDS analysis.

Quantitative EDS analysis can be conducted using standards or employing a standardless technique. Analysis with standards involves EDS analysis of specimens (i.e., standards) that are similar in composition to that of the unknown specimen being analyzed. Spectra obtained from the standard specimen(s) are compared with those obtained from the unknown specimen to determine the concentration of different elements in the unknown specimen. The accuracy of such an analysis is high, but the procedure is cumbersome. Standardless EDS analysis involves testing only unknown specimen and comparing spectra with spectral data stored in the computer to quantify concentrations of elements in the unknown specimen. Such a technique is fast and convenient, but accuracy is less. Both techniques are described in the following sections. It is assumed that specimens tested are flat and polished down to 0.1 μm using standard metallographic techniques. Specimens are stable under the beam and do not undergo beam damage. They are conductive to avoid beam instability and change in x-ray intensity during analysis. It is important that optimum operating conditions are employed to obtain EDS spectra and elements in the specimen have been correctly identified using the qualitative technique as described in the previous sections.

7.5.2 EDS with Standards

7.5.2.1 Castaing's First Approximation

The standard specimen is the one which has known and uniform composition along its analyzed surface at a microscopic level. The standard specimen is selected due to its closeness in composition to that of the unknown specimen. However, this is usually not possible in practice. Therefore, mostly 100% pure elements are used as standards which work equally well for microchemical analysis. In some cases where pure elemental compositions are difficult to achieve, simple compounds such as oxides are employed as standards. Both standards and unknown specimens are prepared to the same level of finish and analyzed in the SEM using the same detector and microscope conditions such as electron beam energy, probe current, EDS detector take-off angle, analysis time, etc. If used, same thickness of the conductive coating should be applied to both standard and unknown specimen. The EDS spectra thus obtained from the standard and unknown specimens are compared. The peak intensity of an element i in the spectrum obtained from the unknown specimen relative to the peak intensity of the same element i in the standard specimen will indicate wt% concentration of element i relative to the wt% concentration of that element in the standard. If the standard is a pure element, same intensity peaks from the standard and the unknown will indicate 100 wt% concentration of element i in the unknown specimen. Similarly, half of the standard intensity will suggest 50 wt% concentration. This can be written as follows:

$$\frac{C_{i\ (\text{unknown})}}{C_{i\ (\text{standard})}} \approx \frac{I_{i\ (\text{unknown})}}{I_{i\ (\text{standard})}} = k_i \qquad (7.5)$$

where

$C_{i\ (\text{unknown})}$ is weight percent concentration of element i in the unknown bulk specimen

$C_{i\ (\text{standard})}$ is weight percent concentration of element i in the standard

$I_{i\ (\text{unknown})}$ is the intensity of the characteristic x-ray peak emanating from element i in the unknown specimen

$I_{i\ (\text{standard})}$ is the intensity of the characteristic x-ray peak emanating from element i in the standard

The ratio between these the two intensities is known as "k-ratio". Equation 7.5 can be written as:

$$C_{i\ (\text{unknown})} = k_i C_{i\ (\text{standard})} \qquad (7.6)$$

For pure standards, $C_{i(\text{standard})}$ equals 1. Since unknown and standard specimens are analyzed under similar conditions, k-ratio is independent of constant factors associated with the instrument and standard and unknown specimens. Weight percent concentrations of other elements in the unknown specimen are determined using k-ratio in a similar manner. k-ratio forms the basis of quantification used in EDS microchemical analysis and is obtained for each element present in the unknown specimen. Peak intensities used in the above equation are net peak intensities obtained by subtracting the peak overlaps and background from the peaks. Various methods such as linear interpolation or extrapolation, filtering, modeling, etc., can be used for background subtraction. Overlapping peaks in the x-ray spectrum need to be separated using de-convolution software programs. Adequate relative intensities should be obtained during analysis since theoretical calculations do not correct for errors in the measurement of x-ray peak intensities. Equation 7.5 above assumes that peak intensities are generated proportional to the respective concentrations of elements. This is called Castaing's first approximation to quantitative analysis.

7.5.2.2 Deviation from Castaing's First Approximation
When multielement specimens are tested, Castaing's assumption fails to hold. It is observed that the ratio of intensities do not vary linearly with the ratio of concentrations as suggested by Eq. 7.5. This is understandable since the matrix of the unknown multielement specimen is not similar to a pure standard. Due to the differences in the matrices, x-rays of a given element in the unknown specimen will undergo absorption more than the corresponding x-rays in the standard sample resulting in lowered intensities, while x-rays of another element in the unknown sample will yield stronger intensities due to a possible fluorescent effect. X-rays emanating from light elements can show strong absorption in the heavy matrix. Due to this phenomenon, measured intensities from unknown specimens need to be corrected for atomic number (Z), absorption (A), and fluorescence (F) effects to arrive at correct intensities and hence generate reliable concentration values. These effects are commonly known as *matrix effects* which can be quite large in certain material systems. The relative intensity of an element does not generally follow a linear relationship with its concentration due to matrix effects. The magnitude of the effect varies with the composition of the material analyzed.

7.5 Quantitative EDS Analysis

The mostly applied correction procedure to counter matrix effects and undertake quantitative microchemical analysis is known as ZAF correction method as shown by the following equation:

$$C_{i\,(\text{unknown})} = [Z_i \times A_i \times F_i] \times C_{i\,(\text{standard})} \times \frac{I_{i\,(\text{unknown})}}{I_{i\,(\text{standard})}} = [Z_i \times A_i \times F_i]\, k_i \quad (7.7)$$

where Z_i, A_i and F_i are correction factors for atomic number, absorption, and fluorescence effects, respectively, for element i in the specimen. Above equation is calculated separately for each element present within the specimen. The aim of matrix corrections is to convert the measured intensity from the sample relative to that from a standard to the actual concentration.

7.5.2.3 Matrix Effects

The intensity of x-rays generated from within a specimen depends on instrumental factors such as accelerating voltage and probe current used as well as specimen factors such as elastic and inelastic scattering processes occurring within a specimen. Measured or detected intensity of x-rays is not equal to generated intensity due to absorption or fluorescence of x-rays generated within the specimen. This variation between generated and detected values of x-ray intensity is governed by the composition of the specimen matrix and is known as matrix effects. Primary phenomena giving rise to matrix effects constitute effect of atomic number (Z), absorption (A) and fluorescence (F), as discussed below.

Atomic Number Effect

Backscattered electrons are those incident beam electrons that undergo elastic scattering upon entering the specimen, get deflected through large angles, and leave the specimen. Backscattered electrons represent a significant proportion of beam electrons that, as they leave the specimen surface, become unavailable to take part in ionization of specimen atoms in order to generate x-rays. Therefore, backscattered electrons do not contribute to x-ray generation within the specimen. The degree of backscattering strongly depends on the atomic number of the specimen material (see Fig. 3.14a). Specimens with high atomic number show a high degree of backscattering. In a multielement specimen, the phase with a high atomic number will eject a large number of backscatter electrons compared to a low atomic number phase. Also, since backscattered electrons constitute a significant proportion of the total energy of the incident beam, a substantial measure of that beam energy is removed from the sample upon their escape.

With regard to microchemical analysis, consider measuring the low concentration of a light element i mixed with a high concentration of a heavy element j in a multielement specimen. The large proportion of beam electrons entering the specimen shall be backscattered by heavy element j and leave the specimen. These electrons shall be unavailable to generate x-rays from within the light element i. In this way, the concentration of i shall be underestimated if its intensity is compared with that originating from a standard with 100% pure element i where a high proportion of beam electrons shall be available to generate x-rays. Therefore, a correction needs to be applied to calculations for this kind of matrix effect in order to use 100% pure standards with multielement specimens to get accurate results.

The fraction of incident beam electrons which do not backscatter and remain available within the specimen to generate x-rays is termed as R. It follows that low atomic number specimens (such as light elements or polymeric/life sciences specimens) produce small degree of backscattered electrons and have a large value of R. Generation of x-rays within a specimen also depends on latter's ability to get ionized, i.e., its critical ionization potential. Specimens with low atomic numbers demonstrate low critical ionization potentials, e.g., they readily ionize compared to heavy elements and generate x-rays. In other words, light elements demonstrate a greater "stopping power" denoted as S (also see Sect. 3.2.3). The higher the value of S of a target specimen, the greater is the rate of energy loss of incident beam energy within that specimen. Thus both backscattering R and stopping power S vary inversely with atomic number Z. Atomic number affects the degree of backscattering and rate of electron energy loss within the specimen and thus influences the degree of x-ray generation at a given depth of specimen. This is especially important when the specimen has light element intermixed within a heavy element matrix. The correction of atomic number effect for a particular element i (Z_i) is obtained by dividing stopping power S for the sample and standard by the backscattering R for the specimen and standard, i.e., $Z_i = S/R$. The S and R factors go in opposite directions and tend to cancel each other out.

It can also be seen that incident beam energy has a similar effect on the values of R and S. Electrons with higher-energy backscatter more and escape specimen surface producing lower values of R. Similarly, stopping power of a specimen is lower for a higher beam energy giving rise to lower S. Therefore, R and S have an inverse proportional relationship to incident beam energy, similar to that with atomic number. Therefore the atomic number effect can be determined by calculating x-ray generation as a function of atomic number and incident beam energy. Study of Monte Carlo simulations reveal that the volume of x-ray generation decreases with increasing atomic number at constant beam energy. This is due to an increase in backscattering in high atomic number specimens which makes a large proportion of electrons unavailable for x-ray generation. In addition, critical excitation energy also increases with an increase in atomic number. The distribution of x-rays within the generated volume is also influenced by atomic number as well as by specimen depth. For convenience, the relative intensity of x-rays generated as a function of mass depth $\varphi(\rho z)$ (where ρ = specimen density, g/cm^3, and z = linear depth, cm) is measured. Mass depth takes into account density of specimens which has a strong effect on the generation of x-rays. In this method, the atomic number correction Z_i can be calculated by taking the ratio $\varphi(\rho z)$ for the standard to $\varphi(\rho z)$ of the element i in a specimen. Calculation of atomic number effect by this method takes into account R and S factors described above.

Absorption Effect

The depth and volume of x-ray generation increase with an increase in incident beam energy. For a given energy, x-ray generation increases with depth from the specimen surface, reaches its highest point quickly, and then falls back to low values at greater depths. This is shown schematically by $\varphi(\rho z)$ curve in Fig. 6.18. As stated previously, not all generated x-rays will reach the detector, and some proportion of it will be absorbed within the specimen matrix. Therefore, the measured intensity I will differ from generated intensity I_0 at a given depth z, and their relationship can be described by Eq. 6.12 given in Sect. 6.4.6:

$$I = I_0 \exp\left[-\left(\frac{\mu}{\rho}\right)(\rho t)\right]$$

where

I is the intensity of x-ray photons when leaving the specimen surface
I_0 is the original intensity of x-ray photons
μ is the absorption coefficient
ρ is the density of the specimen
t is the thickness of specimen traveled
ρt is the area density known also as the mass thickness

Above equation gives a measure of absorption of x-rays within a specimen matrix. It is clear that the greater the depth at which x-rays are generated, the greater is the proportion that is lost due to absorption within the specimen. Mass absorption coefficient depends on the energy of generated x-rays; therefore its value will be different for each characteristic x-ray line. It also depends on the composition of the specimen analyzed. The mass absorption coefficient of a multielement specimen is obtained by multiplying individual absorption coefficients by their mass fractions and adding them up.

Generally, correction for mass absorption is the biggest correction made during quantitative microchemical x-ray analysis. Especially, light elements such as C, N, and O are strongly absorbed in heavy matrices and need to be accounted for during calculations. X-rays might be generated at greater depths within a specimen but might not make it to the surface due to absorption. Only those x-rays that are relatively close to the surface might escape. Absorption can be decreased by using the minimum incident beam energy required to generate characteristic x-rays resulting in lesser beam penetration and lower path lengths (t) that x-rays need to traverse to reach specimen surface. Higher take-off angles also decrease x-ray path lengths and reduce the chance to get absorbed within the specimen.

Fluorescence Effect

Characteristic x-rays generated as a result of the interaction between the electron beam and the specimen can be absorbed within the specimen matrix and cause ionization of atoms resulting in the emission of further characteristic x-rays. This

Fig. 7.17 Plot showing measured k-ratios (curved lines) versus weight fractions (straight lines) of Ni-Fe alloy. Adapted from [8]

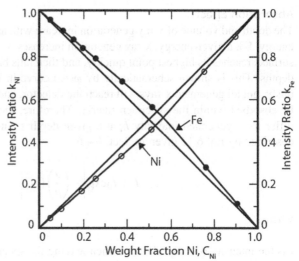

fluorescent effect takes place only if the critical excitation energy of absorbing atoms is less than the energy of generated x-rays. This effect will result in an increase in the measured x-ray intensity by the SEM detector since now both the original x-rays and the x-rays generated due to fluorescence are measured. Intensity will be increased by both continuum and characteristic x-ray; however, the effect of former can be considered negligible. Correction required due to florescence effect is usually smaller compared to that for atomic number and absorption in ZAF corrections. In some cases, fluorescence can result in erroneous peaks in the x-ray spectrum.

As an example, the effect of matrix on the intensity of Fe K_α and Ni K_α characteristic x-rays in Fe-Ni alloy is shown in Fig. 7.17. If Castaing's first approximation holds, the plot should be linear. However, the measured intensity of Fe K_α is higher than theoretical value due to fluorescence induced by Ni K_α, i.e., Fe will be overrepresented. On the other hand, the intensity of Ni K_α is lower than that calculation based on Castaing's first approximation as it is absorbed by the matrix more than in the standard, i.e., Ni will be underrepresented. Such an effect needs to be compensated during calculations. Greatest deviation from a linear relationship is seen in cases where analysis is performed for light elements in a heavy matrix or light elements in a light matrix.

7.5.2.4 ZAF Iterative Process

The aim of quantitative analysis is to determine the composition of an unknown sample. The extent of matrix effects, and its required correction depends on the composition of the unknown sample. Since the composition of the sample is not known in the first instance, the ZAF correction factors are also unknown. Therefore, true concentrations are achieved by an iterative process. In this procedure, k-ratio is calculated from measured intensities and used as an initial estimate of the composition of the unknown sample. Based on this k-ratio, ZAF factors for this composition

7.5 Quantitative EDS Analysis

are calculated. The composition of the unknown sample is calculated by multiplication of k-ratio and corresponding ZAF factors calculated in the previous step. Based on this newly calculated composition, a new set of ZAF factors are calculated. The composition of the unknown sample is again calculated by multiplying the new set of ZAF factors with the original k-ratio. This process continues until concentration does not change appreciably from the previous one as shown below:

First iteration: k-ratio $\rightarrow (ZAF)_1 \rightarrow C_1 = k \times (ZAF)_1$
Second iteration: k-ratio $\rightarrow (ZAF)_2 \rightarrow C_2 = k \times (ZAF)_2$
Third iteration: k-ratio $\rightarrow (ZAF)_3 \rightarrow C_3 = k \times (ZAF)_3$
nth iteration: k-ratio $\rightarrow (ZAF)_n \rightarrow C_n = k \times (ZAF)_n$
The procedure stops when $C_n = C_{n-1}$

7.5.2.5 Phi-Rho-Z Correction Method

Production of x-rays varies with the depth of the specimen. Phi-Rho-Z function $\phi(\rho z)$ was developed to take into account the generation of x-rays as a function of depth and self-absorption. This function uses mass depth ρz parameter instead of simple linear z (see Sects. 6.4.4 and 6.4.5). The $\phi(\rho z)$ function is defined as the x-ray intensity generated in a thin layer at some depth z, relative to the intensity generated in an isolated layer of the same thickness. This is then integrated over the total depth where the incident electrons exceed the binding energy of that particular characteristic x-ray. The curve of $\phi(\rho z)$ versus ρz is generated for each characteristic x-ray (see example in Fig. 6.18). The shape of the curve depends on accelerating voltage, the critical excitation energy of a particular element x-ray line, and mean atomic number of the specimen [9]. These curves are generated using the *tracer method*. $\phi(\rho z)$ is an elemental quantification method based on the matrix and includes fluorescence correction. The common $\phi(\rho z)$ method depends on standardization or reference measurements. Compared to ZAF, the $\phi(\rho z)$ methods have improved the accuracy of microanalysis. They involve complex computations but perform much better with light element analysis.

7.5.3 Examples of ZAF Correction Method

Quantification of the effects of atomic number, absorption, and fluorescence on the concentrations of various elements present within a multielement specimen has been undertaken and refined by various researchers in the past few decades [12–17]. X-ray intensity ratios of the unknown specimen to that of the standard specimen are measured and corrected for Z, A, and F effects for each element to get final values of concentrations, as shown by the Eq. 7.7 ($C_{i\ (unknown)} = [Z_i \times A_i \times F_i]\ k_i$). This section includes an example of ZAF corrections taken from the literature [18].

7.5.3.1 Stainless Steel

The following example is reproduced from reference [18].

Actual composition (wt%): 62.03Fe, 23.72Cr, 13.26Ni, 0.23Mn, and 0.37Si
Standards: Pure Fe, Cr, Ni, Mn, and Si
Accelerating voltage: 25 kV
Detector take-off angle: 30°

Results of ZAF corrections for each element are shown in Table 7.1.

It can be seen from Table 7.1 that if concentrations are calculated based on mere k intensity ratios, the relative error in calculated concentrations would be high. The accuracy of measurements increases when ZAF corrections are incorporated into the measurements. For instance, x-ray intensity of Fe K_α line emanating from the specimen is lower than that suggested by k-ratio due to atomic number and absorption, while it is higher than k-ratio due to fluorescence of Fe K_α by Ni K_α x-rays. The overall effect is lowered intensity which is then corrected by multiplying by $[ZAF]_{Fe}$ factor of higher than 1 (e.g., 1.0235). This compensates for low measured intensity and results in concentration values closer to the actual composition than that suggested by k-ratios. Corrections for other elements are performed in a similar manner. The intensity of Cr k_α is enhanced due to fluorescence by K_α x-rays from Fe and Ni elements present in the specimen which results in an increased overall intensity for Cr. This is corrected by multiplying with $[ZAF]_{Cr}$ factor of less than 1 (e.g., 0.9125). For Ni, there is no fluorescence effect due to other elements present in the specimen thus requiring no correction. For this reason, F_{Ni} factor for Ni K_α is taken as 1. Silicon is heavily absorbed in the specimen and is compensated by a relatively large absorption (A_{Si}) correction. It is also clear from the table that calculations of elements present in the specimen with low concentrations (such as Mn and Si) will yield higher relative errors indicative of the challenges to quantifying such level of concentrations with a high degree of accuracy. For quantitative analysis using standards, the user needs to measure the net intensity of x-ray peaks from the standard and unknown specimen to derive k-ratios. ZAF correction factors are stored in computer memory, and once k-ratio values are entered into the computer, it can calculate elemental concentrations. ZAF factors are obtained from methods developed by various researchers over the years.

7.6 Standardless EDS Analysis

Quantitative EDS analysis with standards is carried out by analyzing unknown specimens and known standard specimens under similar measurement conditions to cancel out any differences arising due to detector efficiency. However, analysis with standards is cumbersome since intensities of all x-ray peaks need to be obtained from both known and unknown specimens in order to calculate intensity (k) ratios. To circumvent this requirement and for the sake of convenience, the microchemical analysis is usually undertaken using *standardless* EDS analysis. In this method,

7.6 Standardless EDS Analysis

Table 7.1 Use of ZAF correction method to analyze concentrations of the stainless steel specimen

Element (K_α lines)	k intensity ratios	Z_i factor	A_i factor	F_i factor	$Z_iA_iF_i$	Measured composition, wt% $C_i = [Z_iA_iF_i]\,k_i$	Relative error, % [a]$(C_{act.} - C_{mea.})/C_{act.} \times 100$
Fe	0.6076	1.0030	1.0316	0.9892	1.0235	0.6219 (62.19 wt%)	0.26
Cr	0.2586	0.9970	1.0070	0.9089	0.9125	0.2360 (23.60 wt%)	0.51
Ni	0.1238	0.9940	1.0804	1	1.0739	0.1330 (13.3 wt%)	0.30
Mn	0.0024	1.018	1.0014	0.9926	1.0118	0.0024 (0.24 wt%)	4.35
Si	0.0024	0.8360	1.8144	1	1.5168	0.0036 (0.36 wt%)	2.70

[a]$C_{act.}$ = Actual composition, $C_{mea.}$ = Measured composition

physical standards (specimens with known composition) are not examined by the user, and only unknown specimen is analyzed. Since the need to analyze standards is eliminated, the whole process of analysis becomes simple and efficient. X-ray peak intensities from standards are still required to estimate elemental concentrations, but these are obtained from theoretical calculations. This procedure is called *first principles standards analysis*. Alternatively, intensities are acquired from standard x-ray spectra stored in the computer. This method is known as *fitted standards standardless analysis*.

7.6.1 First Principles Standardless Analysis

The accuracy with which the intensity can be calculated depends on several critical physical parameters such as the ionization cross section, the x-ray self-absorption inside the target, the fluorescence yield, the backscatter loss, the stopping power, and the detector efficiency. Values of K shell ionization cross sections published in the literature show a variation of more than 25%, especially in the low overvoltage range, $1 \leq U \leq 3$, which is the primary operating range of energy dispersive x-ray spectrometry. Similarly, published K shell fluorescence yield shows a variation of 25% for many elements. Similar data variation is observed for L and M shell transitions. The EDS detector efficiency can also present a major source of error in the standardless analysis because characteristic x-rays of different energies are compared. Due to these reasons, use of standard intensities derived from theoretical calculations generally results in a large relative error in measured elemental concentrations. Due to this reason, this method is seldom employed, and the most commonly used procedure is to derive standard intensities from stored x-ray spectra.

7.6.2 Fitted Standards Standardless Analysis

In this method, the intensity values are obtained from experiments performed on a range of standards consisting of pure elements or binary compounds. A library of intensities for K, L, and M x-ray peaks is obtained for elements ranging from low to high atomic numbers. This results in a database of standard x-ray spectra intensities. Change in x-ray intensities with atomic number is modeled, and mathematical fits are derived to predict the intensity of an element with a specific atomic number. Similarly, the change in x-ray intensities with accelerating voltage can be modeled and dependence of elemental concentration on beam energies is calculated. This modeling is performed at the manufacturers' site and is not undertaken by the SEM user. Intensities of standard spectra obtained at the factory and stored in the computer are adjusted according to the efficiency of individual EDS detector fitted on a particular SEM. The term *standardless* in this procedure to obtain elemental concentrations can be regarded as a misnomer since x-ray intensities used in calculations are actually derived experimentally from physical standards. Perhaps, it came to be known as standardless technique since SEM operators do not measure intensities from standard specimens and only analyze unknown specimens. The accuracy obtained from this procedure can be several orders of magnitude higher than that obtained from standard intensities using theoretical calculations. Accuracy is greater in specimens with similar atomic numbers and for those where only K_α peak is measured. Accuracy decreases with use of L or M peaks for intensity measurements. All results are normalized to 100%, and oxygen is calculated by the direct method or indirectly by the stoichiometric method.

7.7 Low-Voltage EDS

The high voltage and large beam-specimen interacting volume of traditional EDS set-up leads to low spatial resolution of the detecting elements, low detectability of lighter elements (lighter than Be), and the high interaction between beam and specimen may cause beam damage to the sample. More recently, low-voltage EDS (LV-EDS) technique was developed as a microanalysis tool which overcomes the above-stated drawbacks. In this technique, an electron beam energy of <5 keV is used to undertake microanalysis with high spatial resolution, which can be termed as "nano-analysis." LV-EDS can be used for compositional analysis where WDS cannot be applied due to latter's use of high current (tens of nA) during analysis even though the energy resolution of WDS is much better than EDS. The higher current is said to lead to thermal damage and degradation of spatial resolution [19]. In addition, LV-EDS reduces x-ray absorption, thus increasing the accuracy of the quantitative analysis.

More recently, microcalorimeters equipped with enhanced detectors and open windows to extend the sensitivity of the low-voltage characteristic spectrum have been developed. Some of these microcalorimeters are mounted on the SEM itself. In FE-SEM, with the application of LVEDS, elemental mapping of bulk materials at high spatial resolutions has been possible [20]. Commercial μcal EDS detectors have higher peak-to-background ratios, fewer peak overlaps and where peaks do overlap, the detection is improved more than that of conventional EDS detectors [20].

When accelerating voltage used is less than 5 kV, the L and M lines of heavy elements are closely spaced with the K lines of lighter elements. These peaks are then not easily distinguished using the SSD (solid-state detectors) which has a resolution of about 122 eV. To overcome this limitation, an EDS system was developed based on a transition edge x-ray sensor (TES) which has a much higher-energy resolution of less than 20 eV [19]. The best energy resolutions demonstrated by low-voltage EDS technology is said to be 2.0 eV FWHM at 1.5 keV (Al-Kα) and 2.4 eV FWHM at 5.9 keV, i.e., the Mn-Kα using a μcal EDS system [21].

One obvious disadvantage is that elements with high critical ionization potential cannot be detected at low voltages. Peak overlap is another drawback. Moreover, contamination buildup at the specimen surface can seriously hamper accurate quantitative analysis at low kV.

Examples of applications of LV-EDS include composition analysis of interplanetary dust particles [20], oxide surface analysis [20], oxidation-state measurements [21], depth profiling of multilayered films [22], and nano-analysis of semiconductor device [23].

7.8 Minimum Detectability Limit (MDL)

The minimum concentration of an element that can be detected in a specimen is called its *minimum detectability limit* (MDL) or *sensitivity*. For an element to be detected, its characteristic peak needs to visible over its background. Normally, it is assumed that peak is detectable if it is at least two times the mean variation height (standard deviation) of the background, i.e., $2\sqrt{I_b}$ where I_b is the intensity of the background, i.e., mean count level of background and $\sqrt{I_b}$ is the variation or noise about this mean. The minimum detectability limit of an element can be expressed as follows:

$$\mathrm{MDL} \propto \frac{1}{\sqrt{I_p \left(\frac{I_p}{I_b}\right) t}} \tag{7.8}$$

where

I_p is the peak intensity of the x-ray line
I_b is the background intensity
t is the acquisition time

It is clear that better sensitivity can be increased by increasing peak-to-background ratio and acquisition time.

7.9 Wavelength Dispersive X-Ray Spectroscopy (WDS)

The characteristics x-rays produced due to specimen-beam interaction have wavelengths unique to the elements in the specimen. This forms the basis for the qualitative as well as quantitative analysis using wavelength dispersive x-ray spectroscopy (WDS). The wavelength of the x-ray photons provides an avenue for the identification of elements in the sample, and the x-ray peak provides the basis for quantification after taking into account numerous factors that determine peak area.

7.9.1 Instrumentation

The primary instrument used to carry out wavelength dispersive x-ray spectroscopy (WDS) is known as electron probe microanalyzer (EPMA) which is very similar to SEM. EPMA employs electron source (W filament, LaB_6 emitter, or field emission

7.9 Wavelength Dispersive X-Ray Spectroscopy (WDS)

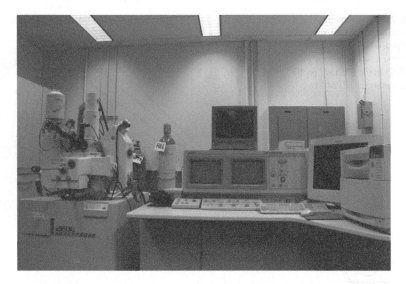

Fig. 7.18 Photograph of electron probe microanalyzer (EPMA) fitted with two wavelength dispersive (WD) x-ray spectrometers

gun), electron column equipped with electromagnetic lenses and apertures, specimen chamber with E-T/BSE detectors and allied vacuum equipment very similar to that in the SEM. The electron beam is generated and accelerated through the column. It is scanned onto the surface of the specimen in the form of a raster. The interaction between specimen and electron beam produces secondary and backscattered electrons as well as characteristic and continuous x-rays. The electrons are used to form SE and BSE images, while the x-rays are used for microchemical analysis. The only difference is that EPMA employs high probe current (typically tens of nanoamperes) and wavelength dispersive x-ray spectrometers instead of energy-dispersive x-ray spectrometer to undertake the chemical analysis. These WD spectrometers are fitted in the ports available in the EPMA specimen chamber. The number of these spectrometers could range from 2 to 4. A photograph of EPMA with 2 spectrometers is shown in Fig. 7.18.

7.9.2 Working Principle

The working principle of EPMA is illustrated in a simplified schematic shown in Fig. 7.19. The electron beam is generated at the top of the electron column. It is accelerated toward the specimen. Typical accelerating voltage used is 20–30 kV. Upon striking the specimen surface, x-ray signals are produced which are directed toward a curved crystal with known interplanar spacing (d-spacing). The x-rays are reflected off the crystal and directed into a detector. The difference here compared to the SEM-EDS technique is that the x-rays do not enter the detector directly upon

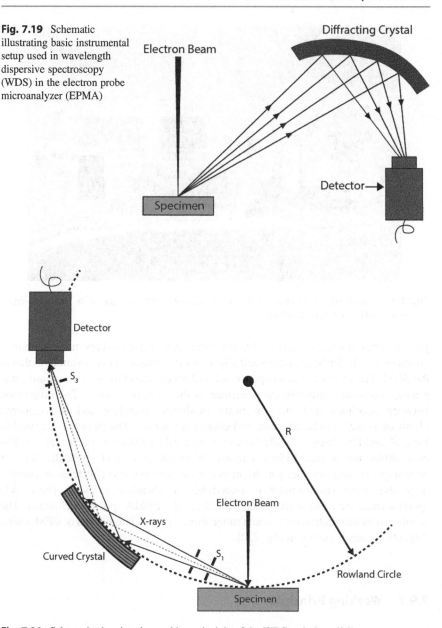

Fig. 7.19 Schematic illustrating basic instrumental setup used in wavelength dispersive spectroscopy (WDS) in the electron probe microanalyzer (EPMA)

Fig. 7.20 Schematic showing the working principle of the WDS technique [24]

ejection from the sample. X-rays are reflected off the surface of a crystal before being directed into the detector.

The sample, crystal, and the detector are positioned on a circle (called Rowland circle) as shown in Fig. 7.20. The sample is located at the bottom of the specimen chamber. The crystal and the detector are made to move on the Rowland circle

7.9 Wavelength Dispersive X-Ray Spectroscopy (WDS)

during analysis. During this movement, the distance between the crystal and the specimen is always kept equal to the distance between the crystal and the detector. The movement of the crystal on the circle serves to continuously change (from low to high) the angle θ at which the x-rays are incident on the crystal. The characteristic x-rays with specific wavelength emanate from the constituent elements of the sample due to orbital transitions and strike the surface of the crystal that has a fixed d-spacing. During the course of crystal movement, it is probable that x-rays (with specific λ) emanating from a particular element in the specimen and upon the striking the crystal (with fixed d) at an angle θ satisfy Bragg's equation (see Eq. 5.3, $n\lambda = 2\,d\sin\theta$) resulting in diffraction.

Upon diffraction, the amplitude of the x-rays will increase manifold resulting in an increase in the intensity of the x-rays at the diffraction angle θ. This increase in intensity is measured by the detector and appears as a peak in the x-ray spectrum recorded in the computer. The movement of the crystal is precisely controlled and monitored thereby providing an exact measure of θ at the time of diffraction. The d value is identified as the crystal used is known. From Bragg's equation, λ is then measured. Working backward, the element that emanates the x-ray with this particular wavelength is identified as the λ is specific to that element. The crystal is moved from low to high θ values, and any diffracted x-rays are detected during this run. In this way, qualitative identification of the elements is carried out. One type of crystal is unable to detect a vast array of possible wavelengths from small to large that encompass the majority of elements in the periodic table. Therefore, two crystals (installed back to back) are used in each spectrometer. Only one crystal can reflect x-rays at a given time. After the first run is completed through all θ angles, the crystals are flipped, and the second crystal with a different d-spacing starts to receive the signals from the specimen. The number of spectrometers coupled to the EPMA can be as high as four. Two crystals installed in two different spectrometers can receive x-ray signals simultaneously. In case of four spectrometers, four crystals (all with different d-spacing) can process the signal at the same time concurrently producing four x-ray spectra on the viewing monitor.

X-ray signal in electron beam instruments is weak. The signal is maximized by using fully focusing x-ray spectrometers with curved analyzing crystals. The two basic types of crystal geometries used in the WDS setup are called *Johann* geometry and *Johansson* geometry. In Johann geometry, the bending curvature of the diffracting crystal is twice the radius of the focusing circle ($2R$), also termed as the Rowland circle as shown in Fig. 7.20. In Johansson geometry, the crystal is bent to have a curvature of radius equivalent to a distance $2R$. It is also surface finished to a distance equivalent to R. This ensures that all x-rays are reflected onto the Rowland circle. In this way, x-rays are deflected off a wider crystal area and are still focused on a single point in the x-ray detector. An optical microscope is mounted to adjust z-position of the sample so that the sample is kept in full focusing condition.

The spectra generated using WDS are normally displayed in the units of $\sin\theta$, wavelength, or millimeters of the crystal movement. This is in contrast to a rather more convenient one of EDS that gives the spectra in intensity as a function of energy. Because of this, some systems allow for the transformation of WDS spectra from the stated units to energy units. WDS is characterized by high-energy resolution typically 10 eV in contrast to 122 eV in EDS.

7.9.3 Analytical Crystals

The range of Bragg angles used in WDS instrument is limited by available space. The maximum achievable θ angle in a WDS system is in the range of 67–73°. Thus, the maximum λ of characteristic x-rays being diffracted is less than $2d$ of the analyzing crystal. Thus, crystals with a wide range of d-spacing values are necessary to cover the entire range of x-ray wavelengths of interest (~0.1 to 2.0 nm). Usually, microprobes employ multiple crystals in each spectrometer to increase analytical flexibility; each spectrometer has a pair of crystals that can be flipped. The most commonly used crystals include lithium fluoride (LiF), thallium acid phthalate (TAP), pentaerythritol (PE), and layered synthetic microstructure (LSM). LiF is an ionic solid, while PET and TAP are organic crystals. X-rays having a long wavelength (low energy) require larger crystal-lattice spacing to obtain optimum diffraction, and hence the layered synthetic microstructure crystal remains a better option. The pseudocrystals are usually produced by depositing vapor on the alternating crystal-lattice layers of the lighter and that of the heavier elements. The selection of the elements is made in order to achieve maximum scattering efficiency; also the width of the alternating layers provides real d-spacing [25].

A given x-ray line can be diffracted by more than one crystal, but the limited space within the spectrometer dictates the most suitable crystal that can be used for this purpose. For example, Fe-K_α with a wavelength of 0.1937 nm is located at θ of 28.75° on LiF, 12.8° on PET, and 4.3° on TAP. Fe-K_α radiation falls at values of θ which are either very near or beyond the mechanical limits of spectrometer's movement for PET and TAP, thus necessitating the use of LiF crystal. Other important characteristics of an analyzing crystal apart from interplanar (d) spacing include stability, reflectivity, spectral range, spectral resolution, and thermal expansion coefficient [26].

7.9.4 Detection of X-Rays

Gas proportional counters are the most common detectors used in WDS for low-energy lines. In this type of detectors, the sample-generated x-rays enter the detector via a collimator and get absorbed by atoms of counter gas in the detector. The atoms of the counter gas, in turn, emit photoelectron on absorbing the x-rays. Finally, the produced photoelectrons are accelerated to a wire located at the center of the tube for further ionization that will generate an electrical pulse having an amplitude proportional to the energy of the x-ray photons generated by the specimen. Theoretically, 16 eV is required to generate one electron pair, but up to about 28 eV is ordinarily required for effective electron pair production. Therefore, the

7.9 Wavelength Dispersive X-Ray Spectroscopy (WDS)

number of electron pair produced should be a value obtained after dividing the energy (eV) by this number.

Another type of detector which is mostly used in high-energy x-ray lines is the sealed proportional counter detector. This type of detector has a thicker window (slit; Mylar) compared to its gas flow counterpart. The ionization gas widely used in sealed proportional counters is Xenon (Xe) gas or Xe-CO_2.

7.9.5 Advantages/Drawbacks of WDS Technique

7.9.5.1 Advantages
Nondestructive technique
High-energy resolution (~5–10 eV)
High peak-to-background ratios (10,000:1)
Good detection efficiency for all x-rays
Fast counting rates
Good detection of light element
Better trace element detection compared to EDS
Suitable for different synthetic and natural solid minerals
The capability of x-ray mapping of elements using *rastered* electron beam

The WDS systems can resolve relative changes in wavelength ($\Delta\lambda/\lambda$) in the range 0.002–0.02 corresponding to the energy range 0.01–0.1 keV. This energy resolution value is >10× better than that of EDS. Modern WDS systems can detect elements from upward of C ($Z = 6$).

7.9.5.2 Disadvantages
Complex mechanical/moving components
Complicated analysis
Operator intensive/time-consuming analysis
Specimen height-dependent focus
Limited solid angles (<0.01 limited solid angles)
Serial detection
Expensive
Low atomic number elements (H, He, Li, and Be) cannot be analyzed using WDS, and hence numerous geologically important materials cannot be measured
Even though WDS has enhanced elemental peaks spectral resolution, some significant peak overlaps still exists (e.g., Vanadium-K_α and Titanium-K_β)
WDS technique does not distinguish different isotopes of elements
The quantification process requires the use of standard reference materials, which makes the method relatively expensive

7.9.6 Qualitative WDS Analysis

In WDS, elements identification in specimens is achieved by obtaining the angles that satisfy the Bragg's law as the scan runs through a range of angles during analysis. The peaks of elements appear at the angles θ where Bragg's law is satisfied. The spectra generated using WDS are given in intensity (y-axis) versus $\sin \theta$, x-ray wavelength, or millimeters of the crystal movement (x-axis). This is in contrast to EDS that gives the spectra in intensity versus x-ray energy. Procedure to undertake qualitative WDS is summarized below:

- Peak acquisition begins at the shortest wavelength end of the crystal spectrum corresponding to the highest photon energy. This enhances the chances of obtaining a first-order peak ($n = 1$).
- The wavelength of the resulting peak is determined by selecting peak with the highest intensity.
- As soon as the element in the specimen is identified, all possible high-order peaks related with each first-order peak throughout the set of crystals are identified.
- After this, the next unidentified high-intensity, the low-wavelength peak is examined. This procedure is repeated for each peak.

References

1. Fitzgerald R, Keil K, Heinrich KFJ (1968) Solid-state energy-dispersion spectrometer for electron-microprobe x-ray analysis. Science 159:528–530
2. Torma P, Sipila H (2013) Ultra-thin silicon nitride X-ray windows. IEEE Trans Nucl Sci 60:1311–1314
3. Torma PT, Kostamo J, Sipila H, Mattila M, Kostamo P, Kostamo E, Lipsanen H, Laubis C, Scholze F, Nelms N, Shortt B, Bavdaz M (2014) Performance and Properties of Ultra-Thin Silicon Nitride X-ray Windows. IEEE Trans Nucl Sci 61:695–699
4. Lee C, Wei X, Kysar JW, Hone J (2008) Measurement of the elastic properties and intrinsic strength of monolayer graphene. Science 321(5887):385–388
5. Huebner S, Miyakawa N, Kapser S, Pahlke A, Kreupl F (2015) High performance X-ray transmission windows based on graphenic carbon. IEEE Trans Nucl Sci 62(2):588–593
6. Williams DB, Carter CB (2009) Transmission electron microscopy: a textbook for materials science, 2nd edn. Springer, New York, USA
7. Williams DB, Goldstein JI, Newbury DE (1995) X-Ray spectrometry in electron beam instruments, 1st edn. Springer, New York, USA
8. Goldstein JI, Newbury DE, Joy DC, Lyman CE, Echlin P, Lifshin E, Swayer L, Michael J (2007) Scanning electron microscopy and X-Ray microanalysis, 3rd edition (Corrected edition), Springer, New York, USA
9. Reed SJB (1993) Electron microprobe analysis and scanning electron microscopy in geology, 2nd edn. Cambridge University Press, Cambrdige, UK
10. Zhou W, Wang ZL (2006) Scanning microscopy for nanotechnology. Springer, New York, USA
11. Hawkes PW, Spence JCH (2008) Science of microscopy, vol 1. (corrected printing. Springer, New York. USA

12. Castaing R (1951) Application of electron probes to local chemical and crystallographic analysis, Ph.D. Thesis, University of Paris, Paris. France. English (trans: Duwez P and Wittry DB) California Institute of technology, 1955
13. Duncumb P, Reed SJB (1968) Progress in the calculation of stopping power and backscatter effects. In: Heinrich KFJ (ed) Quantitative electron probe microanalysis. National Bureau of Standards Special Publication 298, US Government printing office, Washington D.C., p 133
14. Bastin GF, Heijligers HJM, Van Loo FJJ (1986) A further improve- ment in the Gaussian $(pz) approach for matrix correction in quantitative electron probe microanalysis. Scanning 8:45–67
15. Philibert J (1963) In: Pattee HH, Cosslett VE, Engstrom A (eds) Proceedings of the 34th international symposium on X-ray optics and X-ray microanalysis. Academic Press, New York, p 379
16. Duncumb P, Shields PK (1966) Effect of critical excitation potential on the absorption correction. In: McKinley TD, Henrich KFJ, Wittry DB (eds) The electron microprobe. John Wiley & Sons, New York, p 284
17. Reed SJB (1965) Characteristic fluorescence corrections in electron-probe microanalysis. Br J Appl Phys 16(7):913
18. Toya T, Kato A, Jotaki R (1984) Quantitative analysis with electron probe microanalyzer. JEOL Training Center, JEOL Ltd., Tokyo, p 83
19. Tanaka K (2006) A microcalorimeter EDS system suitable for low acceleration voltage analysis. Surf Interface Anal 38:1646–1649. https://doi.org/10.1002/sia.2408
20. Kenik EA, Demers H (2006) Spectrum imaging with a microcalorimeter EDS detector on a FEG-SEM Cr Mn 1 μm. Met Mater 12(Supp 2):140–141. https://doi.org/10.1017/S143192760606658X
21. Cantor R, Croce MP, Havrilla GJ, Carpenter M, McIntosh K, Hall A, Ullom JN (2016) Oxidation state determination from chemical shift measurements using a cryogen-free microcalorimeter X-ray spectrometer on an SEM. Microsc Microanal 22(S3):434–435. https://doi.org/10.1017/S1431927616003020
22. Rickerby DG (1999) Application of low voltage scanning electron microscopy and energy dispersive X-Ray spectroscopy. In: Chapter from book impact of electron and scanning probe microscopy on materials research. Springer, New York, pp 367–385
23. Redfern D, Nicolosi J, Weiland R (2002) The microcalorimeter for industrial applications. J Res Natl Inst Stand Technol 107(6):621–626
24. Goodhew PJ, Humphreys J, Beanland R (2001) Electron microscopy and analysis. Taylor and Francis, London
25. Marco S, Ivan B (2006) An introduction to energy-dispersive and wavelength-dispersive X-ray microanalysis. Microsc Anal 20(2):S5–S8 (UK)
26. René EVG, Andrzej AM (2002) Handbook of X-ray spectrometry, 2nd edn, Revised and expanded. Marcel Dekker, Inc., New York, ISBN: 0-8247-0600-5

Sample Preparation

8

One of the primary reasons why scanning electron microscopy is hugely popular among scientists is that a large variety of specimens can be examined directly with slight or no sample preparation. This allows fast and convenient analysis of surface topography without the possibility of introducing any artifact into the material. While sample preparation is not a requirement, it does become necessary depending on the type of material examined and the nature of the information that needs to be derived from it. Irrespective of the material analyzed, the sample has to be of dimensions that can be accommodated within the specimen chamber of the SEM, and it also has to be sufficiently conductive to enable grounding of the incoming electron beam. The goal of any sample preparation technique is to reveal the fine details of the surface structure of materials without any alteration or introduction of extraneous elements for eventual examination in the SEM. This chapter is organized on the basis of the types of materials that need to be prepared for analysis in the SEM. These include metals, alloys, ceramics, geological (rocks, minerals) and building materials (cement concrete), polymers, and biological materials. Detailed sample preparation techniques are found in the literature [1, 2].

8.1 Metals, Alloys, and Ceramics

8.1.1 Sampling

Sampling involves identifying a suitable area in the sample that adequately represents its structure, morphology, and chemistry. Microstructural variations can appear during the manufacturing or usage of the product. For instance, metal castings, drawn wire, and forged alloys usually display different microstructures in transverse and longitudinal directions. Sintered ceramics may also show structural variations between surface regions and the core. Components exposed to high temperatures during service can lead to the development of oxidation/corrosion products, loss of material, change in original shape and preferential segregation of chemical elements to the grain boundaries, etc. Any sample preparation method should take into account the manufacturing method and usage history of the product under examination. The aim of sample preparation should be clear at the outset. The sample should be transported to the SEM in a box or wrapped up in a dry material. It should be labeled and stored properly.

Preparation of specimens usually starts with selecting and precise cutting of the section that needs to be examined in the SEM. This involves using appropriate techniques that do not damage the sample or contaminate it. In some applications like metallurgical failure analysis, it is usual practice to take photographs of uncut failed components and perform low magnification light microscopy on sectioned parts. This allows correlation of macrographs obtained with a light microscope with the microstructural features seen in the SEM later.

8.1.2 Sectioning

Specimens are usually sectioned with power-driven silicon carbide (Fig. 8.1a) or aluminum oxide (Fig. 8.1b) cutoff wheels that use "plunge-cutting method" as shown in Fig. 8.1c, d. To avoid damage to sensitive materials like coatings on substrates, slow cutters equipped with diamond wheels can be utilized as shown in Fig. 8.1e. It is also ensured that both feed motion and rotation of the wheel proceed from the coating of the sample into the base material. Sufficient quantity of emulsions, hydrous solutions, low-viscosity mineral, or water is used as lubricants or coolants during the cutting process. Manual cutting by saw, chopping off, or dry sectioning is not recommended. Use of cutting fluid saves time, enhances surface quality, and increases the service lifetime of sectioning wheels. Although the SEM chamber can accommodate large-sized samples, the optimum surface area of the sectioned specimens is around 10 mm^2. These are relatively quicker and easier to prepare compared to very large or very small specimens. Large-sized specimens are only inserted into the chamber when it is not appropriate to cut them into smaller sections for some reason.

8.1.3 Cleaning

Cleaning can precede or follow the sectioning process. Cleaning prior to sectioning is required sometimes in order to locate and identify the area of interest. If sampling is accomplished without difficulty, cleaning can be undertaken after the sectioning. Cleaning is usually done in an ultrasonic cleaner with the use of a suitable solvent, such as acetone. In order to remove any remaining surface films, the specimen is washed with methanol. The selected solvent should not modify or damage the surface. If further cleaning is required on a finer scale, low-energy plasma can be utilized. If the surface of the sample contains small particles that are easily removed during cleaning, it is better to examine the specimen without initial preparation.

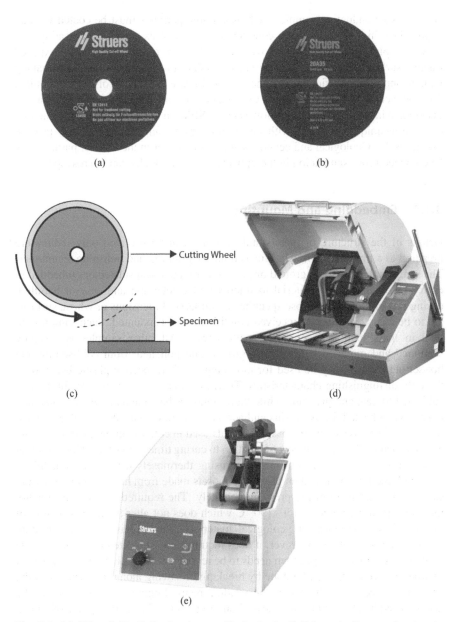

Fig. 8.1 (**a**) SiC and (**b**) Al$_2$O$_3$ abrasive cutoff wheels. (**c, d**) Schematic diagram showing the cutting of specimen using the plunge-cutting method. (**d** and **e**) Photographs of plunge cutter and slow precision cutter, respectively. (Photographs courtesy of Struers)

The cleaned sample must be transferred onto a suitably sized specimen stub. In general, appropriate conductive paint must be used to secure the sample to the stub. The paint also provides an electrically conductive path to ground the specimen current. Nonconducting samples such as ceramic particles must be coated with an appropriate thin and electrically conductive layer such as gold. These coatings are easy to apply but difficult to remove especially from rough samples like exposed fracture surfaces of components. For such samples, it is prudent to carry out imaging without applying a coating layer at the surface. In case application of coating required for conductivity is not possible, examination of an insulating sample can be undertaken in an environmental or low vacuum SEM. In the event of the unavailability of these techniques, imaging with low beam current and low voltage may provide useful results. Geological and ceramic samples that occur in the form of particles can be embedded for examination in the optical and scanning electron microscope.

8.1.4 Embedding and Mounting

Details of the structure of metals and ceramics can be revealed only if they are prepared to a fine surface finish. This is done by grinding and polishing techniques. The samples that are to be ground or polished are first encapsulated or embedded in epoxy or other suitable material to improve edge retention and ease of handling during preparation. Only those specimens with large dimensions or whose surfaces are to be analyzed in the as-received condition are not mounted. Such samples are held in metallic holders equipped with screws. All other samples are usually sectioned and mounted before being ground and polished. Both cold-setting and hot-setting plastics may be used for mounting. Their respective plastic behavior is the only distinguishing characteristics. Thermosets harden (cure irreversibly) when subjected to heat and pressure, while thermoplastics become hard when cooled and plastic when heated. Epoxy resins and acrylates (thermoplastics), as well as polyester resins (thermosets), are the basic materials used in cold mounting media that cure exothermically. The ratio of heat of reaction to curing time is constant. The mounting press can be used to carry out mounting using thermosets with or without fillers. After polymerization has taken place, thermosets made from heating under pressure can be taken away from the press immediately. The required temperature for hot mounting is in the range of 150–190 °C which does not alter the microstructure of metals, alloys, ceramics, or geological materials. Thermoplastics do not require pressure during heating. However, it does require to be subjected to pressure while cooling. If the mounted specimen needs to be recovered, the mounting medium must first be removed. This can be done by breaking or sawing along the surfaces of the embedded specimen. Thereafter, the residual resin clinging to the sample can be decomposed in a laboratory furnace at about 500–600 °C in an atmosphere of air [1].

8.1.5 Grinding, Lapping, and Polishing

Once the specimen is embedded in a mounting material, its surface is mechanically ground using a series of abrasive materials starting from rough to fine grit size e.g., 120-, 240- to 400-, and 600- grit papers. Grinding is usually carried out in a wet condition to avoid overheating and obtain a better surface finish. Grinding involves working with a firmly bonded abrasive for the purpose of flattening the specimen surface and eliminating material rapidly. The applied force on the specimen lies at an angle to the horizontal force. Grinding of specimen reduces the damage caused by the sectioning of the specimen. However, it introduces its own thin layer of damage to the surface regions. Grinding is continued till the grinding lines become finer and finer with each successive grit size and a smooth surface is obtained. By the end of grinding process, any cutting marks or scratches would have disappeared from the surface. The finer is the deformation surface at the end of the grinding procedure, the easier it is to carry out the polishing.

Lapping or polishing process uses loose abrasive between the specimen and hard fabric fixed on a rotating wheel. The rubbing action results in abrasion of the specimen. After the grinding procedure, lapping and/or polishing is undertaken to impart a flat scratch-free mirror-like finish to the surface of the specimen. The polishing abrasives impart a fine surface finish starting from 30 μm and culminating at 1 μm. Polishing is usually carried out in a stepwise manner on a hard synthetic fiber cloth with diamond pastes which has high hardness. Polishing process with fine grain sizes eliminate only a minor amount of material. The optical microscope is used to monitor the condition of the surface at each step. In order to avoid abrasive materials getting embedded in cracks and pores, the specimen should be totally cleaned after each step. Generally, automatic polishers give good reproducibility. Three steps of embedding, grinding, and polishing are outlined schematically in Fig. 8.2.

Ceramic specimens are particularly likely to exhibit cracking, pullout, and fragmentation of grains during grinding procedure. Dislocations and deformation twins can also develop. Lapping process of ceramic involves working with tumbling particles on a lapping disc. Lapping eliminates less material than grinding. The specimen surface shows a dull appearance. During polishing, grains move down and up within the long-napped or short-napped surface layer of the cloth. After polishing, the depth of the pullouts is reduced significantly. The surface of a ceramic specimen will have a faint finish with minor or no scratching [1].

Different types of underlying and abrasive materials are used during grinding, lapping, and polishing. Aluminum oxide, diamond paste, silicon carbide, cubic boron nitride, alumina/zirconium oxide mixtures, and boron carbide are utilized as

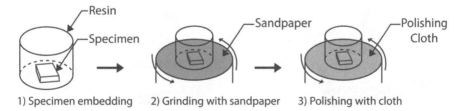

Fig. 8.2 Schematic showing three steps of specimen preparation: (1) embedding of a specimen in a resin, (2) grinding with a series of abrasive sandpapers, and (3) final polishing with polishing cloth made of hard fabric

abrasives. Fluids, water, and carriers containing alcohol and oil are used for loose abrasives in suspensions and pastes. In addition, water can be used as the abrasive medium for all grinding discs.

Use of bonded diamond grinding discs results in the highest volume of material removal. The lowest volume of removal occurs with monocrystalline diamond, polycrystalline diamond suspensions, SiC papers, diamond sprays, alumina slurries, and diamond pastes. As the percentage of material removed increases, the magnitude of damage caused to the specimen surface increases. In the case of relatively porous or soft materials, it is more practical to use an oil-based abrasive medium rather than one based on alcohol or water.

Grinding and lapping/polishing machines allow control of preparation time, direction, and the rotational speed of sample holder and disc rotational speed (see Fig. 8.3a, b). The volume of material removed increases as the pressure on the sample is held constant and the rotation speed of the disc is increased. The grinding speed is usually in the range between 1 and 18 m/s depending on the rotational speed, the position of the specimen, the motion of the specimen induced by the specimen holder, and the disc diameter. Incorrect positioning of the specimen holder results in an unequal removal of material from the specimen resulting in *half-moon effect*. The specimen holder can be rotated in either the opposite direction (*counterrotation*) of the disc or in the same direction (*complementary rotation*). In the complementary rotation, the average disc rotation is in the range of 150–200 rpm which allows the surface to be prepared uniformly. The material removal increases with an increase in the load per unit area. However, the roughness and depth of damage increase by a similar proportion. Specimen grinding is carried out either by hand or with automated equipment. Manual grinding is undertaken using a rotating wheel. After sectioning, it is possible to create a flat sample surface using planar grinding. The degree of surface damage produced using sectioning determines the quality of planar grinding.

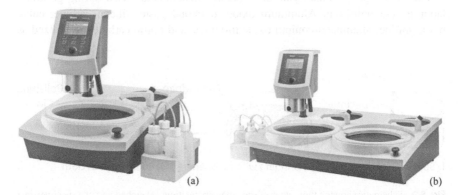

Fig. 8.3 (a and b) A variety of grinding and polishing equipment suitable for all types of materials is available. (Images *courtesy of Struers*)

8.1.6 Impregnation

During the imaging of a ceramic or its composite, it is desirable to distinguish between the closed and open porosity present in the specimen. However, during grinding and polishing, exposed pores in the specimen can serve as initiation points for damage in the surface due to pullouts. This can compromise the accurate representation of porosity of the polished section of the selected ceramic under investigation. Ceramic specimen should, therefore, be impregnated with a synthetic resin of low viscosity in order to ensure a credible representation of its porosity and also to make the polished section considerably easier to prepare. No impregnation is required for ceramic samples with a porosity between 0 and 5%. Ceramic samples with porosity in the range of 5–15% can be prepared more efficiently by impregnating them with epoxy resin. Due to the lack of pore channels, the presence of closed pores, or the narrowness of these channels, thorough impregnation is not always possible. Although ceramic samples with porosity >15% can usually be impregnated effectively, impregnation is impossible for extremely porous specimens with fine pore channels [1].

8.1.7 Etching

Etching may be necessary to reveal microstructural features of the specimen. Etching can be carried out on polished specimens to reveal a contrast in surface features. Exposure to chemical etchants dissolves microstructural constituents of the specimen in a selective manner. The idea behind etching is to reveal the lowest energy surface through chemical or thermal means in order to disclose imperfections such as grain boundaries and enhance the contrast among the various crystallographic orientations or phases. Polishing usually reveals no contrast between the alloy matrix and the grain boundaries. Upon etching, the chemical etchant will corrode the grain boundaries quicker than the grain surfaces. Hence, grooves over the boundary region will develop providing contrast in the microscope. An example of an etched surface is shown in Fig. 8.4.

Etchant should be suitable for the material being etched. For metals and alloys, the etchant is either acid or peroxide solutions. Hydrous nitric acid is frequently used at the beginning, and the additions of a powerful etchant such as (HCl or HF) can be mixed with nitric acid. Sometimes methanol is used as a solvent instead of water. For inert oxides, hot H_3PO_4 is used. The rate of etching must be controlled strictly to avoid over-etching of the surface, and after adequate contrast is observed, the sample is washed using an inert solvent such as acetone and alcohol to prevent additional corrosion at the surface. For inert materials which do not react chemically with an etchant, the surface of the sample is heated to a range where substantial diffusion is possible. This heating effect is comparable to that of chemical etching. Substantial diffusion due to heating will have a tendency to establish an equilibrium state on the surface of the sample which predominantly drives phenomena such as faceting of particular planes and grain boundaries. As a result of these phenomena, contrast is

Fig. 8.4 SEM micrograph of SS304 stainless steel showing *step structure* due to etching with oxalic acid

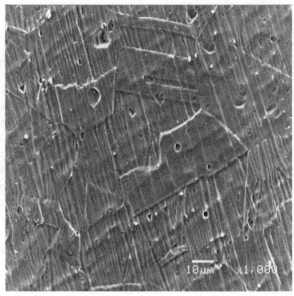

formed. Electrolytic etching can also be undertaken to etch certain materials such as Al alloys. Comprehensive list of etchants can be found online. An alternative to etching is to image the specimen in the backscattered mode that can reveal differences in chemistry between various phases in a specimen. Etching of geological samples and ceramic samples mostly involves HF.

8.1.8 Fixing

Irregularly shaped bulk specimens, powders, and fibers need to be fixed to the specimen stub properly in order to provide a conductive path for the beam electrons to prevent charging. The schematics in Fig. 8.5a–c show the correct fixing procedure for each of these three types of materials.

8.1.9 Fracturing

During metallurgical failure analysis, fracture surfaces of metals and alloys are routinely studied using optical and scanning electron microscopy. This investigation is known as *fractography*. Study of fractures reveals important information about the origin, propagation and timescale of failure, the magnitude of stress, and mode and cause of failure. The fracture surface may belong to a component that had failed in service, or a sample may be fractured in the lab to study its microstructural features. In the latter case, the piece of metal or alloy is scratched to introduce a notch at the center of the specimen from one edge to the other. The sample is then dipped in

8.1 Metals, Alloys, and Ceramics

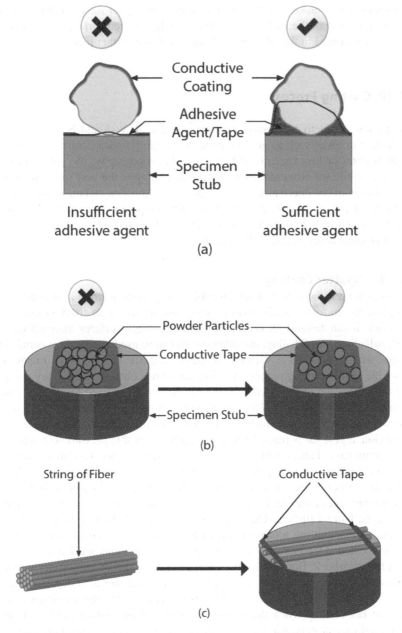

Fig. 8.5 Schematic showing correct methods to fix various forms of materials to the specimen holder (**a**) bulk, (**b**) powder, and (**c**) fiber

liquid nitrogen to make it brittle. Application of an impact force on the other side results in fracture of the metal piece initiated at the notched surface. These fracture surfaces are studied in the SEM after cleaning without further preparation.

8.1.10 Coating Process

Due to non-conductivity of most ceramic samples, they should be coated with a conductive material such as gold to avoid charging up under electron beam unless environmental or low vacuum SEM is used. For topography, the gold coating is preferred, while for elemental analysis carbon coating is the best solution because carbon has minimal interference in EDS analysis. However, carbon is not preferred to use for imaging since its uniformity is poor compared to other metal coatings. Chromium and iridium can be used for topography, but they are less suitable for EDS elemental analysis and BSE imaging.

8.1.10.1 Sputter Coating

Sputter coating is the most efficient, reliable, and popular method as it results in an exceptionally uniform distribution of coating on the specimen from all directions. Therefore, it can be used for complex shapes and rough surfaces more effectively. Gold, palladium, Au-Pd alloy, and other metals can be used as coating material. This method works under lower vacuum compared to thermal evaporation. Figure 8.6a shows the sputter coating process. The low vacuum is obtained by a rotary pump. It is essential that the chamber contains no molecules of H_2O and O_2 as these might damage the sample surface. Without adequate vacuum, the instrument will not be able to introduce argon gas inside the chamber to continue the coating process. After evacuation, argon gas is purged into the chamber to enable pressure to reach about 1 Pa. Argon gas is ionized when free electrons in the chamber spiral under magnetic field and strike argon molecules. This generates more electrons which ionize Ar molecules further giving rise to a cascading effect. A magnet located at the center of the chamber forces the electrons to move away from the specimen so as not to damage the latter's surface. The high voltage applied (1–3 kV) between the target (cathode) located at the top of the evacuated chamber and the specimen platform (anode) located at the bottom forms a gas plasma. Argon ions accelerate toward the target foil to strike it. Bombardment of a metal target by argon ions erodes the metal target and extracts atoms from the thin target foil which are then deposited on the specimen surface. Target atoms hit Ar ions on their way to the specimen and are scattered before they reach the specimen surface. This random motion of target atoms produces the deposition in all directions on the specimen and enables irregular surfaces to be uniformly coated. The thickness of the coating depends on the applied current and the coating duration. In order to undertake high-resolution microscopy, a turbo-pumped sputter coater is used with Cr, Pt, W, or Ta target. Furthermore, the sample stage can be cooled down, and pulse voltage can be used to reduce the temperature and save the specimen from thermal damage.

8.1 Metals, Alloys, and Ceramics

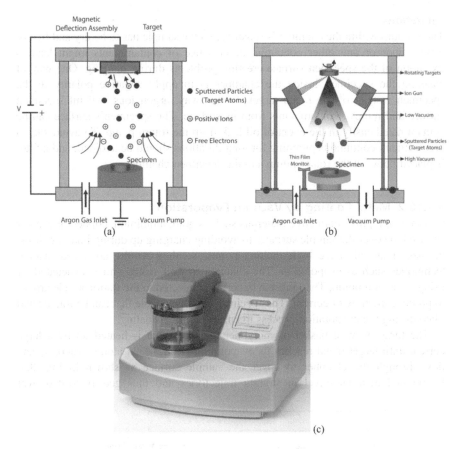

Fig. 8.6 Schematic representation of plasma magnetron sputter coating process. (**a**) Ar gas is ionized within the chamber, and (**b**) ion guns are used to bombard Ar ions onto the metal target. (**c**) Photograph of a typical sputter coater. (*Courtesy of Quorum Technologies*)

In another variation of the process (Fig. 8.6b), the chamber is equipped with Ar^+ ion guns which generate and bombard Ar ions (at 8 keV) onto the metal target located at the topside of the chamber. The target is rotated to generate target atoms uniformly throughout its surface. Apart from plasma and ion beam, sputtering can be undertaken by generating active atoms using radio frequency, magnetron, and penning sputtering method. Photograph of ion sputter coating machine is shown in Fig. 8.6c.

Advantages
The sputter coating is a common process and suitable for a wide variety of specimens. The most important advantage is that the metallic target atoms are scattered within the chamber to strike the sample surface from all directions. This results in a uniformly deposited coating that covers all nooks and corners of the specimen.

Limitations

The vacuum within the chamber is sensitive; so the sample has to be dry and free of water vapor or any other contaminant. Presence of contaminants might lead to oxidation of the specimen surface creating problems during imaging. One way of cleaning the specimen inside the chamber is to apply negative polarity to the specimen instead of the metal target. In this way, argon ions will hit specimen surface and remove contaminants from its surface. The specimen's surface can be contaminated easily in the event of oil leak from the rotary pump. To avoid such a predicament, continuous pumping for long durations should be avoided, and a filter is placed between the rotary pump and the specimen chamber.

8.1.10.2 Metal Coating by Vacuum Evaporation

Coatings of highly conductive materials such as gold, aluminum, copper, and silver are used to cover the sample surfaces to avoiding charging up during imaging. Most of these materials have high melting points, so they require advanced heating techniques such as evaporation. The chamber is well sealed and is evacuated by using a vacuum pump. The tungsten basket is connected to the stationary electrodes to produce the flow of current. The specimen rests on a table that can be rotated and tilted during the evaporation process to ensure uniform coating.

The tungsten wire basket or molybdenum sheet boat is heated up by a large current until target metal reaches its evaporation point. Metal atoms evaporate and flow through the chamber to cover the sample surface as shown in Fig. 8.7. Evaporated material travel in straight trajectories and thus gets deposited over

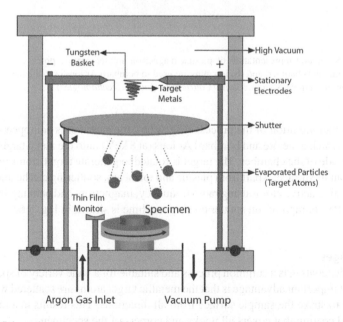

Fig. 8.7 Schematic representation of vacuum evaporation method for coating samples [1]

8.1 Metals, Alloys, and Ceramics

chamber walls and other parts of instrumentation also. Once metal particles reach the specimen, the latter will dissipate heat, and the particles diffuse at the surface of the specimen. The coating rate depends on many factors, such as the rate of evaporation, specimen topography, and target material. The coating continues to cover the entire surface until the required thickness for the intended application is reached. The coating material used is preferred to be an alloy instead of a pure metal, because dissimilar atoms will not pile over each other on the sample surface. Instead, they will form thin layers, as they will grow near to each other. For backscattered electron imaging, coating with the metal of high density is preferable.

Advantages
The advantage of this technique is that different metallic target materials can be used to coat using vacuum evaporation. Also, the instrumentation available for this technique can cope with metals that have a wide range of melting points. Moreover, this common coating method can be used repeatedly for all SEM samples with reproducible results. Finally, the coating layer produced with this method is relatively thin and has an adequately small grain size that helps imaging specimens at high resolution.

Disadvantages
Disadvantages of this technique include longer preparation time compared to a sputter deposition method. The difference in evaporation temperature can cause problems with an alloy target, as the difference in melting points of alloying additions can reach up to 600 °C. Also, the rough surfaces may not be coated properly, due to the straight line trajectories of the evaporated metallic target particles. This problem is mitigated by continuous rotation and tilting of the specimen to enhance its exposure to the evaporated material. Lastly, the high temperature may damage the specimen, but this concern can be addressed by inserting a shutter over the specimen or by using pulsed power source.

8.1.10.3 Coating by Carbon Evaporation
Carbon coating is helpful for EDS elemental analysis as it does not interfere with the detection of other elements. Evaporated carbon also has a fine grain size which makes it suitable for use at high magnifications. Granular morphology of the coating does not appear as an artifact during high-resolution imaging. Figure 8.8a depicts this C evaporation process where the specimen is put in a vacuum chamber that contains a carbon evaporation source consisting of two connected carbon rods (3–6 mm diameter). The photograph of a carbon evaporator is shown in Fig. 8.8b. The rods are resistively heated by passing a current (100 A) through them for a few seconds, and the carbon evaporates from the contact position. The pressure inside the chamber is kept below 10^{-2} Pascal. Evaporated carbon atoms travel in straight lines. Therefore, the sample is rotated to cover all areas. However, this technique is only suitable for flat surfaces. The optimal thickness of carbon layer on the sample surface is around 20 nm; the increase in thickness could be accomplished by increasing the process time as well as by the magnitude of current. The thickness of the coating can be monitored during deposition. This is undertaken by using quartz crystal as part of

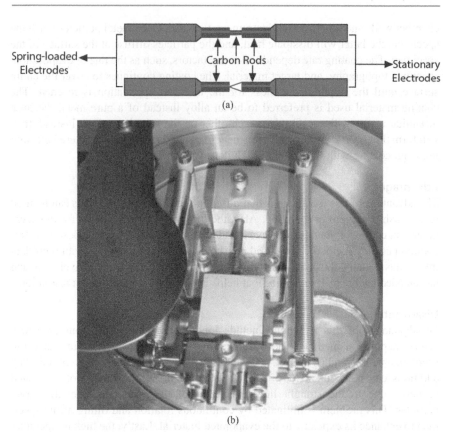

Fig. 8.8 (a) Schematic showing carbon evaporation method. (b) Inside chamber of carbon evaporator depicting head with carbon rods. (*Courtesy of Quorum Technologies*)

an electronic oscillator circuit in which the frequency of oscillation is used to determine the thickness. Other monitoring methods depend on either electrical resistance or optical density of the layer of coating. The color of the coating also can be used to determine the thickness of the coating. For example, a polished metal such as brass shows that orange indicates 15 nm, red is around 20 nm, blue is 25 nm, and green is about 30 nm [1].

8.1.10.4 Imaging of Coated Specimens

Powder or bulk specimens are fixed on specimen stub. If the specimen is conductive, it is not necessary to apply a layer of coating at its surface. In this manner, the original surface features can be examined (Fig. 8.9a). However, use of high accelerating voltage can result in charging of the specimen. Therefore, the use of low beam energy during imaging might become necessary. For a non-conductive specimen, a thin coating will allow imaging of surface features as well as the use of high beam energy (Fig. 8.9b). For the same non-conductive specimen, application of

8.1 Metals, Alloys, and Ceramics

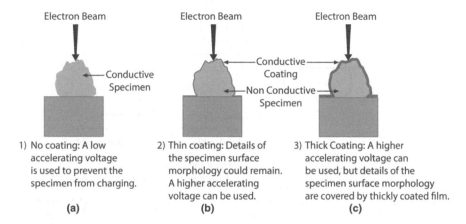

Fig. 8.9 Schematic illustrations of various imaging and coating conditions used depending on the type of specimens. (**a**) For conductive specimen, imaging can be performed without applying any coating. (**b, c**) For non-conductive specimen, a thin coating will allow imaging of surface features, while a thick coating will hide them

a thick coating layer will allow usage of high accelerating voltage, but fine surface details might get concealed under the thick coating (Fig. 8.9c). Therefore, coating thickness should not be higher than a minimum value that prevents charging.

8.1.11 Marking Specimens

For identification purposes, the samples should be marked. Aluminum SEM stubs can be inscribed upon or written on with pens. The label can be embedded with the sample. The label can be written on the back by a diamond point. Occasionally, the area of interest is marked to make it easier to locate during the SEM examination. Currently, SEMs incorporate optical imaging which can be used to locate the area of interest at low magnification and immediately magnify the same area on the SEM imaging screen.

8.1.12 Specimen Handling and Storage

Specimens could be stored, if necessary, in a dust-free area. Moreover, sensitive samples need a vacuum to avoid moisture. SEM stub is kept in special plastic boxes. Handing samples and stub with gloves to avoid carbon contaminants and also cleaning by a solvent are required in some cases. However, the solvent should not damage the specimen, mounting materials, and marking ink. Dust may be cleaned by an air jet. Standard samples are also stored in clean places for frequent use. Sample preparation and imaging flowchart based on solid and nonsolid samples is given in Fig. 8.10.

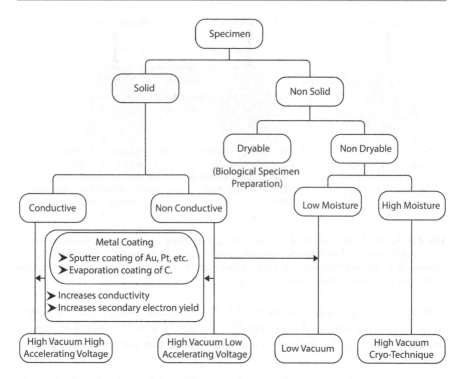

Fig. 8.10 Flow chart illustrating sample preparation procedures adopted for solid and nonsolid specimens. (Adapted from [2])

8.2 Geological Materials

8.2.1 Preliminary Preparation

The initial process consists of removing the contaminants and cleaning the sample. Sediments and soils need to be dried first while porous materials sometimes require impregnation. The samples like rocks should also be cut into small pieces for preparation as per the dimensions of the specimen chamber in the SEM.

8.2.2 Cleaning

Geological samples normally contain undesirable contaminants. These components should be cleaned well because they hinder the examination of features of interest. Sediments are washed with distilled water to remove chlorides and other salts. In order to clean the minerals, a gentle agitation is utilized; ultrasonic cleaning could damage the mineral grains. Hydrochloric acid, iron oxides with stannous chloride, and organic matter with potassium permanganate or hydrogen peroxide can be used to remove unwanted carbonate. Hydrocarbons are preferably removed with soaking in solvents like trichloroethane.

8.2.3 Drying

Drying is used to remove the moisture in wet samples. This is accomplished by heating in air at temperatures greater than 50 °C. When the feature of interest is a fragile structure, the moisture is removed by volatile liquid with lower surface tension such as amyl acetate before the drying process is undertaken. For drying fragile biological materials, the water could be removed by sublimation in a vacuum. To guarantee prevention of growth of ice, the samples need to be frozen quickly; liquid nitrogen could be beneficial for this process.

The nondestructive method (with least damage) is critical point drying; however, it is a slow process. It depends on converting the liquid to gas at a temperature greater than the critical point. This can be achieved by using methanol or liquid carbon dioxide to replace water and then heating it up above critical point (38 °C for CO_2 liquid).

8.2.4 Impregnation

Impregnation with low viscous resin is used to provide the necessary mechanical strength to fragile materials in order to withstand preparation procedure. Furthermore, the porous material is filled to avoid entrapment of polishing materials which then degas in a vacuum. Dilution of the medium with solvent as toluene or acetone is preferred in some cases. The air inside pores can be removed by vacuum and by applying a medium in atmospheric pressure to enable insertion into pores. A schematic in Fig. 8.11 shows the impregnation process.

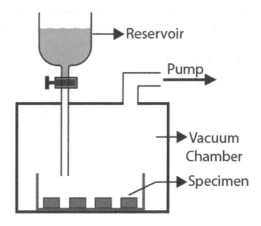

Fig. 8.11 Schematic showing vacuum impregnation process [3]

8.2.5 Replicas and Casts

Replicas and casts are usually used when the pore structure is the target of investigation in spectroscopy. Impregnation process mentioned above is applied here followed with the dissolution of the sample material by hydrochloric acid (for carbonate materials) and hydrofluoric acid (for silicate materials). For SEM, latex rubber casts of fossil plant impressions could be utilized.

8.2.6 Rock Sample Cutting

Few millimeter thick samples are obtained by sectioning the rock using a circular diamond saw. Fragile materials need impregnation before cutting. For materials such as calcium oxide, water-sensitive materials, or sintered magnetite, use is made of mineral oils or alcohol during sectioning.

8.2.7 Mounting the Sample into the SEM Holder

8.2.7.1 Using Stub

Stub is a disc onto which the sample is mounted as seen in Fig. 8.12. It has a pin in the center which is held in the holder. Its size varies between 1 and 3 cm in diameter so as to accommodate specimens of various dimensions. It is made of Al. In some cases, graphite stub is used to reduce the background of x-rays during EDS analysis. Glue or double-sided sticky tape (with low vacuum pressure) is used to fix the sample onto the stub. Conductive double-sided adhesive carbon tabs are convenient to use for this purpose. The aim is for the sample to connect to the stub for electrical conductivity.

8.2.7.2 Embedding Media to the Sample

In this process, the sample is placed into a nonstick mold, and then the embedding medium (such as epoxy resin) is poured as shown in Fig. 8.13. The bubbles formed in the process can be eliminated by carrying out this procedure under vacuum.

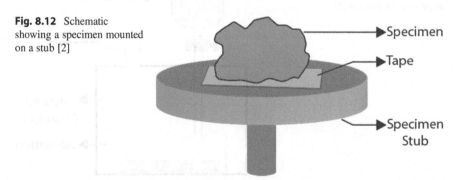

Fig. 8.12 Schematic showing a specimen mounted on a stub [2]

8.2 Geological Materials

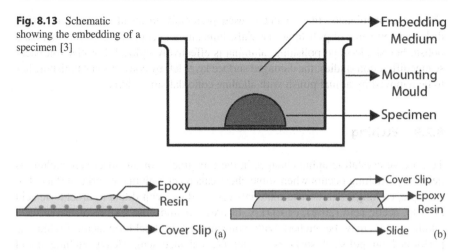

Fig. 8.13 Schematic showing the embedding of a specimen [3]

Fig. 8.14 Schematic showing methods to mount small grains in geological materials

8.2.7.3 Grain Mounts

Grains or powders are mixed with embedding medium and put in a mold, or small layer of grains is put on epoxy resin, and glass cover is put above it. Then the glass cover with the sample is inverted, and the cover is ground away exposing the grains for testing as shown in Fig. 8.14. Furthermore, grains can be put onto a sticker or on top of carbon paint.

8.2.7.4 Mounting Standards

Standards are ready samples with known chemical composition and phase constitution. These are usually supplied by the manufacturer and can be used for calibration or comparing the results with unknowns that have a similar constitution. Standard samples are mounted as described above. The standard samples are usually small, so the user can mount more than one sample on the holder at the same time. Standards can also be prepared by the user in-house. Some of these standards can be assigned permanently for specific applications.

8.2.8 Polishing

For rocks which contain silicates, it is preferred to prepare flat and well-polished samples for EDS analysis as well as for backscattered electron imaging and to avoid topographic effects. The sample surface polishing is carried out with progressively finer grades of abrasive, such as carborundum or emery for the coarser grades and diamond or alumina for later use. Paper and woven nylon laps are used which tend to produce surface relief between minerals of different hardness. Lap could be used by rotating or vibrating motion. After each process, the sample should be cleaned to remove abrasive materials. To get the fine surface, alumina may be used. For cleaning abrasive contaminants, the solvent is used preferably in an ultrasonic bath which dislodges remnants of polishing materials. Fine alumina could be used for final hand polish. For single-stage polishing, alumina is effectively utilized. For electron backscatter diffraction studies, the damaged surface layer left by conventional polishing has to be removed by a final polish with alkaline colloidal silica slurry.

8.2.9 Etching

To observe crystallographic changes in the topography of the specimen, etching is necessary. Etching occurs when some chemicals react with the material surface. For example, etching of carbonate materials can be done with dilute hydrochloric acid (1–5%), acetic acid (20%), or EDTA. In order to study the texture of the surface, quartz grains may be etched with concentrated hydrochloric acid. Etching is performed on polished surfaces for texture enhancement. Heavy etching could remove carbonate cement leaving grains of quartz exposed. Polished section etched by hydrofluoric acid can be used to reveal textures of quartzite sandstones. In some cases, etching by fume is used by suspending the sections above an acid bath.

8.2.10 Coating

Geological materials are mostly non-conductive and require the application of coatings similar to that discussed in Sect. 8.1.10. Figure 8.15 shows a sputter-coated surface of carbonate rock mineral sample as an example. Gold-palladium coating is popular for imaging while C coating is used for EDS analysis. In some cases, the gold coating needs to be removed such as for elemental analysis. In such cases, fine polishing is used for flat surfaces. Gold can be removed from non-flat surfaces by special treatment using 10% aqueous solution of sodium cyanide while silver can be removed easily by photographic Farmer's reducer [3].

Fig. 8.15 Secondary electron SEM image of a coated carbonate rock sample

8.3 Building Materials

Scanning electron microscopy is used to examine different types of building materials including cement materials, cement powder, and cement clinker. Furthermore, cement pastes and hardened concrete can also be analyzed [4]. Microstructural details of cement materials and hardened concrete can be obtained from SEM in fine detail. Different techniques of sample preparation for cement and concrete are discussed in the literature in detail [5, 6].

8.3.1 Preparation of Cement Paste, Mortar, and Concrete Samples

Two methods can be used to prepare cement pastes and concrete samples:

8.3.1.1 Dry Potting
This method can be used for specimens that are dried before preparation. In this way, cracks caused by drying shrinkage no longer remain a concern. Moreover, this method is used when specimens need to be prepared in a short period of time. In dry potting, cut sections or blocks of material are dried at a temperature less than 65 °C. Water is removed from specimens because it can interpose with polymerized epoxy. After that, the top surface of the specimen is exposed to air, and the remaining sides of the specimen are coated by epoxy. Through capillary suction, epoxy is drawn into the microstructure. To make infiltration faster, the specimen is immersed completely in epoxy. Also, a vacuum is used to extract the remaining air. The pore system is filled with epoxy depending on the release of vacuum. After curing the epoxy at 65 °C, cutting and polishing of the specimen is undertaken [4, 5].

8.3.1.2 Wet Potting
This method is used to prepare sections after polishing when the material is not dry. In this way, cracking due to drying shrinkage does not occur because the material is still wet. If cracks are observed in the material, it will be due to physical or chemical processes. Wet potting of specimen consists of three steps:

(a) Replacing the pore solution alcohol (200 proof ethanol).
(b) Replacing the ethanol with a low viscosity epoxy
(c) Curing of epoxy

Firstly, lubricating of plate and strips of saws with propylene glycol or isopropyl alcohol during cutting is carried out to maintain specimen dryness. The cut section is placed in a container filled with 200 proof ethanol for replacement stage of the alcohol-pore solution. Use of a companion specimen is necessary to measure the required time for replacing the alcohol-pore solution. After that, the specimen is placed in a container filled with a deep red ethanol dye. After a period of time, the companion specimen is split or sawed to see the replacement depth due to dye coloration. If this depth is equal to half of the section depth, this indicates that the pore solution in the section is replaced by alcohol. Next, the section is placed in a container with the low-viscosity epoxy. Epoxy replacement takes at least the same time required for the first replacement stage. The alcohol-pore solution replacement time is usually 1.5 times for that of the epoxy-alcohol replacement. The time required for each stage is shorter for the thinner section. The specimen is placed in a container with fresh epoxy. Finally, curing of the specimen at low temperatures is done according to the manufacturer's specifications. Next, the cutting and polishing of the specimen is undertaken [4, 5].

8.3.2 Cutting and Grinding

In all preparations, the specimens are cut and ground to expose a fresh surface for testing. These steps are required to examine the microstructure of the specimens. To expose the surface of the specimen in the fresh state, blade plates of diamond or strips of saws are lubricated with propylene glycol. Then, smoothing of the surface is done by grinding. Materials by grinding are removed using abrasive papers of 110, 200, 300, 410, and 610 grit size (silicon carbide paper) that should be dry. Also, the damage produced by earlier grit is removed using finer grades of abrasive papers. The smoothness of surface required for polishing can be accomplished with diamond blades after the 600 grit grind. The removed layer of material using grit is indicated by grinding striations on the specimen surface. To ensure grinding of the entire surface, alternating grinding directions by 90 degrees should be carried out [4].

8.3.3 Polishing

The damage caused by sawing and grinding is removed by polishing. At this stage, diamond pastes are used for polishing with particle size of 6–0.25 µm. Also, lap wheels are used for polishing. Both polishing pastes and lap wheels are covered with cloth of low relief, and they are used in a sequence. This can be done manually or by using automated polisher for larger specimens [4, 5]. The clarity of microstructure is increased by removing the damage from grinding. This can be performed using diamond paste with the size of 6 µm.

The clarity of the clinker surface is increased with continuity of polishing. This is very important to examine the microstructure of cementitious materials. Next, polishing is carried out with pastes of size 3 µm, 1 µm, and 0.25 µm to remove fine pits remaining after the 6 µm polishing.

8.3.4 Impregnation Techniques

8.3.4.1 Epoxy Impregnation
The epoxy impregnation is done for the pore system to serve two purposes:

1. To fill the pores after curing the cement paste or concrete specimen. The aim is to support the microstructure to resist shrinkage cracking.
2. To enhance the contrast between the hydrated products and cement materials.

Epoxy with low viscosity is used for cement pastes or concrete specimens that are highly permeable. Also, it is used for powders of Portland cement. However, for cement pastes and concrete that are less permeable, epoxy with ultra-low viscosity is used. This leads to fill voids in the structure quickly [4].

8.3.4.2 Dye Impregnation Method
Dye impregnation method is used to decrease the time required to prepare the specimen. In this technique, the water-soluble red powder is dissolved in 100 ml of ethanol solution. The specimen is kept in the dyed solution for 5 min. The second impregnation is carried out in the dye solution. Then the specimen is polished using 6 μm diamond paste to remove excess dye. Finally, the specimen is polished again with 3 μm and 1 μm of diamond paste to achieve the proper polishing finish [5].

8.3.4.3 Impregnation by Wood's Metal
Wood's metal can be used instead of epoxy. The aim of impregnating the cement pastes or concrete with Wood's metal is to provide stability and better contrast for identifying microcracks in the specimens. Also, this technique can be used for concrete specimens. A thin slice of cement paste or concrete is cut and washed to remove attached contaminates. After that, the cleaned sample is kept in an oven at 65 °C for 24 h to extract the water present in the specimen. To impregnate the sample with Wood's metal, it is placed in a steel mold to facilitate the impregnation process. The specimen in the steel mold is subjected to vacuum with a pressure of 7000 Pa for half an hour. After keeping the temperature of the oven at 90 °C for 2 h, the vacuum is replaced by nitrogen pressure of 2 MPa for 3 h. Finally, the specimen is allowed to cool down to be followed by cutting [6].

8.3.4.4 High-Pressure Epoxy Impregnation Method
In this technique, the cement paste or concrete specimen is placed at high pressure of approx. 2 MPa. After drying, the specimen is kept in a steel cylinder filled with epoxy resin at least 5 cm higher than the upper surface of the specimen for 2 days at 65 °C. This cylinder is blocked from the top and bottom by Teflon, and the desired pressure is applied through the hydraulic piston. Then, the curing of epoxy is done by heating it up to 50 °C for 1 day. This procedure results in uniform distribution of epoxy throughout the material structure. The main advantage of this method is that the effective penetration of epoxy in the material is high.

8.3.5 Drying the Specimen

The scanning electron microscope operates with a vacuum. Therefore, the cement pastes or concrete specimens should be dry during SEM analysis. Otherwise, the results of the analysis will not be accurate, and the images will not be clear [5].

8.3.6 Coating the Specimen

Samples are coated to increase the conductivity of the cement pastes and concrete specimens in the SEM. The coating will prevent charge accumulation on the specimen surface by conducting it to the ground [5].

8.3.7 Cleaning the Surface of the Specimen

The cleaning of cement pastes, mortars, and concrete surface specimens is very important to remove the unwanted deposits such as dust, silt, or other contaminants. The cleaning of the specimen surface is usually carried for specimens that are used to investigate the effect of the environmental conditions on the microstructure of cement pastes, mortars, or concrete surface [4].

For backscattered electron (BSE) imaging, the surface of cement pastes or concrete samples should be highly polished to get optimum images of the microstructure. The surface with a low degree of polishing will be rough in texture, and the quality of the image will be unclear. The rough surface reduces the quality of the image by decreasing contrast and loss of feature definition. Further, if the polishing of specimen surface is not good, this leads to inaccurate quantitative estimate.

If the cut surface of cement pastes or concrete specimens is not impregnated with epoxy, it does not present clear microstructure, and the examination without bias will be difficult. The cutting operation damages the microstructure of the cement pastes or concrete specimen surface by creating a series of fractures that are increased with drying shrinkage. For secondary electron (SEI) imaging, the damage caused by cutting controls the topographic features, and the resulting plates of cutting deposited on the surface may disturb the analysis of the material. These difficulties can be avoided by polishing and impregnating of surfaces with epoxy [4, 7]. Figure 8.16 shows SEM image of a cement sample.

8.4 Polymers

Polymer molecules have high molar masses (are called macromolecules). These macromolecules are formed by combining together a large number of small molecules, or small repeated units called monomers, in the chain. Monomer units can be repeated linearly, in branched fashion, or in an interconnected network. The broad forms of polymers are homopolymer, composed of a single repeating

Fig. 8.16 SEM image of cementitious sample clearly showing multi-phase structure

monomer, and heteropolymers [8]. Polymers can exist naturally like proteins, cellulose, starches, and latex, or can be synthesized. Synthetic polymers are usually manufactured on large scale with a wide spectrum of properties, for instance, plastics. Polymers are used extensively in everyday life, such as in housing materials, clothing, automotive parts, aerospace industry, and in communication. As for metals, material science is applied to the polymers to study the relationship between the processing of polymers, produced structures, and the resulting properties. Besides their low processing cost, polymers have low weight, low toughness, and optical properties such as transparency which, in some application, give them an advantage over metals and ceramics [8, 9].

8.4.1 Types of Polymers

One way to classify polymers is through their end use application. Another important method for polymer classification is according to the behavior or response of polymers to rising temperature. Within this scheme, the following types are discussed.

8.4.1.1 Thermoplastics
Thermoplastics are the class of polymers that, when heated, soften and harden when cooled without burning. As the temperature is raised, molecular motion increases, resulting in a consequent diminishing in the forces of the secondary bonding, which facilitates the relative movement of adjacent chains when a stress is applied. Thermoplastics are usually found with linear or branched chains. These materials are manufactured by concurrent application of pressure and heat. Poly(vinyl chloride), poly(ethylene terephthalate), polyethylene, and polystyrene are all thermoplastic polymers.

8.4.1.2 Thermosets

These polymers become permanently hard after formation and do not soften when heated. The structure of thermosets is usually three-dimensional networks with chains that are cross-linked with each other. In most cases, bonding between these chains is covalent. When the polymer is subjected to a high temperature, these bonds tend to prevent both rotational and vibrational motion of the chains, and therefore, softening of the material is prevented. Thermoset polymers are stronger and harder than thermoplastics. Epoxies, vulcanized rubbers, phenolic, and unsaturated polyester are examples of thermoset polymers.

8.4.1.3 Rubbers and Elastomers

These polymers have long flexible chains between cross-links, and when heated to high temperatures, they do not soften, and therefore cannot be melted. The cross-links are in the form of three-dimensional networks. These polymers are stretchable and have the capability to restore to the original form when applied loads are removed. Examples of elastomers are polybutadiene, styrene-butadiene rubber, ethylene-propylene copolymers, and ethylene-vinyl acetate.

8.4.2 Morphology of Polymers

In general, morphology is the study of size, shape, texture, and phase distribution of objects. For polymer science, morphology discusses the organization and form of the polymer structure on a size scale that is smaller than the size of the sample but above the atomic arrangement. The size and distribution of the structural units and the shape of filler and additives are examples of polymer morphology. As far as morphology is concerned, polymers are classified as amorphous or crystalline or more correctly semicrystalline.

8.4.2.1 Amorphous Polymers

These are polymers which do not exhibit any crystallinity in their structure when they are examined by x-ray diffraction. The molecules are randomly oriented and intertwined, and the polymer has transparent glass-like nature. There are a lot of polymer materials that fall under the umbrella of amorphous polymers including glassy brittle polymers such as polystyrene, poly(methyl methacrylate), styrene-acrylonitrile, and cyclic olefin copolymer and ductile polymers such as polycarbonate PC and polyvinyl chloride PVC.

When loads are applied to amorphous polymers, they tend to deform in localized zones, such as shear bands or craze bands. Especially in brittle amorphous polymers, plastic deformation occurs due to crazing, which is the formation of network of small cracks perpendicular to the tensile stress direction. Crazing is enhanced by the presence of rubber inclusions, which reduce the brittle fracture of the polymer by termination crazes, which can be seen with the naked eye, at the rubber particle. On the other hand, shear banding is a phenomenon at which a localized deformation is formed in a way that a high degree of the chain orientation appears at a plane that is oriented at 45° to the stress direction. Crazes and shear bands play an important role in the mechanical properties of amorphous polymers.

8.4.2.2 Semicrystalline Polymers

These polymers, when scanned with x-ray diffraction or electron microscopy, show a crystalline order or periodic arrangement of atoms to form structures that range from nanometers all the way to millimeters. Besides the crystalline order, they also exhibit melting transition temperature and glass temperature. The atomic arrangements in polymers are more complicated than that in metals and ceramics, because of the fact that the structure is based on molecules rather than atoms, and therefore unit cells in polymers are quite complex. Usually, crystalline orders in polymers exist only in certain regions within an amorphous material; this is why polymers are called semicrystalline materials. However, up to 95% crystallinity exists in polymers. Usually, amorphous polymers have a lower density than crystalline polymers, because crystalline polymer chains are closely packed. Cooling rate during polymer solidification process along with chain configuration dictates the degree of crystallinity in a polymer. The ability of a polymer to crystallize is also influenced by the chain configuration and the molecular chemistry. Usually, crystallization is unlikely to happen in chemically sophisticated repeat units like polyisoprene; on the other hand, crystallization is almost inevitable in simple polymers like polyethylene. Polymer structure also plays an important role in crystallization. Linear polymers easily form crystalline material because there are no hindrances to prevent chain arrangement. Excessive branching is likely to prevent crystallization, while most of cross-linked and network polymers are amorphous. As in metals and ceramics, the degree of crystallinity, to some extent, influences the physical properties of polymers. Crystalline polymers are stronger than amorphous ones. Also, they don't soften easily by heat.

One of the models that tried to explain crystalline polymers assumes that a semicrystalline polymer consists of small regions that exhibit crystallinity (crystallites), which are interspersed with regions that are composed of randomly orientated molecules. These crystals are shaped in the form of platelets or lamellae. Usually, lamellae are of 10–20 nm thickness and 10 µm long. Spherulite is a common type of polymeric structure that is formed by bulk polymers. It consists of many ribbon-like lamellae that radiate outward from a common nucleation point in the center [9].

8.4.3 Problems Associated with the SEM of Polymers

In general, there are four fundamental issues regarding the characterization of polymers.

8.4.3.1 Radiation Sensitivity of Polymers

When the incident electron beam interacts with the organic material, usually inelastic scattering takes place. This will result, primarily, in breaking chemical bonds of the polymer. It also produces some secondary effects like mass loss, heat generation, degrading of crystallinity, and charging. Mass loss is mainly caused by bond-breaking processes, where bubbles and cracks are formed in thick samples. The image appears to have uniform depression in the surface. Bond-breaking also results in loss of crystallinity which often leads to major changes in the lamellae and spherulitic textures. Irradiation may also distort the sample and induce some dimensional changes that greatly affect the accuracy of the image. All of these artifacts may lead to misinterpretation of images formed by electron microscopy.

The process of irradiation happens quickly even before focusing and saving images and often goes undetected because, in many times, it cannot be seen by the naked eye. Sensitivity to radiation decreases as the content of carbon increase in the polymer [10]. Oxygen can greatly increase the radiation effects because it tends to form peroxides. Polymer morphology also has an effect on the degree of sample radiation damage. Crystalline polymers, usually, experience less radiation damage than amorphous polymers.

There are a lot of techniques used to minimize damage caused by beam irradiation. The standard procedure is to reduce the beam current; however the signal-to-noise ratio will be degraded, and result in poor image quality. Other techniques include [10]:

1. Focusing the beam into one location on the specimen and taking images from the different non-irradiated location. This process is called *low-dose* techniques.
2. Cryo-microscopy, or imaging at very low temperatures. This will greatly reduce the mobility of the molecules, and thus all of the secondary process (e.g., cross-linking, loss of crystallinity, etc.) will be reduced. However, cooling the sample to such lower temperatures is quite difficult.
3. Dispersing thin layers of evaporated carbon on both sides of the sample to improve the conductivity and hinder sample movements and volatilization of molecular fragments.

Although the irradiation sensitivity of polymers may have detrimental effects on the image quality, it has been used for contrast development. In a mixture of polymers, the different irradiation sensitivities of the components may result in less mass loss in one component than in the other, giving rise to contrast at the beginning of the electron microscopy. It can also give rise to contrast in semicrystalline polymers between the crystalline and the amorphous portions. This is because the cross-linking process caused by irradiation is stronger in amorphous region than that in the crystalline regions.

8.4.3.2 Low Contrast of Polymers
Polymers, when imaged, exhibit a very low contrast between the structural details. This is because polymers consist of the same elements, carbon, hydrogen, and oxygen. These elements are light, and they weakly interact with the incident electron beam, giving rise to weak contrast.

Fig. 8.17 Secondary electron SEM image showing charging of ultrahigh molecular weight polyethylene (UHMWPE) sample. The surface of the specimen exhibits high brightness regions hiding topographic details

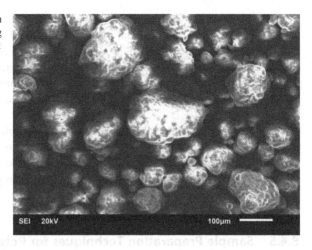

8.4.3.3 Charging

Charging occurs when electrons in the beam accumulate at the surface of a polymer sample, preventing the normal emission of the secondary electron. Charging happens in polymers because they are not conductive. A wide range of behaviors are believed to be caused by charging, including small imperfections in the image like darkening/brightening in some areas, abnormal contrast, and it can reach to an extreme level where the image is completely degraded because the beam is displaced by the charging spot. An example of charging is illustrated by the SEM image shown in Fig. 8.17.

8.4.3.4 Degraded EDS or WDS Spectrum

Analyzing polymers with EDS or WDS involves some difficulties, mainly, the same difficulties polymers face in SEM imaging. The fact that EDS and WDS analyses require a relatively high beam current to generate x-rays makes the damage caused by the incident beam more severe. Moreover, the rapid mass loss for some polymers combined with differential mass loss takes place in a mixture of polymer which makes the quantitative analysis a difficult task to achieve.

8.4.4 General Aspects in Polymers Preparation for SEM

Issues in electron microscopy of polymers (i.e., beam sensitivity, low atomic number, etc.) also cause difficulties in sample preparation of polymers. In general, there are three stages in polymeric sample preparation:

1. Sampling, dimensioning, or forming a sample from the bulk material. Simple cutting usually comes first, where the specimen is cut from the bulk using conventional mechanical techniques like diamond cutting. Often, the surface is

further prepared by grinding, sectioning, and polishing especially when the internal structure of the material is to be investigated.
2. Contrast enhancement. Polymers consist of light elements and therefore exhibit a low contrast in the SEM. There are a lot of techniques used to enhance the contrast including but not limited to etching, staining, and replication. All of these processes might introduce some degree of damage or change to the original material.
3. Minimizing beam damage. This can be done by application of conductive tape and conductive coating, which allow for electrons dissipation and thus minimize beam damage. Replication is also used to completely avoid beam damage especially for highly sensitive samples, whereby a replica is scanned by SEM instead of the sensitive sample.

8.4.5 Sample Preparation Techniques for Polymers

8.4.5.1 Cutting and Sectioning

In SEM machines, specimen size is always limited by chamber size, door and stage size, objective length design, CCD cameras, detectors, and other components inside the chamber. Therefore, the size of large specimens has to be reduced in order to fit into the SEM chamber. Simple cutting is used to extract a small sample from a bulk material which can be easily accommodated inside the SEM chamber. Rotating cutters made of diamond or some hard metals are used to cut polymeric samples.

Sectioning is usually conducted to generate a cross section of the material to be characterized or to specifically cut an interior portion of the material to create an SEM sample out of it. A lot of techniques are used for sectioning including but not limited to abrasive cutting, hand saws, and cutting blades in case of soft materials. Both sectioning and cutting are usually undertaken with very low cutting speed to minimize specimen damage that might be introduced with high-speed cutting.

8.4.5.2 Microtomy of Polymers

Similar to sectioning, microtomy is a method used to prepare semi-thin, thin, and ultrathin flat sections of polymers and biomedical and biological samples that are almost completely free from artifacts. Microtomy was first introduced to biological samples examination, and later, it has been adopted for polymers.

As for polymers, there are three useful microtomy methods: thin sectioning, ultrathin sectioning, and peel-back method. For normal thin sectioning, the sample is cut with a glass or steel knives to a thickness of approximately few microns (1–40 µm), whereas in ultramicrotomy, the thickness of specimen may be brought down to nanometer scales, usually between 30 and 100 nm. Schematic showing general sectioning procedure for polymers is shown in Fig. 8.18. Some of the polymers are soft to an extent that they cannot be sectioned at room temperature. Therefore, to cut such polymers, cryo-microtomy or cryo-ultramicrotomy is used, whereby the specimen is cooled down to a very low temperature during microtomy process.

8.4 Polymers

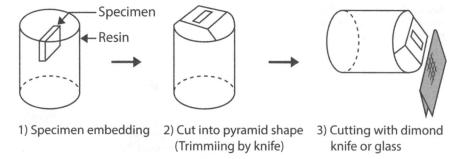

1) Specimen embedding 2) Cut into pyramid shape 3) Cutting with dimond
                         (Trimmiing by knife)   knife or glass

Fig. 8.18 Schematic illustrating sample preparation procedure for polymeric samples

8.4.5.3 Peel-Back Method
Peel back is a microtomy method used to section or to produce a longitudinal splitting of synthetic fibers. The basic idea of peel-back microtomy is to open a specific fiber with a minimum possible disruption through what is called *orientation splitting*. In this technique, the fiber is cut at an oblique angle up to halfway using a razor blade, and then it is cut along its axis. The second cutting process is to peel back with forceps and give a thin section that is aligned with the longitudinal axis of the fiber.

8.4.6 Devices Used in Microtomy

8.4.6.1 Microtome
Microtome is the device used to cut a thin slice of materials. Inside the machine, diamond, steel, or glass cutting blades are used, depending on many factors like material to be sliced and desired thickness. Microtome is capable of producing samples as thin as <5 μm in thickness. There are several types of microtomes including sledge microtome, rotary microtome, saw microtome, vibrating microtome, and laser microtome. Figure 8.19 shows a photograph of microtome used for preparing polymer slices.

8.4.6.2 Ultramicrotome
Ultramicrotome is similar to microtome except that it is capable of producing a thinner slice than microtome. A specimen of thickness less than 200 nm can be made with ultramicrotome. The knife of ultramicrotome is often made of diamond. Sample preparation techniques like embedding, chemical modification, and trimming may be needed for the sample to be cut by ultramicrotome. A photograph of an ultramicrotome is shown in Fig. 8.20.

8.4.6.3 Cryo-microtome and Cryo-ultramicrotome
Cryo-microtome and cryo-ultramicrotome are same equipment as microtome and ultramicrotome except they work at low temperature. As mentioned before, soft

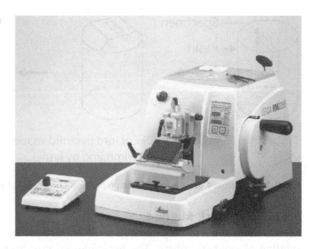

Fig. 8.19 Photograph of the rotary automated microtome. (Courtesy of Leica)

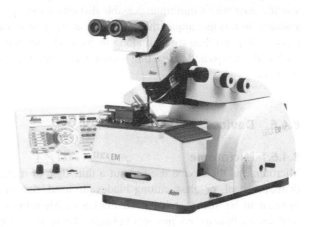

Fig. 8.20 Photograph of an ultramicrotome. (Courtesy of Leica)

polymers can be damaged if sliced at room temperature. The working temperature of cryo-microtome and cryo-ultramicrotome can be lowered down to −185 °C. Figure 8.21 shows a photograph of cryo-ultramicrotome.

8.4.7 Sample Preparation Procedure for Polymers

8.4.7.1 Mounting of Polymer Samples

A small and irregular specimen which cannot be held easily during preparation and those which are difficult to fit into specimen holder have to be embedded in supporting (mounting) medium such as epoxy, paraffin, acrylic or Bakelite [11]. Currently, epoxy is a commonly used as the mounting medium for hard polymeric specimens because of the fact that epoxies are unmixable with water. Paraffin is

Fig. 8.21 Photograph of cryo-ultramicrotome. (Courtesy of Leica)

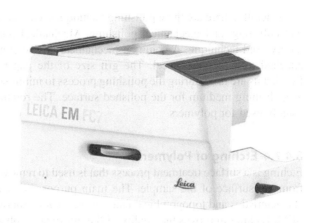

rarely used because it provides a poor cut at room temperature, and also maintaining hot paraffin is time-consuming [8].

The specimen is placed inside the chamber of the pneumatic mounting press machine, and epoxy powders are put over the specimen. Pressure is then applied along with a 200 °C temperature, and the chamber is held closed for 20 min so that the epoxy powder melts and forms the shape of the chamber. The mixture is then cooled down to room temperature where the epoxy solidifies and completely holds the sample.

8.4.7.2 Grinding of Polymers

The general purpose of grinding and polishing is to make the surface of the polymeric specimen flat and clean and remove any topographical feature irrelevant to the specimen's microstructure. Polishing and grinding normally utilized for metals and ceramics also find a use for polymers and polymer composites. Fiber glass and machined semicrystalline polymers were shown to have some artifacts, most probably generated by machining [8].

Grinding is done in several stages to remove artifacts that might have been produced at previous preparation stages like cutting and sectioning, or remove damaged layers of the specimen which may exist at the top surface. Grinding is usually done by using abrasive papers with different grit sizes, starting from coarser to finer. Water is usually used during grinding to reduce the friction, cool the surface of the specimen, and clean it from contaminants. Sometimes glycerin and kerosene are used as lubricants. Grinding is performed under gently applied pressure on the mount to avoid damaging the feature of the specimen.

8.4.7.3 Polishing of Polymers

Similar to grinding, polishing is utilized to create a smooth surface and remove scratches and damages which might have been caused by grinding itself. The main difference between grinding and polishing is the surface roughness; polishing produces a much smoother mirror-like surface.

Generally, there are three polishing techniques: chemical polishing, electrochemical polishing, and mechanical polishing. Mechanical polishing is the most commonly used technique. A common mechanical device consists of a fine grid paper attached to a rotating wheel. The grit size of the paper is approximately 1 μm. Lubricants are used during the polishing process to minimize friction as well as to act as a cleaning medium for the polished surface. The rotational speed of 600 rpm is usually used for polymers.

8.4.7.4 Etching of Polymers

Etching is a surface treatment process that is used to remove the unwanted substance from the surface of the sample. The main purpose of etching is to reveal surface characteristics and topographical features that would not otherwise be detectable by SEM. Etching improves the quality of the information obtained by a microscope as shown in Fig. 8.22. However, it comes with the expense of the possibility of introducing many artifacts. Although it improves the image quality, the methods are not reproducible, and it is difficult to interpret the images. To use an etched sample in the SEM, a conductive coating is applied to it. Preparing a sample using etching can be useful for the purpose of comparison with other samples prepared using other methods.

Etching can be divided into two broad categories: chemical treatments or bombardment with charged particles. In the former, fragments from the specimen are removed from the specimen as a result of the chemical attack. Chemical treatments can further be divided into dissolution, acid treatment, and acid treatment with permanganate. Plasma and ion beam etching are examples of the bombardment with charged particles in which atoms and molecules at the surface of the sample are removed by hitting with a plasma beam [8].

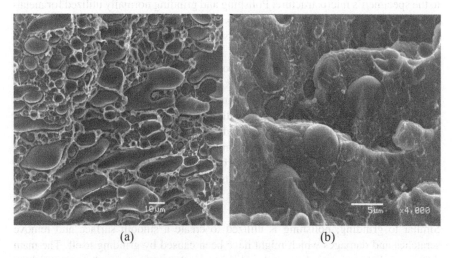

Fig. 8.22 Secondary electron SEM images of etched samples of (**a**) polyethylene polypropylene and (**b**) polyethylene polypropylene royalene

Image quality can be markedly enhanced by etching. Some of the polymer etching techniques are briefly described below.

Solvent and Chemical Etching
This is a useful and attractive way to observe the microstructure. As the material is dissolved, molecules are removed from it. This method, known as solvent extraction, has a tendency to give incomplete results as some material is often left on the specimen surface.

Acid Etching
In this method, chemically resistant semicrystalline polymers (e.g., polyolefins) are etched by using oxidizing acids which diffuse in and attack the amorphous regions in the specimen. After etching, crystalline lamellae will stick out of the specimen's surface and are observed in the SEM after being coated with a conductive metal. Large-scale features (e.g., spherulites) may be visible even with reflected light microscopy.

Permanganate Etching
Permanganate acid is the most widely used etchant to study polymers today [12]. Being weaker than the nitric acid, the permanganate acid selectively attacks amorphous regions in crystalline materials like polyolefins, and hence reveals the internal lamellar structure and organization. Table 8.1 shows common chemical etching media for common polymers.

Plasma and Ion Etching
Leaving residue on the specimen surface is highly unwanted, and in order to avoid it, the etching time is kept as short as possible (in the order of 5–30 min). Aside from the fact that a smoother surface means lesser artifacts, target and specimen rotation and ion gun cooling will help reduce the likelihood of artifact formation. In general, compared to ion etching, plasma etching has a lesser possibility of artifacts formation. In this etching technique, directional plasma such as low-temperature RF plasma oxidizes the organic matter and minimizes the formation of artifacts.

Focused Ion Beam Etching
The focused ion beam is a physical process that is used to prepare specimens with a finely focused ion beam (probe size less than 10 nm). The ion beam is usually composed of positive gallium ions that are used to scan and gradually etch the surface of the specimen away. The sample surface is coated with a conductive coating for the purpose of charge dissipation. The artifacts formation, in this case, is due to the high-energy ion beam resulting in changes in polymer materials like chain shrinkage, chain scission and/or cross-linking, surface chemistry, and crystallinity changes. The focused ion beam can be used in both sample preparation and imaging.

Table 8.1 Etching techniques for common polymers [12]

Polymer			Etching agent	Action/result
Polyolefins		PE	Permanganic Solvent (xylol)	Removes amorphous phase/ spherulitic texture
		COC	Permanganic	
Copolymers of polyolefins		EOC		Degrades amorphous matrix/ visible crystalline lamelle
		PP		Removes amorphous phase/ spherulitic texture, lamellar structure
Blends of polyolefins				Removes amorphous or less crystalline phases
Polyesters		PHA	Methylamine, alcohols	Removes amorphous phase/ spherulitic texture, lamellar structure
		PA	Acids	
		PET	Hydrolysis, nitric acid	Fibrillar structure of oriented fibers
		PLA	Solvents (xylol, heptane, benzol)	Removes amorphous phase/ spherulitic texture
		POM	Alcohols	
		PC	Diethylamine, ethanol	Removes amorphous phase/ spherulitic texture, lamellar structure (if applicable)
Blends		PP/ PEO	Permanganic	Removes PEO, PET phases, and amorphous phase of PP/spherulitic texture, lamellar structure
		PET/ PP		
		PCL/ PS		–
		PE/ NR		
		ABS	Chromic acid, permanganic	Removes rubber phase
		HIPS		
Block copolymers		SBS	Permanganic	Removes butadiene phase
Others		sPS		Removes amorphous phase/spherulitic texture, lamellar structure
		PEEK		
		PVDF	Permanganic, chromium trioxide, phosphorus pentoxide	

Replication of Polymers

Replication is conducted to prepare specimens for SEM and optical microscopy as well as producing thin specimens for TEM. The produced replicas have surface characteristics identical to that of the original specimen, and hence eliminate the need to image sensitive specimens directly. SEM sample replicas can be made for a variety of different sizes and properties of samples. In order to reveal their internal structures, the specimens to be replicated are often pretreated by preparation methods like etching. Disadvantages include time consumption and difficulty in image interpretation.

8.4 Polymers

The polymer replication process for SEM starts with the specimen being placed onto a thick tape to allow the extraction of the replica then pouring a mixture of silicone and a curing agent over the polymer which is left to cure and then peeled off. Examples of the mixtures used for replication include: Dow Corning RTV silicone rubber with Silastic curing agent, Dow Corning RTV 3112 encapsulant and catalyst F (fast cure), and Xantopren (Unitek). Next step is to use an ultrasonic bath for cleaning the specimen. A low viscosity embedding resin is then dropped into the molded silicone (negative) replica, and the mold is then turned over onto an SEM stub and left to cure overnight. Finally, the silicone is removed from the resin replica and positive replica is metal coated for SEM [13].

Staining of Polymers

Due to the fact that they are formed of low-atomic-number elements, polymers experience very small electron energy variations. Hence staining that involves the introduction of heavier atoms into the polymeric sample is used to enhance the density and accordingly the contrast. Staining can involve physical or chemical incorporation of heavier atoms. The first application of staining was to natural polymer fibers like cellulosic textiles [12].

Staining can be divided into positive and negative staining. In positive staining, which is the most commonly used type, the desired zone is stained black by chemical interaction or physical adsorption. On the other hand, in negative staining, small particles mounted on the smooth substrate are revealed by staining the zones around the particles instead of the particles themselves. Staining can be applied to improve the SEM contrast to reveal details in multiphase polymers.

Conductive Coatings

Because they are typically non-conductive specimens, polymers do not allow the dissipation of beam electrons thus resulting in charging. Polymers are generally metal coated to minimize the polymer surface charge, to minimize radiation damage, to reduce the heating effects, and to enhance the electron emissions provided by the electrically conductive layer at the sample surface. However, non-conductive polymers without conductive coating can be examined by using techniques like field emission gun (FEG) and variable pressure SEM; the latter comes at the expense of some loss of resolution.

Deposition of conductive coating can be done using methods like vacuum evaporation and sputtering. Vacuum evaporator thermally evaporates materials, like metals, to obtain a conductive layer on the specimen when the evaporated material deposits on it. On the other hand, sputter coating removes the metal on the target by inert gas ions and directs it to the specimen. A rotary vacuum pump is used to evacuate the device's chamber. After the desired vacuum is reached, the whole system is flushed with argon gas, and the gas flow is controlled during the process. Sputter coating process has a drawback of heating the specimen, but it can be minimized.

Methods other than vacuum evaporation and sputtering include plasma magnetron coating, ion beam sputtering, and Penning sputtering. Unlike ordinary sputter coating, the specimen in an ion beam sputter coating is not heated or bombarded with plasma. A fine, collimated, and water-cooled ion source is used to sputter the target's metal onto the specimen. All the aforementioned are more expensive and take longer time and require more effort than the first two, but they have the advantage of imparting a fine-grained coating which is suitable for high-resolution SEM imaging.

Applying a conductive layer on the sample surface helps to minimize surface charging, minimize damage caused by radiation, and enhance electron emission. A good coating has characteristics as being thin and continuous following sample surface topography.

Artifacts are better identified by experience; however, charging and beam damage are common in polymer specimens. The purpose of coating a polymer sample is to make it conductive, so a poor conductive coating will still lead to charging in the polymer specimen. Charging can cause effects like poor signal output, bright spots, image shift, and snowy images. Another artifact that is taken into account is the damage caused by the beam and also distinguished from damage or artifacts induced by conductive coatings, as similar symptoms may be observed in both cases.

8.4.7.5 Cryogenic and Drying Methods

In order to provide a solid matrix that is suitable for microscopy, liquid or flexible materials are converted to a solid state. The disadvantage of this approach is represented in the formation of crystalline materials during the phase change from liquid to solid. Many polymeric specimens are ready for direct examination. However, a diversity of materials including polymer blends and ductile polymers are not directly ready for examination and hence require preparation before SEM.

8.4.7.6 Simple Freezing Methods

Liquids, colloidal suspensions, and microemulsions may be cooled by simply placing thin layers of these materials on a cold stage. Polymers may become rigid when cooled to low temperatures. Polymers can be immersed in liquid nitrogen for 20–30 min, fractured, and then mounted and coated for observation. Multiphase polymers are notched prior to cooling and fractured in an impact testing machine.

8.4.7.7 Freeze-Drying

Due to the detrimental effects of solvents on polymers, water is not generally replaced with a solvent that has a lower surface tension. Freeze-drying is used to prepare film-forming latex and emulsions which have glass transition slightly below room temperature. Freeze-drying works because at certain temperatures and pressures, solidified liquids will sublime (go from solid to gas phase directly without becoming liquid). Almost all the freeze-drying work involves water. Using freeze-drying is not just about removing water from the sample as water plays a role in providing structural support for macromolecules preventing them from piling up on top of each other, and hence the water removal can cause shrinkage and molecular collapse. Photograph of freeze dryer is shown in Fig. 8.23.

Fig. 8.23 Photograph of a liquid nitrogen cooled freeze dryer. (Courtesy of Quorum Technologies)

8.4.7.8 Critical Point Drying

The critical point drying is done using transitional fluids which go from liquid to gas state through the critical point. For water, the critical point (above 300 °C and above 21 MPa) is too high to be used; hence water is replaced with a transitional fluid. To avoid damaging the specimen when conducting critical point drying, a suitable series of fluids and conditions must be selected carefully. To best identify any artifacts introduced by the preparation method, a microscopic comparison between samples prepared by different methods is done. Photograph of critical point dryer is shown in Fig. 8.24.

8.4.7.9 Yielding and Fracture

Plastic flow and fracture of polymers can be studied using microscopy, with SEM being the most commonly used technique. SEM is mostly used because of its superior depth of field which is suitable for the rough surface examined. Fractured sample preparation is better done by using an easy and reproducible physical test method as in tensile or impact machine at standard settings so that the deformation is reproducible.

Bulk fractured surface can come from three sources: sample deformation in a standard mechanical testing device (like in a tensile, tear, or impact tests), fracture deformation by in situ testing device, or yielding or fracture during service or testing (see Fig. 8.25). The idea behind conducting fractured sample microscopy is to specify failure propagation mode and a link between the obtained mechanical data and the microstructure of the sample.

SEM of fractured surfaces gives the opportunity to compare the observations with the standards available for similar materials. The second fracture source, in situ

Fig. 8.24 Photograph of critical point dryer. (*Courtesy of Quorum Technologies*)

Fig. 8.25 Schematic illustrating one method of inducing a fracture in the polymer in the lab

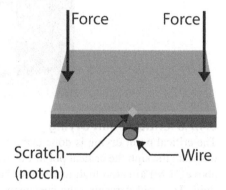

deformation, is almost always used for SEM polymeric samples at low magnifications and usually for fabrics and fibers. The third failure and fracture source occur during testing or service (outside the laboratory), and most often, the microscopy is used to determine the mode of failure or to help in products development.

8.5 Biological Materials

Identification and quantification of infections, diseases, pollutants, etc., are very efficient through the examination of biological samples. Biological samples have long been utilized in research that aimed at the development of cure or prevention of hazards caused by these living organisms. Water, air, food, and human contact through inhaling and body absorptions are different forms of the medium where these living organisms are found and may affect one's health. However, development of antidote or drugs can counter their effect. Accumulation of these minute organisms in small concentrations in the liver, kidney, blood, and various tissues and in the habitual environment can adversely affect human body metabolism functionality which can lead to several health issues if a certain threshold of the contaminant is exceeded. In this case, proper monitoring of biological samples by

8.5 Biological Materials

using different technological advancements have been a promising option available to efficiently measure the extent of exposure of human and other living things to various hazardous substances [14–16].

High depth of field, resolution, and 0.1–0.2 weight percentage (wt. %) elemental detection sensitivity make SEM a very effective tool to examine biological samples [17]. Ease of identifying changes and modification in physical properties of a wide variety of samples in dry and wet states, relatively easy preparation prior to analysis, and prompt results are responsible for its utilization in life science applications [18, 19].

The types of biological samples for SEM analysis dictate the level of their treatment before the examination. Urines, serum, skin, hair, nail, organs (liver, kidney, spleen), and tissue samples, as well as small living organism samples, are expected to have different special treatments and preservation techniques (storage temperature and conditions) to avoid deterioration, degradation, and deviation from their original features. The nature of these samples (mostly wet) causes electron beam damage, charging, and many effects that hinder characterization [20–22].

Biological sample preparation for SEM analysis is critical due to the very nature of the samples which are characterized by their nonuniform structure, organic makeup, and water content. Biological samples should be well collected to conserve their features and prepared to avoid any distortion of the image during analysis. The main reason for image distortion is associated with wet nature of biological samples which might release gas in the vacuum chamber. Radiation damage can also occur in the sample. Therefore, a biological sample should be well dried and prepared without damaging their structural features. Due to the poor electrical and thermal conduction of biological samples, the necessary sample preparation procedure to avoid their susceptibility to radiation should be adopted. The common preparation techniques include sample fixation, drying, and coating to ensure better SEM analysis [22, 23].

8.5.1 Fixation

Fixation is an important stage after sample collection of a biological nature. Fixation is used to stabilize and prevent sample structure from decomposition in order to preserve the nature and structure of the sample during analysis. This process involves several important stages such as prevention of further decomposition of a sample by enzyme activities and removal of the water content of a sample through dehydration processes. Fixation itself should not change any natural structural features of a biological sample. Common fixation methods have been divided into two major groups including *chemical fixation* and *physical fixation* [22, 23].

8.5.1.1 Chemical Fixation

Chemical fixation is commonly used for biological sample preparation in the laboratory. This process involves the use of a certain chemical fixative agent, whose selection depends on the nature of the sample and on the objectives of the analysis. The type of fixation for a certain sample is not specified; however, it can be determined by trial and error.

Chemical fixation consists of the use of several fixatives such as acetone, alcohol, tannic acid formaldehyde (FA), glutaraldehyde (GA), osmium tetroxide (OT), protein cross-linking reagents, and acrolein. Some of the chemical fixatives such as alcohols and acidic fixatives are coagulative and denaturing which can cause changes to biological macromolecules, e.g., protein in the sample. Most of these fixatives are used for samples examined by light microscope where the protein cannot be visualized. However, other chemical fixatives such as formaldehyde (FA) and glutaraldehyde (GA) are non-coagulative which preserve a sample by cross-linking and preserve the protein in the sample. Some fixative agents penetrate rapidly and others slowly into the tissue and can be removed easily. These fixatives preserve the nature of biological samples such as cells and tissues [23].

Various chemical fixation agents used in the preparation of biological samples for analysis in the SEM are discussed as follows.

Formaldehyde (FA)

Formaldehyde is among the most commonly used compounds in chemical fixation of biological samples with chemical formula CH_2O. Formaldehyde tends to react with amino groups of proteins in biological samples and form a cross-link with protein molecules. This chemical fixation method is always slow compared to the glutaraldehyde compound. Time for fixation may take 2 h or more. For SEM analysis, the use of formaldehyde is followed by application of glutaraldehyde compound [24].

Glutaraldehyde (GA)

Glutaraldehyde is a compound made of dialdehyde which can react with protein in the biological sample during the fixation process and result in cross-links of molecular substances. It is a protein cross-linking compound or agent which can also form cross-links with its molecules which will result in a long chain. Glutaraldehyde has the ability to react with a protein and form a neutral pH because of its high reactivity. Also, its reaction to form cross-linking is irreversible which changes the nature of cell cytoplasm to be in the form of a molecular gel. Glutaraldehyde is a good preservative for SEM samples which requires a well-preserved structure at the macromolecular level resolution and tends to preserve the sample throughout the following preparation procedure. Due to this attribute, it is a preferred choice for chemical fixation of a biological sample. However, glutaraldehyde lacks the ability to fix other compounds such as lipids [25].

Osmium Tetroxide (OT)

Osmium tetroxide is also among the popular reagents for chemical fixation of biological samples for SEM analysis. It is usually used as a secondary fixative which follows after glutaraldehyde. Osmium tetroxide tends to cross-link most of the lipids and gives them an ability to resist any further dehydration thus promoting lipid retention in biological samples. The preservation through osmium tetroxide tends to make a cell inactive osmotically and provide permeability of cell in water. Osmium tetroxide can also perform the function of staining a biological sample for SEM analysis. However, it has been reported that osmium tetroxide can cause damage to protein, therefore, it is advisable to use this compound only for a specific purpose [22].

Protein Cross-linking Reagents

Protein cross-linking reagents are compounds whose function resembles that of glutaraldehyde. These compounds form inter- and intramolecular cross-links between amino groups such as proteins. The formation of cross-link forms depends on the functional groups at the reactive end of the molecule which also determines the distance between cross-links [26].

8.5.1.2 Physical Fixation

Physical fixation is one way of fixating the biological samples before analysis. Physical fixation can be done either by heating or freezing. Physical fixation using heating is limited. It might denature proteins which may affect the structure of the sample [22].

Freezing (cryo-fixation) is the most important technique to physically fixate the biological specimen to preserve its structure. Cryo-fixation provides clear morphological details in samples compared to chemical fixation. Cryo-fixation is undertaken in four steps. Initially, microbiopsy is carried out using a very small needle. The sample is then directly dipped into 1-hexadecene (oil substance) to stop the moisture loss from the sample. The needle is mounted and the sample is transferred to the cavity of biopsy platelet using a transfer tool. Platelet is fixed and loaded into the rapid loader and then introduced into high-pressure freezing. Cryo-fixation rapid chills the specimen to a state of vitrification so that the sample can be fractured and micro-features observed using SEM [22, 27].

8.5.2 Examples of Biological Sample Preparation

During the preparation of the biological sample for SEM, there is a possibility of changing the natural structure of the sample; thus wrong conclusions will be reached. In addition, unclear images of the specimen are obtained during SEM. Also, the information obtained might not be satisfactory due to various reasons that might have caused this phenomenon. For biological sample preparation, common problems that contribute to these phenomena include evaporation, surface charging, and dissolving of the sample. To a great extent, these particular problems play a significant role in biological sample preparation. Flowchart given in Fig. 8.26 summarizes the sample preparation procedure for biological samples.

8.5.2.1 Bone Tissue

Bone samples are prepared for SEM examination after drying at 80 °C for 24 h and converted into powder form. SEM sample preparation begins with fixation with 3%

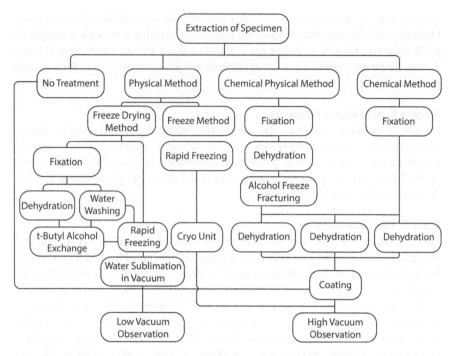

Fig. 8.26 Flow chart showing sample preparation procedure for biological samples. Adapted from [2]

glutaraldehyde in phosphate buffer solution, dehydrated in various grades (30–100%) of acetone in ascending order for 30 min each and dried for morphological study using SEM [28].

8.5.2.2 Heart Tissue

SEM can be used to study the cardiac tissues present in the heart of human, cow, and pig. The larger portion from the left ventricles of the heart muscles for cow and pig are excised immediately after slaughtering and placed in a jar containing 96% ethanol for sterilization (chemical fixation). A block of human tissue is chemically fixed with ethanol (96%). All samples then undergo critical point drying (cryofixation) and made to about 1 mm^3 blocks of tissue with the aid of a scalpel. Samples are viewed with the light microscope to check the orientations and positions of the tissue blocks before finally sputtered coated with gold for SEM analysis [29].

8.5.2.3 Stem Cells

Stem cells have emerged to be an important subject in biology due to its ability to generate different tissues in the body of living organisms. Stem cells are generally used in the cure and treatment of many diseases such as different type of diabetes, cartilage disorder, tendon and ligament injuries, as well as neuron-associated disorders and cardiac failure disease. Stem cells found in different tissues can be extracted and used for different purposes. Most commonly advanced technique which has been used in examining stem cell is the SEM. Through this technique, it has been possible to study clearly the nature and characteristics of stem cells in biology for different treatments. However, stem cells are required to be prepared in a way that preserves its features that need to be studied using SEM. Therefore, proper procedures are followed in the preparation of stem cell for SEM analysis [30].

Stem cells coated with 0.2% gelatin are put in 35 mm dimension dishes that contain coverslips with a mark of differentiation. Another dish may be assigned as control and can be kept in induction for the same duration. The dishes which contain the coverslips with differentiated cells are used for SEM study after 10–14 days. Differentiated adipocytes are then removed in the experiment, and by following fixation and dehydrating procedures, lipids are extracted. A procedure is developed for handling and fixing of the specimen as well as the dehydrating process of stem cells. Treatment medium solution which was in the dishes is decanted, and the cells are fixed with a fresh solution of prepared 2.5% glutaraldehyde in Dulbecco's phosphate-buffered saline at a temperature of 4 °C for around 3 h. After that, the dishes which contain fixed cells are then cleaned twice with this buffer solution, and dehydration is carried out with methanol [30].

Dehydration is conducted in sequence in the dishes by using a solution of methanol which has percentage concentrations of 20, 40, and 60 for 5–6 min each. It is washed with 80% methanol for 3 min and 100% methanol for 30 s, repeated 5 times. Coverslips with dishes undergo drying in a vacuum desiccator for around 10 h and are stored at room temperature until the SEM analysis is carried out. For coating, the surfaces of the coverslips are sputtered coated in a vacuum with an electrically conductive 5-nm-thick layer of gold-palladium alloy. Samples are then taken for SEM analysis where images are recorded. The scanning electron microscope is used at a low voltage (1 kV) and low vacuum mode with a specimen tilt of 30° [31].

8.5.2.4 Bacteria

For bacteria specimen, 2% osmium tetroxide and saturated mercuric chloride at a ratio of 6:1 can be used as a fixative. Samples are held in fixative at the temperature of 4 °C for the duration of 12–18 h. Subsequently, the specimens are kept in reservoirs of either glass or Perspex tube with dimensions of internal diameter of 1 cm and are cut into 1 cm sections. The membrane filters are then attached to their base with the help of a silicone glue. In this simple and cost-effective technique, there is no chance of mechanical damage or loss of specimen cells during processing. Uni-pore polycarbonate filters with 0.4 micron pore size are selected over the cellulose membranes which are commonly used. These are selected based on their properties to resist solvents and a comparative unsculptured surface [32].

After fixation, the bacteria specimens are then taken to small reservoirs followed by properly cleaning in water for around 10 min. Sequential dehydration by ethanol or acetone solution instead of a single dehydration step using methoxyethanol is implemented. After treatment with methoxyethanol, specimens finally undergo acetone treatment. The solutions may then be substituted by gentle draining into the absorbent tissue through the membranes used.

Cell specimens are then critical point dried while still in reservoirs in which carbon dioxide is used as the transitional fluid. The membranes are mounted with double-sided tape to the aluminum stubs and later gold sputter coated before examination in a scanning electron microscope working at 20 kV. In other case, specimens could be taken out of the reservoir gently to the double-sided tape on the aluminum stub to help adherence and decrease subsequent loss [32].

8.5.2.5 Insect

The most common procedure usually followed during the preparation of insect specimens for SEM analysis is dehydrating, clearing, and embedding. Dehydrating the sample is very important which can be done using several methods. Acetone is an excellent solution for dehydrating an insect for SEM analysis. The procedure is outlined as follows:

1. Fixed tissue is placed in 50% acetone for 15 min.
2. It is transferred to 70% acetone for 15 min.
3. It is then placed in 90% acetone for 15 min.
4. It is immersed in 100% acetone 3 times for 10 min.

Clearing is an important step to remove the dehydrating material or compound with a substance that will mix well with the embedding material such as paraffin. There are a lot of clearing agents used for preparation. The type of clearing agent selected depends on the choice of dehydrating agent and mounting material. One of the potential clearing agents that can be used is carbol-xylol (carbolic crystals 1 pt, xylene 1 pt). The mosquito larva is fixed in 70% alcohol and cleared using the solvent density gradient of Pantin. The solution layers used are alcohol 70% glycerin lacto-glycerol [33].

Bayberry paraffin is one way that is recommended as a general embedding medium with the formula shown in Table 8.2.

Use of bayberry wax imparts fine grain and does not change the paraffin melting point, and it mixes better than paraffin. The sample is fixated using glutaraldehyde solution with the formula shown in Table 8.3 [33].

SEM images of various biological samples are shown in Fig. 8.27a–f.

Table 8.2 Different solutions in bayberry paraffin [33]

Compound	Amount (%)
Bayberry wax	1
Hard paraffin (52–56 °C)	50

Table 8.3 Different solutions with glutaraldehyde used for fixation [33]

Compound	Amount (%)
Sodium dihydrogen phosphate (anhydrous) 3.5	3.5
Disodium hydrogen phosphate (anhydrous) 6.5	6.5
40% Formaldehyde 100	40
25% Glutaraldehyde 100	25
Distilled water 800	

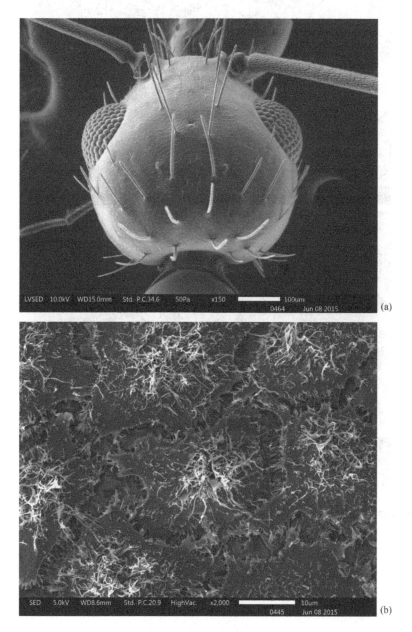

Fig. 8.27 SEM images of (**a**) ant, (**b**) hela cells, (**c**) kidney, (**d**) microvilli, (**e**) pollen grains, and (**f**) leaf with fungal infection images at LN_2 temperature

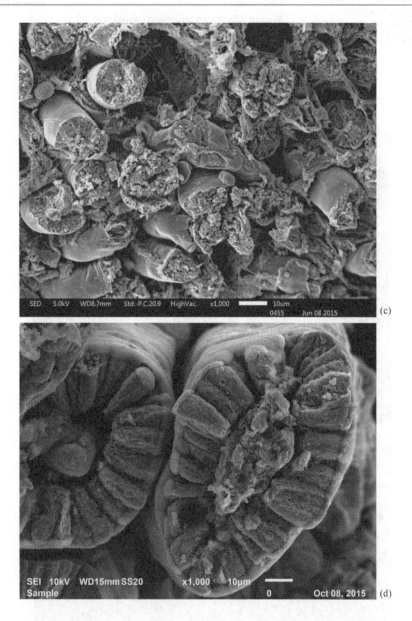

Fig. 8.27 (continued)

8.5 Biological Materials

Fig. 8.27 (continued)

Questions

8.1 What is meant by sampling?
8.2 Why are certain liquids used during sectioning?
8.3 How is grinding different than polishing?
8.4 What is the purpose of impregnation during sample preparation of ceramics and geological materials?
8.5 Why is sputter coating process preferred for imaging? Why is C coating suitable for elemental analysis?
8.6 What is the difference between dry and wet potting?
8.7 Why are polymers difficult to image? What can be done to overcome these difficulties?
8.8 What is the difference between microtome and ultramicrotome?
8.9 How is a polymer replica prepared?
8.10 What is staining of polymers?
8.11 Why is it difficult to examine biological materials in the SEM?
8.12 Explain fixation in biological samples.

References

1. Elssner G, Helmut H, Gonde K, Wellner P (1999) Ceramics and ceramic composites: materialographic preparation. Elsevier, Amsterdam, p 98
2. Invitation to the SEM world: for people who are using the SEM for the first time, JEOL Ltd. Tokyo, Japan. https://www.jeolusa.com/RESOURCES/Electron-Optics/Documents-Downloads/EntryId/257
3. Reed SJB (2005) Electron microprobe analysis and scanning electron microscopy in geology. Cambridge University Press, Cambridge
4. FHWA, Sample preparation for and examination with the scanning electron microscope, Federal Highway Administration Research and Technology. Coordinating, Developing, and Delivering Highway Transportation Innovations, 2006
5. Marusin SEA (1995) Sample Preparation - the Key to SEM Studies of Failed Concrete. Cement Concrete Comp 17(4):311–318
6. Shah VS (2004) Detection of microcracks in concrete cured at elevated temperature, MS Thesis. University of Florida, School of Building Construction, Florida, USA
7. Sas LEA (1990) A note on the preparation of clinker and cement paste samples for scanning electron microscopy. Cem Concr Res 20:159–116
8. Sawyer L, Grubb D, Meyers G (2008) Polymer microscopy, 3rd edn. Springer, New York
9. William D. Callister and David G. Rethwisch (2009), Materials science and engineering: an introduction, Eighth Edition, John Wiley and Sons, NJ, USA
10. Michler GH (2008) Electron microscopy of polymers. Springer, Berlin
11. Sulchek T, Yaralioglu GG, Quate CF, Minne SC (2002) Characterization and optimization of scan speed for tapping-mode atomic force microscopy. Rev Sci Instrum 73(8):2928
12. Goldstein J, Newbury D, Joy D, Lyman C, Echlin P, Lifshin E, Sawyer L, Michael J (2003) Scanning electron microscopy and x-ray microanalysis, vol 1. Springer, New York
13. Sawyer LC, Grubb DT, Meyers GF (2008) Polymer microscopy, 3rd edn. Springer, New York
14. Kaláb M, Yang A, Chabot D (2008) Conventional scanning electron microscopy of bacteria. Infocus (10):44–61

15. Stimpfl T, Muller K, Gergov M, LeBeau M, Polettini A, Sporkert F, Weinmann W (2012) Recommendations on sample preparation of biological specimens for systematic toxicological analysis. In: Committee of systematic toxicological analysis,– the international association of forensic toxicologist TIAFT bulletin XLI, vol 2. Seville, Spain, pp 1–11
16. Kashi AM, Tahermanesh K, Chaichian S, Joghataei MT, Moradi F, Tavangar SM, Lotfibakshaiesh N, Beyranvand SP, Anvari-Yazdi AF, Abed SM (2014) How to prepare biological samples and live tissues for scanning electron microscopy (SEM). Galen Medical Journal 3(2):63–80
17. Goldstein J et al (2003) Scanning electron microscopy and x-ray microanalysis, 3rd edn. Kluwer, New York
18. http://www.geosci.ipfw.edu/sem/semedx.html
19. Leng Y (2008) Materials characterization: introduction to microscopic and spectroscopic methods. Wiley, Hoboken
20. Egerton RF, Li P, Malac M (2004) Radiation damage in the TEM and SEM. Micron 35(6):399–409. https://doi.org/10.1016/j.micron.2004.02.003
21. Glaeser RM (1971) Limitations to significant information in biological electron microscopy as a result of radiation damages. J Ultrastruct Res 36:466–482
22. Moran P, Coats B (2012) Biological sample preparation for SEM imaging of porcine retina. Micros Today 20(02):28–31. https://doi.org/10.1017/S1551929511001374
23. Bozzola JJ, Russell LD (2002) Electron microscopy. Jones and Bartlett Publishers, Sudbury
24. Fox CH, Johnson FB, Whiting J, Roller PP (1985) Formaldehyde fixation. J Histochem Cytochem 33(8):845–853. https://doi.org/10.1152/ajpgi.00048.2011
25. Migneault I, Dartiguenave C, Bertrand MJ, Waldron KC (2004) Glutaraldehyde: behavior in aqueous solution, reaction with proteins, and application to enzyme crosslinking. BioTechniques 37(5):790–802. https://doi.org/10.2144/3705A0790
26. Sinz A (2006) Chemical cross-linking and mass spectrometry to map three-dimensional protein structures and protein–protein interactions. Mass Spectrom Rev 25(4):663–682. https://doi.org/10.1002/mas.20082
27. Vanhecke D, Graber W, Studer D (2010) Rapidly excised and cryofixed rat tissue. In: Methods in cell biology, vol 96. Elsevier. https://doi.org/10.1016/S0091-679X(10)96021-4
28. Sivakumar S, Khatiwada CP, Sivasubramanian J (2014) Studies the alterations of biochemical and mineral contents in bone tissue of mus musculus due to aluminum toxicity and the protective action of desferrioxamine and deferiprone by FTIR, ICP-OES, SEM and XRD techniques. Spectrochim Acta A 126:59–68. https://doi.org/10.1016/j.saa.2014.01.136
29. Saunders R, Amoroso M (2010) SEM investigation of heart tissue samples. J Phys Conf Ser 241:012023. https://doi.org/10.1088/1742-6596/241/1/012023
30. Ozen A, Sancak IG, Tiryaki M, Ceylan A, Alparslan Pinarli F, Delibaşi T (2014) Mesenchymal stem cells (Mscs) in scanning electron microscopy (SEM) world. Niche Journal 2(2):22–24. https://doi.org/10.5152/niche.2013.162
31. Parameswaran S, Verma RS (2011) Scanning electron microscopy preparation protocol for differentiated stem cells. Anal Biochem 416(2):186–190. https://doi.org/10.1016/j.ab.2011.05.032
32. Seviour RJ, Pethica LM, McClure S (1984) A simple modified procedure for preparing microbial cells for scanning electron microscopy. J Microbiol Methods 3:1–5. https://doi.org/10.1016/0167-7012(84)90038-1
33. Barbosa P (2015) Insect histology: practical laboratory techniques. Wiley-Blackwell, Chichester, p 2015

ns/Answers

Chapter 1

1.1 What are the advantages of the SEM over optical microscopy?
Answer:
Advantages: Higher resolution and greater depth of field and microchemical analysis
Disadvantages: Expensive, images lack color, difficult to image wet/live samples, analysis takes more time, and equipment maintenance is relatively tedious.

1.2 What are the different types of samples imaged in the SEM?
Answer:
Metals, alloys, semiconductors, polymers, coatings, ceramics, rocks, sand, corrosion products, catalysts, membranes, carbon nanotubes, nanopowders tissues, cells, insects, leaves

1.3 List various applications of the SEM.
Answer:
Materials identification, materials science, forensic science, metallurgical and electronic materials failure analysis, corrosion science, rock mineralogy, geosciences, nanodevices, polymer science, catalysis, semiconductor design, desalination, life sciences, and oil and gas mining

1.4 Name the industries where SEM is used.
Answer:
Academic and research, oil and gas, power generation, metals and alloys, industrial manufacturing, automobile, aero, aerospace, petrochemical, geosciences, nanotechnology, semiconductor, computer, chemical process industry, mining

1.5 What was the imaging resolution achieved in the earliest SEM? What is the routinely achieved resolution nowadays?
Answer:
50 nm, 1 nm

1.6 Why the scanning electron microscope is so named?
Answer:
It is an instrument that is used to magnify small objects using an electron probe that is scanned across the surface of the object.

1.7 Define resolution limit.
Answer:
It is the smallest distinguishable distance separating two objects.

1.8 What is a useful magnification? Why simply magnifying objects is not enough to keep revealing fine details?
Answer:
It is the maximum magnification beyond which the image becomes blurred and the objects are not resolved clearly.
It is not possible to keep revealing newer details in an object by simply increasing the magnification. Fine details in an image cannot be resolved beyond a certain magnification. This is due to limitations imposed by the resolving power of the imaging technique as well as that of the human eye.

1.9 Who is credited with the invention of the first scanning electron microscope? State the approximate resolution achieved by the SEMs constructed in the early days.
Answer:
German physicist Max Knoll introduced the concept of a scanning electron microscope in 1935. He proposed that an image can be produced by scanning the surface of a sample with a finely focused electron beam. Another German physicist Manfred von Ardenne went on to produce the earliest scanning electron microscope in 1937.

1.10 Can SEM produce color images? State the reason for your answer.
Answer:
No. The probe used in the SEM is electrons whose wavelength does not fall within the visible light spectrum.

1.11 What capability of the SEM enables it to produce 3-D-like images?
Answer:
Large depth of field

Chapter 2

2.1 Calculate the maximum theoretical brightness for W filament operated at 20 kV accelerating voltage and a temperature of 2700 °K. Take current density J_c of W cathode as 3.4 A/cm².
Answer:
$$\beta_{max} = \frac{J_c eV_0}{\pi kT}$$
9.2×10^4 A · cm^{-2} · sr^{-1}

2.2 Determine the brightness of an electron beam that has a current of 3 mA, probe diameter of 0.5 mm, and convergence angle of 0.04 radians.
Answer:
$$\beta = \frac{4i_p}{\pi^2 d_p^2 \alpha_p^2}$$
$\beta = 3.04 \times 10^9$ A · cm^{-2} · sr^{-1}

2.3 What is the brightness β of a probe that has a current density of 8×10^3 A/cm² and a convergence angle of 0.05 radians?
Answer:
$$\beta = \frac{j_c}{\pi \alpha^2}$$
$\beta = 1 \times 10^6$ A · cm^{-2} · sr^{-1}

2.4 What are the four main types of electron guns used in the SEM?
Answer:
Tungsten filament thermionic emission gun
LaB6 thermionic emission gun
Schottky field emission gun
Cold field emission gun

2.5 Name the three main parts of the thermionic emission electron gun.
Answer:
Emitter (cathode, negative electrode, filament)
Grounded plate (anode, positive electrode)
Surrounding grid cover (Wehnelt cup, control electrode)

2.6 List at least three required characteristics of the electron gun.
Answer:
High brightness
Small source size
Low energy spread
Beam stability

2.7 List three advantages and disadvantages of W filament.
Answer:
Advantages:
Low cost.
Reliable.
Replacement is fairly straightforward.
Disadvantages:
Large energy spread, and ΔE is 3eV.
High work function of 4.7eV.
Insufficient beam current.
Short service lifetime.

2.8 Describe the working of thermionic emission W filament.
Answer:
The filament is heated resistively to 2400 °C to exceed work function of W.
Negative bias exists between the filament and Wehnelt cylinder.
Wehnelt cylinder focuses the beam and controls electron emission.
Electrons make a crossover of diameter d_0 (50 microns) between Wehnelt cylinder and anode.
Gun is at high negative potential, and anode is at ground (zero) potential resulting in electron acceleration.
Anode allows only a fraction of electrons to pass through.
Bias resistor keeps the negative bias at the Wehnelt cylinder at an optimum level resulting in a stable beam.

2.9 What is the difference between filament, emission, probe, and beam current?
Answer:
The current provided to the filament is called "filament heating current" i_f.
Total emitted current (100 mA) from the filament is "emission current" i_e.
The fraction of emission current leaving thru the anode is "beam current" i_b.
Current decreases at every lens and aperture; the final current measured at specimen is called "probe current" i_p.

2.10 List three advantages of LaB_6 filaments over W filaments.
Answer:
Five to ten times brighter than W filaments
Improved image quality
Longer service life ($\approx$1000 h)

2.11 List three disadvantages of LaB_6 filaments over W filaments.
Answer:
About ten times more expensive
Require higher vacuum to work
Chemically reactive when it gets hot

2.12 Describe shortly the working of the field emission electron gun.
Answer:
The field emission cathode is usually a wire of single-crystal tungsten fashioned into a sharp point of about 100 nm or less. If the tip is held at negative 3–5 kV relative to the anode, the applied electric field at the tip is so strong that the potential barrier for electrons becomes narrow in width. This narrow barrier allows electrons to "tunnel" directly through the barrier which leave the cathode without requiring any thermal energy to lift them over the work function barrier.

2.13 List three advantages of field emission gun over the LaB_6 emitter.
Answer:
Field emission gun is 100 times brighter than a LaB_6 emitter.
Service life is up to 2000 h.
Field emission guns produce fine probe sizes with large probe currents allowing for high spatial resolution of microscopes.

2.14 List two disadvantages of field emission gun over W or LaB$_6$ guns.
Answer:
High cost
High vacuum requirement, i.e., 10^{-9} to 10^{-10} Pa

2.15 What is an aperture disc, and what is its function?
Answer:
Apertures: It is a strip of metal (Mo or 95%Pt-5%Ir alloy) with precision-drilled holes of various sizes, e.g., 10, 50, 100, and 500 µm.
Function: The aperture discs are located at critical positions within the SEM column and serve to block off-axis beam electrons from reaching the specimen. It limits or controls the number of electrons passing through the column.

2.16 List two types of electromagnetic lenses in the SEM and their functions. How are they controlled?
Answer:
Condenser lenses: Demagnify the electron beam by regulating the current in the condenser lens coils. It is controlled by the knob labeled as "spot size."
Objective lenses: To focus the beam onto the specimen surface and form the image. The objective lens is controlled by adjusting the "focus" control knob during the SEM operation.

2.17 What does strong and weak condenser lens signify? What effect it has on the spot size?
Answer:
Weak condenser strength results in higher current passing through, i.e., bigger probe size (large spot size).
Strong condenser strength results in lower current passing through, i.e., smaller probe size (small spot size).

2.18 What is spherical aberration, and how can it be corrected?
Answer:
The ability of electromagnetic lenses to focus the beam into a fine symmetrical probe is limited by defects called lens aberrations. Electrons near the edge of the lens are bent more strongly than the ones away from the edge resulting different points of focus. It can be corrected by using a small aperture.

2.19 What is chromatic aberration? Is it higher at low kV or high kV?
Answer:
Electrons of different energies focus at different focal points. Less energetic electrons will be bent more strongly by the lens. It is enhanced at low kV.

2.20 How does an astigmated image appear? How can it be corrected?
Answer:
The image appeared stretched in one direction due to astigmatism, which can be removed using astigmator coils.

2.21 Where are scan coils located, and what are their function/purpose?
Answer:
Scan coils are located within the objective lens assembly in the electron column. They scan the electron beam from left to right across the surface of the specimen. They are used to control the magnification of the image.

2.22 Calculate the probe diameter of the electron beam with an energy of 20 keV as it passes through a condenser lens set at a magnetic field strength (B) of 1.0 tesla. Take electron mass (m) and charge (e) as 9.109×10^{-31} kg and 1.602×10^{-19} Coulombs, respectively.
Answer:
$$d_0 = \frac{1}{2B} \sqrt{\frac{2mV_0}{e}}$$
0.24 μm

2.23 What unit dimension is used to express spherical aberration coefficient, C_s? Calculate the beam diameter formed at the Gaussian image plane in an SEM equipped with an immersion lens that has a focal length of 5 mm. Take convergence angle to be 0.005 radians, and ignore other lens aberrations.
Answer:
Millimeters.
The beam diameter at the Gaussian image plane is measured as $2\,C_s\alpha^3$
$2 \times 5 \times (0.005)^3 = 1.25$ nm.

2.24 Calculate the minimum probe size of an electron beam with an energy of 20 keV, probe current of 3 mA, the brightness of 3.04×10^9 A · cm^{-2} · sr^{-1}, and a spherical aberration of 2 mm (take $K = 1$ and $\lambda = 0.008$ nm).
Answer:
$$d_{p,\min} = KC_s^{1/4} \lambda^{3/4} \left(\frac{i_p}{\beta\lambda^2} + 1\right)^{3/8}$$
125.46 nm

2.25 What is the solid angle of a detector? How does it affect the efficiency of the signal collection?
Answer:
The size of the detector is described by the solid angle (Ω – omega) which is the ratio of the area of the face of the detector to the square of the radial distance to the beam impact point, $\Omega = A/r^2$.

2.26 Name different components of Everhart-Thornley detector (ETD) and describe their functions.
Answer:
Faraday cage: Secondary electrons are pulled toward the detector by applying a bias (-50V to $+250$ V) to F placed at the front end of the detector.
Scintillator: The scintillator (S) surface is doped plastic/glass or crystalline CaF_2 doped with europium with a $+10$–12kV. The electrons are converted into light (photons) upon striking the scintillator surface.
Light guide: The light then passes through a total internal reflection plastic or glass pipe (light guide, LG).
Photomultiplier: The first photocathode of PM converts the photons into an electric signal (electrons) that is amplified (up to $\times 10^6$ gain) to levels suitable for image formation on a CRT tube or computer monitor.

2.27 What are secondary and backscattered electrons, and what is their origin?
Answer:
SE = secondary electrons (low-energy electrons belonging to sample)
B = backscattered electrons (high-energy electrons originally belonging to the electron beam)

2.28 Where is backscattered electron detector usually located?
Answer:
It is located directly above the sample (below the objective lens cone).

2.29 Why can't we use rapid scan rates with a solid-state BSE detector?
Answer:
The detector response time is slow for BSE solid-state detector. Rapid scan rates can be employed with a scintillator-type BSE detector.

Chapter 3

3.1 Define the picture element and pixels. What parameters are stored in the computer for each pixel? How can we compare the size of the picture element and pixel?
Answer:
Discrete locations on the specimen where the electron beam dwells to generate signals are called picture elements. Each picture element will have a specific value of location (x, y) and intensity I. The signals are then processed by the detector and are used to form an image on a screen at corresponding locations called pixels. A pixel in an image is larger than the corresponding picture element on a specimen by a magnitude that equals the magnification used to take that image.

3.2 State the benefits of digital imaging.
Answer:
- Digital imaging technology is efficient and cost-effective and easily lends itself to storage and further handling or processing.
- Several images or frames can be acquired and averaged to reduce noise or charging effects.
- Digital images are stored in memory and can be displayed on the screen without having to continuously scan the beam on the specimen surface thus reducing the probability of beam-induced contamination or damage in sensitive specimens.

3.3 How is the size of the picture element in the specimen determined?
Answer:
Pixel length is determined by dividing the horizontal length of the image with the number of image pixels present in the horizontal direction. The picture element is calculated by dividing the pixel size by the value of magnification.
$$L_{\text{picture element}} = \frac{L_{\text{pixel}}}{M}$$

3.4 How does the size of the interaction volume affect the image resolution?
Answer:
Interaction of the probe with the specimen gives rise to large interaction volume. If the diameter of the projected area of interaction is comparable to the size of one picture element in the specimen, the image will be in sharp focus. If the area extends to several picture elements, the image will become blurred since information from several locations will overlap.

3.5 What practical steps can be taken to increase the signal-to-noise ratio?
Answer:
Increase probe current, reduce image scan rate, perform frame averaging, increase solid angle of electron collection, and carry out post-acquisition smoothing.

3.6 What is the contrast formed by a particle emitting 500 secondary electrons compared to 100 SEs ejected from the background?
Answer:
$C = \frac{S_A - S_B}{S_A}$
$C = 0.8 = 80\%$

3.7 Why are secondary electrons suitable to study surface features?
Answer:
Secondary electrons are produced from near the surface of the specimen and hold information about the surface features. They are suitable to study the topography of a material.

3.8 What happens when +200 V and at −50 V bias voltage is applied on the Faraday cage of an E-T detector?
Answer:
At +200 V setting, the E-T detector attracts all SEs and only line-of-sight BSEs. At −50 V, the E-T detector rejects all SEs and receives line-of-sight BSEs.

3.9 What is edge effect? Why does it occur?
Answer:
Higher incidence angle allows greater penetration of the beam into surface regions such that the escape distance toward one side of the beam decreases and the number of secondary electrons emitted from the specimen increases. Higher secondary electron emission results in a brighter contrast. This is why steep surfaces, protrusions, and edges within a specimen tend to appear brighter compared to the flat surfaces. This is called *edge effect*.

3.10 Why is an E-T detector commonly known as SE detector?
Answer:
Images formed by the E-T detector have a strong SE component due to the positively biased Faraday cage.

3.11 What is interaction volume and why is it important?
Answer:
Elastic and inelastic scattering events make electrons penetrate into the depth and spread laterally across the breadth of the specimen resulting in a relatively large "interaction volume." This volume encompasses most of the scattering events. This is the volume of the specimen within from all imaging, and the microchemical information is extracted directly affecting the spatial and analytical resolution of the SEM.

3.12 Describe the effect of beam energy and atomic number of the specimen on the shape and size of interaction volume.
Answer:
The higher is the beam energy (E_0), the greater is the depth and breadth to which the electrons can travel into the specimen as they lose energy at a lower rate which is proportional to $\frac{1}{E_0}$. Increasing beam energy also reduces its probability to scatter elastically (as a function of $\frac{1}{E_0^2}$) thus penetrating deeper into the specimen. The trajectories of the electrons near the specimen surface are straight resulting in widening of the interaction volume away from the surface. High beam energy is more likely to form a teardrop interaction volume, and low beam energy tends to form a hemispherical shape.
For specimens with high Z, the elastic scattering is greater resulting in a deviation of the electrons from their original path more quickly and reducing the distance that they travel into the specimen. This results in a hemispherical shape of the interaction volume. On the other hand, elastic scattering and the rate of energy loss per unit length are lower in low atomic number targets due to which beam electrons manage to maintain their straight trajectories for larger depths in the specimen. This results in the formation of a large interaction volume taking the form of a *teardrop*.

3.13 How does beam energy and Z affect electron range?
Answer:
Electron range or depth of beam penetration increases with increasing beam energy and decreasing Z.

3.14 What is the difference between BSE and SE?
Answer:
SE are low-energy (<50 eV) electrons that belong to the target material. They are ejected out due to inelastic scattering of the electron beam close to the surface of the specimen. BSE are high-energy (several keV) electrons belonging to the original beam that enters the specimen and gets elastically scattered at large angles to emit out.

3.15 What is the origin of compositional contrast?
Answer:
Compositional contrast is formed by BSE. Scattering of BSE strongly depends on the atomic number of the specimen material. Target materials with high atomic number show a high degree of elastic scattering resulting in high angles of deflection and large backscattering effect. Two phases with different atomic numbers present within a specimen will exhibit different values of backscatter coefficient. This will result in a contrast where the phase with a high atomic number will appear relatively brighter (due to a larger number of backscatter electrons ejecting out of this phase), while phase with a low atomic number will appear relatively dark. This forms the basis for a compositional or atomic number or Z contrast.

3.16 Calculate the natural BSE contrast formed between two adjacent phases of pure Au and Al.
Answer:
Backscatter coefficient is given as:

$$\eta = -0.0254 + 0.016Z - 1.86 \times 10^{-4}Z^2 + 8.3 \times 10^{-7}Z^3$$

$\eta_{Au} = 0.48$
$\eta_{Al} = 0.15$
From Eq. 3.3, contrast $= \frac{\eta_{Au} - \eta_{Al}}{\eta_{Au}} = 0.69$ or 69%.

3.17 Why does backscatter coefficient increase with specimen tilt?
Answer:
As the angle of inclination between the beam and specimen is narrowed (i.e., tilt, θ increases), a greater number of electrons can escape the specimen surface as backscattered electrons. The surface, when tilted, lies closer to the forward scattering direction of electrons.

3.18 What can be done to maximize BSE signal if the detector is placed (a) directly above and (b) to the side of the specimen?
Answer:
(a) Image the specimen at zero-degree tilt.
(b) Image the specimen by tilting it toward the detector.

3.19 How does the variation of spatial distribution of BSE with the atomic number of the target material affect the image?
Answer:
Low atomic number targets will exhibit larger depth and lateral dimensions from which the BSE are derived. This will add to the noise component of the image leading to the degradation of its resolution.

3.20 How much of a backscattered contrast between two phases is appropriate?
Answer:
Contrast between 1% and 10% is appropriate. Less than 1% is difficult to interpret. More than 10% is considered strong.

3.21 What is the origin of trajectory component in BSE image?
Answer:
Features facing the detector will appear bright in the BSE image. The features facing away from the detector will appear relatively dark even though they have the same density. This is due to the fact that BSE emitted from the specimen have straight trajectories and continue their travel in the direction they are originally emitted. Apart from the atomic number Z component, this characteristic adds a trajectory component to the BSE image. If the specimen is flat, the BSE trajectories will not vary, and the contrast will solely depend on the Z component. If the surface is rough, the BSE trajectories will vary as per orientation of features, and the image contrast will be composed of Z and trajectory components. Separation of Z and trajectory components present in the BSE image can generate BSE compositional or topography images independently.

3.22 Why is the spatial resolution of the BSE image worse than that of SE image?
Answer:
The spatial resolution of backscattered electron images varies between 50 and 100 nm for beam energies of 10–20 keV that are employed during routine imaging. This is directly related to the comparably large volume within the specimen from where the backscattered electrons are derived to form the image. At low beam energy of 1 keV, the information volume of SE and BSE becomes

comparable. For high-resolution microscopy, low-loss BSE are used which are ejected from the area immediately surrounding the point of beam incidence. These electrons undergo single or lesser number of scattering events and represent high-resolution signal.

3.23 Summarize the factors that influence BSE contrast formation.
Answer:
By convention, the energy of the BSE is generally defined as 50 eV $< E_{BSE} \leq E_0$. Specimens with large Z will have high BSE energy distribution. Light elements will show low-energy distribution.

The depth and lateral dimensions of interaction volume formed in a given target material increase with an increase in beam energy.

Specimens with a high Z show a high degree of elastic scattering resulting in high angles of deflection and large backscattering coefficient η. Light elements exhibit small η. Due to this reason, a phase with a high Z will appear bright, while a phase with a low Z will appear dark. This is called compositional or atomic number or Z contrast.

Backscattered signal from the specimen can be increased by tilting the specimen in such a way that the incidence angle of the beam with the specimen surface is small, i.e., at large tilt angle θ.

At zero-degree specimen tilt, the maximum backscattered electrons are emitted along the beam. BSE emitted at small angles relative to the specimen are small in number, following a cosine law.

The collection efficiency of a detector for an untilted specimen can be maximized by placing it directly above the specimen.

BSE topographic contrast can be obtained by collecting BSE signals separately by two segments A and B which are located at opposite sides of the specimen. The image formed by these segments will differ as per the orientation of the features. Features normal to segment A will appear bright in A and dark in B. Features normal to B will appear bight in B and dark in A. Subtraction (A-B) of these images will reveal topographic contrast.

3.24 An electron beam with primary current i_B of 2 µA gives SE current i_{SE} of 0.6 µA and BSE current i_{BSE} of 0.4 µA. Calculate the specimen current i_{sp} and SE yield δ and BSE yield η for this beam.
Answer:
Primary beam current, $i_B = i_{SE} + i_{BSE} + i_{sp}$
Specimen current, $i_{sp} = i_B - i_{SE} - i_{BSE}$
$i_{sp} = 2 - 0.6 - 0.4 = 1.0$ µA
SE yield, $\delta = \frac{i_{SE}}{i_B} = 0.3$
BSE yield, $\eta = \frac{i_{BSE}}{i_B} = 0.2$

3.25 Why is escape depth of SE smaller in metals compared to insulators?
Answer:
The escape depth depends on the type of material. It is smaller (around 1 nm) in metals and up to 20 nm in insulators. This is due to the fact that SE generated within the specimen are inelastically scattered due to the presence of a large number of conduction electrons in metals. This scattering prevents the SE generated within greater depths of metals to escape the surface. Due to a lack of electrons in insulators, inelastic scattering of generated SE is not significant thus allowing them to reach and escape the specimen surface from greater depths.

3.26 Compare escape depth of BSE and SE.
Answer:
The probability of escape of secondary electrons as a function of specimen depth Z decreases sharply with depth. In the range of 10–30 keV, secondary electron escape depth is around 100× smaller than the backscattered electron escape depth.

3.27 What is the energy distribution of SE?
Answer:
Although, the secondary electron energy can be up to 50 eV, 90% of secondary electrons have an energy <10 eV. The distribution of the secondary electron energy is generally peaked in the range (2–5 eV).

3.28 Which signal dominates in an image, SE_1 or SE_2, and why?
Answer:
The contribution of the SE_1 signal in secondary electron emission is greater than SE_2 signal in light elements. This is due to the fact that backscattering is low in light elements due to their small atomic size. Therefore, the main component of the total SE signal is generated by the primary beam incident on the specimen surface. In heavy elements, SE_2 signal dominates since backscattering is prominent due to large atomic size.

3.29 How does SE_3 signal represent BSE contrast?
Answer:
The number of SE_3 signal emanated from the hardware of the SEM specimen chamber depends on the number of BSE originating from the specimen. SE_3 increases when BSE increases and SE_3 decreases when BSE decreases. Thus, SE_3 is an indication of atomic number contrast in the image.

3.30 What is the effect of beam energy on SE yield?
Answer:
Secondary electron emission is higher at lower incident beam energy. There is a significant increase in secondary electron emission below 5 keV. This is due to the fact that secondary electrons have low energy and can only escape the specimen if they are generated near its surface. At low keV, the penetration of incident beam is shallow, and most of the secondary electrons are generated near the specimen surface. This enables them to escape resulting in a higher secondary electron coefficient (δ) at lower beam energy.

3.31 How low beam energy be used to obtain high-resolution images?
Answer:
At low beam energy, the SE_2 electrons are generated closer to the beam impact point decreasing the range of SE_2 signal. SE_2 spatial distribution approaches that of SE_1 under such conditions. Due to their close proximity to the beam impact point, both types of signals now serve to make up the high-resolution image. The use of low beam energy, therefore, allows exerting better control on interaction volume and image contrast and facilitates the acquisition of high-resolution images rich in near-surface information.

3.32 Does SE yield change with Z?
Answer:
For practical considerations, the secondary electron emission can be considered to be independent of atomic number and does not change significantly from light to heavy elements.

3.33 What is the effect of specimen tilt on SE yield?
Answer:
When the specimen is tilted at increasing angles θ, secondary electron coefficient δ increases obeying a secant relationship.

3.34 List the factors that affect topographic contrast.
Answer:
Sharp edges and tilted features in a specimen yield more SE as secondary coefficient δ increases as a secant function of tilt θ. Sharp edges, therefore, appear bright in SE image. Tilted surfaces in a specimen will also generate a higher number of BSE as backscatter coefficient η increases significantly with the angle of tilt. This

will, in turn, increase the SE_2 signal also adding a number component to the topographic contrast formed in the SE image. Moreover, titled features will generate BSE in a direction defined by the beam direction and surface normal of the feature. This adds a trajectory component to the SE image.

The yield of SE from tilted surfaces is not significantly different in various directions, so the trajectory component in this regard is negligible.

The Everhart-Thornley detector receives line-of-sight BSE from features of specimen facing the detector. These BSE will also form part of the SE image.

Elements with high atomic number have a greater yield of backscattered electrons compared to elements with lower Z. This adds a number component to the topographic contrast, and high Z elements, therefore, may appear brighter in the SE image.

Contrast in SE image is enhanced at lower accelerating voltage. Charge accumulation or buildup in partly coated areas of the specimen can also result in an increase in contrast.

Magnetic areas in a specimen may either deflect or attract the beam to affect the yield of SE.

3.35 Fracture surface is best imaged using SE or BSE?
Answer:
Fracture surface is usually imaged to observe the topography of material. Therefore SE signal is more adequate for fracture surface imaging. SE gives better image resolution and can be collected efficiently from surfaces facing away from the E-T detector. Various features within the fracture surface such as fatigue striations, dimples, ridges, beach marks, crevices, etc. can be identified unambiguously.

3.36 What imaging mode suits a specimen with two phases widely separated by Z?
Answer:
It is more appropriate to use BSE imaging to observe the contrast between two phases with different Z as SE yield does not differ appreciably with Z.

Chapter 4

4.1 What is the resolution of the SEM at 30, 20, 10, 5, and 1 keV if convergence angle is 0.01 radian?
Answer:

$$d = \frac{0.753}{\alpha\sqrt{V}} \text{ where } d \text{ is the resolution limit}$$

At 30 keV, $d = 0.435$ nm
At 20 keV, $d = 0.532$ nm
At 10 keV, $d = 0.753$ nm
At 5 keV, $d = 1.065$ nm
At 1 keV, $d = 2.38$ nm

4.2 Summarize the operating conditions that favor high-resolution microscopy.
Answer:
Short working distance produces a small probe size and reduces lens aberrations that serve to improve resolution. High accelerating voltage should be used as it produces a small probe size, unless surface features are to be imaged in which case low accelerating voltage is suitable as it limits the size of the excitation volume. Smallest probe current that gives adequate signal-to-noise ratio combined with frame averaging should be used. Intermediate final aperture is suitable to balance out aberration and diffraction effects.

4.3 Calculate the wavelength of the electrons that have an energy of 10 keV.
Answer:
$\lambda = \sqrt{\frac{1.5}{V}}$ nm
$\lambda = 0.0122$ nm

4.4 Summarize the operating conditions that favor high depth of field.
Answer:
Large working distance, small final aperture (e.g., small convergence angle), and small magnification favor images with large depth of field.

4.5 Why depth of field in the SEM is far larger than in the optical microscope?
Answer:
The depth of field is large because the convergence/divergence angle formed in the SEM is small (in milliradians). Due to this reason, the probe size remains small for longer depth (z-d axis).

4.6 Find the depth of focus of a beam when the angle of convergence is 0.2 radian and magnification is 10,000×.
Answer:
$D = \frac{200}{\alpha M}$ μm
$D = 0.1$ μm

4.7 Describe the relationship between the probe current and size.
Answer:
Probe diameter increases with probe current at all accelerating voltages. At high accelerating voltage, the probe diameter is smaller compared to low accelerating voltage.

4.8 Summarize the advantages/disadvantages of the use of high/low probe current.
Answer:
High probe current results in smooth images but degraded image resolution. It can also induce beam damage. Low probe current can realize high image resolution due to small probe size, and the specimen is susceptible to less beam damage. Very low probe currents give rise to grainy images which tend to hide surface details. A critical level of probe current is required to achieve an acceptable contrast in the image.

4.9 What is the advantage of FESEM for high-resolution use?
Answer:
Field emission electron guns can concentrate a large amount of current in small probe resulting in high brightness and small spot size.

4.10 State the benefits/drawbacks of using large working distance during microscopy.
Answer:
Large WD increases the depth of field due to smaller convergence angle. It also allows the specimen to be observed at small magnifications. Large WD lowers the spatial resolution due to increased probe diameter. The signal strength at large WD decreases, and the image can appear relatively noisy.

4.11 What is the positive charging effect on the specimen surface?
Answer:
A charge balance is achieved when incident electrons impinging upon the specimen (i_B) are equal to those leaving the sample ($i_{sp} + \eta + \delta$). If the number of incident electrons is higher than the emitted electrons, the sample will charge negatively

which is the usual case. If the number of emitted electrons become more than the incident electrons at localized specimen surface, then positive charging takes place at that region. This does not pose much difficulty as the positive charge created at the specimen surface is neutralized by the SE emitted from the specimen and pulled back toward the surface.

4.12 How is charging recognized?
Answer:
Charging effect may present itself in many forms such as unusual contrast (fluctuation in image intensity such as excessive brightness/darkness in images), horizontal lines on images, beam shift and image distortion (spherical objects appear flat), etc.

4.13 What is the most effective and common used method to combat charging effects?
Answer:
Sputter coat with a thin metal layer.

4.14 What means can be adopted to prevent contamination during imaging?
Answer:
Contamination from the instrument is reduced by employing dry pumps or installing a vapor trap in the pump backing line that can control hydrocarbon contamination originating from vacuum pumps. In addition, cold traps can be employed to seize contaminants, and the SEM chamber is purged with dry nitrogen gas during specimen exchange. Contamination from the specimens can be reduced by properly handling (e.g., use of gloves and completely dry specimens) the specimens and using minimum amount of adhesive tapes, conductive paints, embedding agents, and resins. Size of outgassing biological or hydrocarbon volatile specimens should be kept to a minimum.

4.15 How to combat beam radiation damage during microscopy of sensitive specimens?
Answer:
Use low accelerating voltage, decrease probe current, reduce exposure time, use low magnifications/large scan areas, and apply conductive coatings such as of gold, carbon, etc. at the specimen surface to improve thermal conductivity.

4.16 How does EMI present itself? How can it be reduced?
Answer:
EMI results in image distortion usually at high magnifications. Edges of objects appear jagged. EMI is reduced by installing screening cage and EMI cancellers that generate electromagnetic radiation of the similar magnitude but in the opposite direction.

Chapter 5

5.1 Why is low-voltage imaging beneficial?
Answer:
As the beam energy is lowered, the specimen interaction volume decreases sharply resulting in a high-resolution high-contrast SE and BSE signal emanating from fine features close to the specimen surface giving rise to images with greater surface detail. Due to the small interaction volume generated at low kV, both SE and BSE signals produced are of a high spatial resolution giving rise to stronger image contrast. A high-resolution image is obtained by separating the high-resolution SE and BSE signals from low-resolution SE and BSE signals, respectively. Production of low-resolution signals such as SE_2, SE_3, and BSE farther away from the probe is eliminated. Effects of specimen charging and edge brightness are also reduced at low beam energies. This imaging technique is also suitable for beam-sensitive specimens as it minimizes radiation damage.

5.2 State the challenges to low-voltage microscopy. What can be done to encounter these?
Answer:
Boersch effect (defocusing at crossover), decreased gun brightness, increased chromatic aberration, increased diffraction at the aperture, and contamination buildup relative to the low depths from which the signals are generated. Low energy beams are also susceptible to electromagnetic interference effects. If the beam current and gun brightness are kept constant, operation at low kV results in a significantly larger spot size resulting in decreased resolution.
Microscopes equipped with high brightness source with low energy spread and an immersion lens help to maintain reasonable image contrast. It is advisable to employ short working distance during imaging at low voltages to mitigate the effects of lens aberration and any extraneous electromagnetic field present in the work environment. The rate of contamination buildup can be reduced by avoiding high magnification, focusing and removing astigmatism in an area other than that used for imaging and not using spot or reduced area raster mode. Clean specimen chamber with a high-quality vacuum system, stable and vibration-free platform,

and proper shielding from electromagnetic influences has enabled imaging at a few tens of volts.

5.3 What is energy filtering and how is it useful?
Answer:
Electron energy filtering separates the low-energy SE from the high-energy SE and BSE. The final image can be selected to be composed of mainly SE or BSE or combination of both depending on the detected signal. Low-energy SE can be rejected or its detection can be controlled in combination with BSE by regulating the extent of negative bias on the control electrode. Low-energy SE are primarily responsible for charging effects. The ability to filter low-energy from the high-energy signal enables better control during low voltage imaging.

5.4 How does beam deceleration work and what is its benefit?
Answer:
In beam deceleration technique, the electron beam is kept at high energy as it passes through the SEM column. Once it exits the final lens, the beam is decelerated before it strikes the specimen surface. Beam deceleration is accomplished by applying a negative bias (up to -4 kV) to the stage which sets up an electrical field between the specimen and the detector, acting as an additional electrostatic lens working to retard the beam accelerating voltage immediately before it hits the specimen. The energy with which the beam lands onto the specimen surface is known as landing energy and is equal to accelerating voltage minus stage bias.

By maintaining the beam at high energy during its movement through the column and lenses, large energy spread, *Boersch* effect, and chromatic aberrations are avoided. The beam lands on the specimen surface with lesser energy which serves to reduce beam penetration and interaction volume. With this technique, greater flexibility in the selection of beam voltages becomes available. It enables detection of electrons scattered at shallow depths emphasizing its surface features. It improves microscope resolution and contrast at low accelerating voltages. Beam deceleration is a relatively simple technique that can be incorporated within the existing electron sources and columns eliminating the need for a separate SEM system.

5.5 What purpose does imaging at low vacuum mode serve?
Answer:
SEM in low vacuum mode serves to analyze damp, dirty, or insulating samples. High vacuum is unsuitable for these sample types since they tend to charge up or degas.

5.6 Describe the working principle of low vacuum mode.
Answer:
In low vacuum mode, gas or water vapor is injected into the specimen chamber around the specimen surface area. High-energy electron beam penetrates the water vapor with some scatter and interacts with the specimen surface. Secondary and backscattered electrons emanating from the specimen strike the water molecules and produce secondary electrons which in turn produce more secondary electrons upon interaction with the surrounding water molecules. Water molecules are changed into positive ions as a result of this interaction with incident beam and secondary/backscattered electrons emerging from the specimen. Positive bias applied to a detector accelerates secondary electrons toward the detector, while positive ions are pushed toward the negatively charged areas on the specimen. The water vapor thus serves to produce positive ions and also increase the number of secondary electrons resulting in gas amplification. Generation of positive ions and their movement toward the negatively charged areas of the sample neutralizes the negative charge accumulated at the specimen surface.

5.7 Why is SE detector not used in low vacuum mode generally?
Answer:
Secondary electron image is usually not available in low vacuum mode because SE interact with the water vapors immediately upon emitting from the specimen. Also, conventional E-T detector relies on a high bias (+10 kV) applied upon the scintillator to enable electron-to-photon conversion. Such high bias can easily ionize water vapor and arc the detector to ground. Thus, the use of E-T detector for SE imaging is ruled out, and BSD is generally employed to undertake BS imaging.

5.8 Is there a way to undertake SE imaging in low vacuum mode?
Answer:
Gaseous secondary electron detector (GSED) is used in low vacuum mode. GSED is a positively biased electrode which is mounted on the objective pole piece and can be dismounted after use. The positive bias of up to +600 V is applied on the GSED to attract secondary electrons. Due to gas amplification caused by collisions with water vapor, the current collected by the detector is hundreds or even thousands time greater than the original signal. The detector is also placed closer (a few mm) to the specimen compared to an E-T detector, thus collecting SE efficiently. Positive bias on GSED also drives positively ionized water molecules toward the specimen to effectively neutralize the accumulated negative charge at specimen surface.

5.9 How does degraded vacuum in the chamber (in low vacuum mode) not affect the vacuum in the column?

Answer:

A conical pressure limiting aperture is provided at the center of the detector to sustain low vacuum in the chamber while maintaining a high vacuum in the electron column. The smaller the bore size in the center of the GSED, the higher the pressure that can be maintained in the specimen chamber. For instance, one-half of a millimeter can support a pressure of 1.3 kPa.

5.10 What is the beam skirt and how its effect can be diminished?

Answer:

Electron beam on its way to the specimen gets scattered due to the presence of water vapors in low vacuum mode. The scattered electrons move away from the focused beam and strike the specimen surface at a point away from the probe. This results in broadening of the electron beam which takes on a *skirt-like* form.

This scattering effect can be reduced by employing an extension tube with pressure-limiting aperture mounted at the end. This long tube is fitted to the objective pole piece. Electrons enter this tube after emanating from the objective lens assembly. In this manner, the distance (gas path length) that the electrons have to travel in gas vapor is reduced, resulting in less scatter.

5.11 How is FIB used to sputter and deposit materials? How is imaging performed using FIB?

Answer:

The focused ion beam (FIB) is an instrument that uses positively charged heavy ions (such as gallium instead of electrons) to raster the specimen surface. Ions are heavier than electrons and carry a greater momentum. The use of heavy ions makes it easier to remove material from the specimen. Therefore, FIB is used for sputtering, etching, or micromachining of materials. It is also useful for milling, deposition, and ablation of materials.

Different gases can be injected into the system near the surface of the specimen to deposit required materials on the sample surface.

When the focused ion beam interacts with the surface of the material, it results in the generation of secondary ions, secondary electrons, and neutral atoms. Information from secondary electrons and secondary ions help in the formation of an image in the same manner as that in the SEM. The resolution of the image can be as high as 5 nm.

5.12 What is the difference between the bright field and HAADF contrast in the STEM?

Answer:

The bright field image is formed by collecting electrons that are scattered at small angles and are centered on the optic axis of the microscope while passing through the specimen (e.g., *direct beam*). The *incoherent* dark-field image is formed by (off-axis) electrons scattered at high angles and shows atomic number and mass-thickness contrast. The detector that collects the strongly scattered electrons to form a high STEM image contrast is called high-angle annular dark-field (HAADF) detector.

5.13 Which detector in the SEM can be used for structural analysis?
Answer:
Electron backscatter diffraction (EBSD) detector

5.14 How is EBSD used to obtain structural information?
Answer:
When electron beam strikes its surface, specimen acts as a divergent source of high-energy backscattered electrons which are incident upon sets of parallel lattice planes present within the crystal and are scattered in a manner that satisfies the Bragg's equation. This type of scattering is termed as electron diffraction. For each set of lattice planes for which the above Bragg condition is fulfilled, the diffracted beams emerge out of the specimen in all directions in the form of a cone. Two cones are formed for each set of lattice planes, one at the front and the second at the rear of the lattice plane. These cones intersect the phosphor screen as two dark lines bordering a bright band which represent a family of parallel planes with a specific value of d-spacing. The distance between two lines is inversely proportional to the d-spacing for that specific plane. From this information, the lattice parameter of the specimen material can be determined.

5.15 State the applications of E-beam lithography.
Answer:
Electron beam lithography is the commonly used method for precise patterning in nanotechnology. Generally, it could be used in the nanoelectromechanical system (NEMS), quantum structures, magnetic devices, solid-state physics, biotechnology, and transport mechanisms. It is used in the fabrication of many functional devices and products such as IC fabrication mask, nano-transistors, nano-sensors, and biological applications such as biomolecular motor-powered devices.

5.16 How does EBID work?
Answer:
Electron beam-induced deposition (EBID) is a process of decomposing gaseous molecules by an electron beam leading to deposition of nonvolatile fragments onto a nearby substrate. The electron beam is typically provided by scanning electron microscope (SEM), which brings about high spatial resolution and most likely to generate free-standing, three-dimensional structures. The electron beam

interacts with the material resulting in the emission of secondary electrons which in turn decompose molecular bonds of precursor gaseous materials resulting in deposition.

5.17 What is cathodoluminescence?
Answer:
A particular class of materials can emit light (photons of characteristic wavelengths in ultraviolet, visible, and infrared ranges) when bombarded with an electron beam in the SEM. This phenomenon is known as cathodoluminescence (CL) which occurs when atoms in a material excited by high-energy electrons in the beam return to their ground state thus emitting light.

Chapter 6

6.1 Distinguish between characteristic and continuous x-rays.
Answer:
Primary electron beam ejects inner shell electron of specimen atom creating a vacancy in the orbital and turns the atom into an ion of the excited state. This vacancy is filled when an outer shell electron is transferred to the inner shell, which brings the atom to its ground state with an accompanying release of energy equal to the difference in the binding energy of the two shells. This excessive energy is released in the form of an x-ray photon. Characteristic x-rays have sharply defined energy values and occupy distinct energy positions in the x-ray spectrum which are unique to the element they emanate from. Distinct energy positions of x-ray lines form the basis for microchemical analysis where different elements in a specimen material are identified based on unique orbital transition energy.
Primary electron beam also decelerates (brakes) due to interaction with atomic nuclei of the specimen. The energy loss due to deceleration is emitted as photons of energy ranging from zero to the maximum energy supplied by incident electrons thereby forming a continuous electromagnetic spectrum called *continuum* or *white radiation or bremsstrahlung (braking radiation)*. Continuum is generated due to a combination of all atoms in a specimen and appears as background in an x-ray spectrum. Since it is not unique to a particular element, it is devoid of any unique feature and cannot be used for microchemical analysis.

6.2 What is the notation used for characteristic x-rays?
Answer:
X-ray lines are denoted by the shell from where the electron was originally ejected (i.e., shell of innermost vacancy) such as K, L, M, etc. This is followed by a line

group written as α, β, etc. If the transition of electrons is from L to K shell, transition line is designated as K_α. If the transition is from M to K shell, it is designated as K_β. Since the energy difference between K and M is larger than that between K and L, K_β is of higher energy than K_α. Lastly, a number is written to signify the intensity of the line in descending order such as 1, 2, etc. Therefore, the most intense K line is written as $K_{\alpha 1}$, and the most intense L line is denoted as $L_{\alpha 1}$.

6.3 What is the significance of Moseley's law?
Answer:
In 1913, the English physicist Henry Moseley discovered that when the atomic number changes by one, the energy difference between the shells varies in a regular step. The energy of photon can be given by Moseley's law below:

$$E = A(Z - C)^2$$

where E is the energy of the x-ray line, Z is the atomic number, and A and C are constants with specific values for K, L, M, etc., shells. This forms the basis for identification of elements in materials using x-rays. The above relationship describes energy required to excite any series of transition lines. For instance, x-ray photons of the highest energy in an atom are emitted from K_α shells. This energy equals the binding energy of 1s electron which in turn is proportional to Z^2 as described above. This energy will be different for each element (depending on its atomic number) and can be used to identify it.

6.4 What is the excitation potential?
Answer:
The minimum energy required to eject an electron from an atomic shell is known as excitation potential (E_c). As the size of the atom increases (e.g., from light to heavy elements), the energy required to excite any particular transition line also increases. For instance, E_c for Ni K_α is much higher than that for Al K_α. The excitation potential of K shell is higher than other shells. In addition, the excitation potential of K shell increases extensively with a small increase in the atomic number.

6.5 Define cross-section of inner shell ionization.
Answer:
It is the probability for an incident beam electron to be scattered inelastically by an atom per unit solid angle Ω. It is denoted by σ or Q. The cross section decreases as the primary electron energy E_0 increases. Also, it is lower for elements with the higher atomic number since the critical excitation energy increases with Z.

6.6 What is the x-ray range?
Answer:

X-ray range is the depth of x-ray production within the interaction volume. It mainly depends on the beam energy, critical excitation energy, and the specimen density.

6.7 What is x-ray spatial resolution?
Answer:

X-ray spatial resolution is defined as the maximum width of the interaction volume generated by electrons or x-rays projected up to the specimen surface. The low atomic number and low-density specimens allow deeper and wider electron beam penetration and generation of x-ray lines which degrades the x-ray spatial resolution achieved.

6.8 Why is mass depth used instead of linear depth?
Answer:

The use of the mass depth term ρz is more common than the use of linear depth term z because the mass depth eliminates the need for distinguishing different materials because of their different densities when illustrating the relation with the depth distribution $\varphi(\rho z)$.

6.9 What is the primary mechanism of x-ray absorption?
Answer:

X-rays are primarily absorbed in specimen material by photoelectric absorption. An x-ray photon loses all its energy to an orbital electron, which is ejected with a kinetic energy equal to the difference in photon energy and critical ionization energy required to eject the electron.

6.10 An EDS detector has beryllium (Be) window which is 8-μm in thickness. Calculate how much fluorine K_α radiation can pass through this window onward to be detected by the detector. The mass absorption coefficient for F K_α radiation in Be is 2039 cm²/g [1].
Answer:

$$I = I_0 \exp\left[-\left(\frac{\mu}{\rho}\right)(\rho t)\right]$$

$$\frac{I}{I_0} = \exp\left[-\left(2039\,\frac{\text{cm}^2}{\text{g}}\right)\left(1.848\,\frac{\text{g}}{\text{cm}^3}\right)\left(8 \times 10^{-4}\,\text{cm}\right)\right]$$

$$\frac{I}{I_0} = 0.05 = 5\%$$

Only 5% of the F K_α radiation will pass through the Be window used in this example. It is therefore not practical to employ such a window in an EDS detector to detect fluorine K_α radiation.

6.11 What is secondary fluorescence and what is its significance?
Answer:
When primary electron beam penetrates a specimen, it ionizes atoms to generate characteristic x-ray photons. These photons, while they are out of the specimen, may interact with other specimen atoms to cause secondary ionization, resulting in the generation of additional characteristic x-rays or Auger electrons. The process by which x-rays are emitted as a result of interaction with other x-rays is called *secondary x-ray fluorescence*.

X-ray fluorescence can complicate quantification of elemental concentrations present within specimen material. For example, the K_α x-ray of Cu element has a value of 8.05 keV, and it can be generated by K_α x-ray of Zn that exists in a brass sample. In 70Cu-30Zn alloy, more than expected Cu K_α and less than anticipated Zn K_α x-rays will be generated due to the fluorescence effects. In this way, Cu will be overrepresented, and Zn will be underreported unless corrections are made to the calculations. X-ray fluorescence acquires importance in alloys that have elements with similar Z because it affects the relative amount of characteristic x-rays emanating from compounds.

Chapter 7

7.1 List advantages of an EDS detector.
Answer:
An EDS detector is simple, robust, versatile, and easy to use and does not take up a large amount of space. Its functionality is seamlessly integrated into SEM operation. It undertakes a simultaneous analysis of all elements.

7.2 What is the energy resolution of the EDS detector? How is it measured?
Answer:
The ability to distinguish between peaks in the EDS spectrum is the energy resolution of the EDS detector. It is usually measured at FWHM using MnK_α (5.9 keV) as the reference peak. The energy resolution of modern EDS detector is approx. 122 eV.

7.3 What can be done to improve x-ray collection efficiency for EDS?
Answer:
The detector is placed close to the sample. Detector with a large surface area is used.

7.4 What is the purpose of a window in the EDS detector?
Answer:
The window is used to protect the detector from the SEM environment. It has the adequate mechanical strength to withstand pressure variations inside the SEM chamber during specimen exchange. It also protects the detector crystal surface from visible radiation. It also prevents any contaminants to move into the detector which functions in vacuum at low temperatures.

7.5 Describe shortly the history of the development of the EDS window.
Answer:
Until 1982, the only available window was made of beryllium usually around 8 µm in thickness. This window would absorb x-rays of energy less than 1 keV, thus preventing the detection of light elements such as boron, oxygen, nitrogen, carbon, etc. To enable light element detection, an "ultrathin" window (UTW) made of thin (tens to a few hundred nanometers) organic film Formvar coated with gold was used instead of beryllium. This window is unable to withstand atmospheric pressure, and the detector assembly is kept under vacuum. The window can be removed altogether, and the detector can be used in a "windowless" mode. However, this leaves the detector exposed to contamination. In this situation, if the SEM chamber is vented, hydrocarbon condensation and ice formation will occur on the detector surface. The light will also be transmitted onto the semiconductor surface.

Presently, the ultrathin window of polymer covered with a thin layer of evaporated Al and supported with Si grid at the detector side is used as a standard. Due to grid support, the window is able to withstand the pressure of 1 bar in the SEM chamber. Support structure blocks part of the low-energy radiation thus reducing the detector efficiency to some extent. The grids are therefore designed to have up to 80% area available for x-ray transmission. This type of window can transmit low-energy ($\approx$100 eV) x-rays and is a preferred choice for light element analysis. Evaporated Al coating serves to restrict the passage of light through the polymeric material which otherwise exhibits high optical transparency. Modern EDS detectors routinely detect elements from beryllium to uranium.

7.6 Explain how EDS detector works.
Answer:
X-ray photons striking the detector surface ionizes Si atom through photoelectric effect creating electron-hole pairs. Upon application of a bias voltage between the

thin gold contacts present at opposite ends of the semiconductor, these electrons and holes move in opposite directions toward the collection electrodes creating a charge pulse. The mean energy required to create one electron-hole pair (one electric pulse) in undoped Si is taken as 3.86 eV. The number of charge pulses generated in the detector can be counted, and the x-ray photon energy responsible for this pulse output is calculated by multiplying this number by 3.86. For instance, if the pulse output count is 1642, the x-ray energy that would produce such a number will be $1659 \times 3.86 = 6403$ eV or 6.4 keV. This energy corresponds to K_α x-ray line which is emitted when an electron transitions from L to K shell in the Fe atom. The energy value is fixed for this particular transition, and thus whenever a magnitude of pulse equaling the number of 1659 is measured, Fe is identified as a possible constituent of the specimen under examination. The greater the number of times this particular value of pulse count is generated, the higher is the elemental concentration of Fe in the material.

7.7 What is the dead time?
Answer:
EDS detector's capacity to receive and process x-ray photons is not unlimited. While one x-ray event is received and processed, other simultaneous incoming x-rays are not processed. The duration for which these x-ray signals are not processed is known as "dead time."

7.8 List four overlapping peaks in the EDS spectrum.
Answer:
$SK_{\alpha,\beta}$-MoL_α, CrK_β-FeK_α, MnK_α-CrK_β, and $TaM_{\alpha,\beta}$-$SiK_{\alpha,\beta}$

7.9 How does escape peak form?
Answer:
X-ray photons emanating from specimen enter the detector and ionize Si releasing K-type x-ray photons. If this transition occurs closer to the detector surface, the photons can escape the detector. This will decrease the energy of the x-rays emanating from the specimen by an amount equal to that required for Si K transition event. Due to this event, an escape peak is generated in the x-ray spectrum at energy $E_{(specimen)} - E_{SiK\alpha}$. For instance, if Cu is the specimen material tested, an escape peak at 8.04 $(E_{CuK\alpha}) - 1.74$ $(E_{SiK\alpha}) = 6.3$ keV can form in the x-ray spectrum

7.10 What is Castaing's first approximation?
Answer:
Castaing's first approximation assumes that peak intensities in eds spectrum are generated proportional to the respective concentrations of elements.

$$\frac{C_{i\,\text{(unknown)}}}{C_{i\,\text{(standard)}}} \approx \frac{I_{i\,\text{(unknown)}}}{I_{i\,\text{(standard)}}} = k_i$$

7.11 What is matrix effects?

Answer:

Measured or detected intensity of x-rays is not equal to generated intensity due to absorption or fluorescence of x-rays generated within the specimen. This variation between generated and detected values of x-ray intensity is governed by the composition of the specimen matrix and is known as matrix effects.

Atomic Number Effect

Consider measuring the low concentration of a light element i mixed with a high concentration of a heavy element j in a multielement specimen. A large proportion of beam electrons entering the specimen shall be backscattered by heavy element j and leave the specimen. These electrons shall be unavailable to generate x-rays from within the light element i. In this way, the concentration of i shall be underestimated if its intensity is compared with that originating from a standard with 100% pure element i. This is called the atomic number effect and needs to be corrected.

Absorption Effect

The greater the depth at which x-rays are generated, the greater is the proportion that is lost due to absorption within the specimen. Mass absorption coefficient depends on the composition of specimen analyzed. Generally, correction for mass absorption is the biggest correction made during quantitative microchemical x-ray analysis. Especially, light elements such as C, N, and O are strongly absorbed in heavy matrices and need to be accounted for in calculations. Absorption can be decreased by using the minimum incident beam energy required to generate characteristic x-rays resulting in lesser beam penetration and lower path lengths (t) that x-rays need to traverse to reach specimen surface.

Fluorescence Effect

Characteristic x-rays generated as a result of the interaction between the electron beam and the specimen can be absorbed within the specimen matrix and cause ionization of atoms resulting in the emission of further characteristic x-rays. This fluorescent effect takes place only if the critical excitation energy of absorbing atoms is less than the energy of generated x-rays. This effect will result in an increase in the measured x-ray intensity by the SEM detector since now both the original x-rays as well as the x-rays generated due to fluorescence are measured. Correction required due to fluorescence effect is usually smaller compared to that for atomic number and absorption in ZAF corrections. In some cases, fluorescence can result in erroneous peaks in the x-ray spectrum.

7.12 State the working principle of WDS technique.
Answer:
The sample, crystal, and the detector are positioned on a circle (called Rowland circle). The sample is located at the bottom of the specimen chamber. The crystal and the detector are made to move on the Rowland circle during analysis. During this movement, the distance between the crystal and the specimen is always kept equal to the distance between the crystal and the detector. The characteristic x-rays with specific wavelength emanate from the constituent elements of the sample due to orbital transitions and strike the surface of the crystal that has a fixed d-spacing. During the course of crystal movement, it is probable that x-rays (with specific λ) emanating from a particular element in the specimen and upon the striking the crystal (with fixed d) at an angle θ satisfy Bragg's equation resulting in diffraction. Upon diffraction, the amplitude of the x-rays will increase manifold resulting in an increase in the intensity of the x-rays at the diffraction angle θ. This increase in intensity is measured by the detector and appears as a peak in the x-ray spectrum recorded in the computer. The movement of the crystal is precisely controlled and monitored thereby providing an exact measure of θ at the time of diffraction. The d value is identified as the crystal used is known. From Bragg's equation, λ is then measured. Working backward, element emanating x-ray with this particular wavelength is identified as the latter is specific to that element.

Chapter 8

8.1 What is meant by sampling?
Answer:
Sampling involves identifying and extracting a suitable area in the sample that adequately represents its structure, morphology, and chemistry. Any sample preparation method should not damage the sample or contaminate it. The sample should be transported to the SEM in a box or wrapped up in a dry material. It should be labeled and stored properly.

8.2 Why certain liquids are used during sectioning?
Answer:
Liquids are used for cooling and lubrication during sectioning.

8.3 How is grinding different than polishing?
Answer:
Mechanical grinding is carried out using a series of abrasive materials such as SiC from rough to fine grit size (e.g., 120-, 240- to 400-, and 600-grit papers).

Grinding eliminates material rapidly and reduces the damage caused by the sectioning of the specimen. By the end of grinding process, any cutting marks or scratches would have disappeared from the surface. However, grinding introduces its own thin layer of damage to the surface regions.

The polishing process uses loose abrasive (e.g., diamond paste, Al_2O_3, SiC, c-BN, etc.) between the specimen and hard fabric (e.g., nylon) fixed on a rotating wheel. The rubbing action results in abrasion of the specimen. Polishing is undertaken to impart a flat scratch-free mirrorlike (up to 1 mm) finish to the surface of the specimen. Polishing process with fine grain sizes eliminates only a minor amount of material. After polishing, the sample can be observed in the SEM.

8.4 What is the purpose of impregnation during sample preparation of ceramics and geological materials?

Answer:

The aim of impregnation is to reduce pullouts and ease sample preparation in ceramics. This way the degree of porosity in a sample can be accurately determined. In geological materials, impregnation with low viscous resin is used to provide the necessary mechanical strength to fragile materials in order to withstand preparation procedure. Furthermore, the porous material is filled to avoid entrapment of polishing materials which degas in vacuum later.

8.5 Why is sputter coating process preferred for imaging? Why is C coating suitable for elemental analysis?

Answer:

Sputter coating results in uniformly distributed fine-grained thin coatings. Carbon coating is preferred during chemical analysis since it does not interfere with different elemental peaks in an EDS spectrum.

8.6 What is the difference between dry and wet potting?

Answer:

Dry potting is used for specimens that are dried before preparation. In this way, cracks caused by drying shrinkage no longer remain a concern. Moreover, this method is used when specimens need to be prepared in a short period of time. Wet potting is used to prepare sections after polishing when the material is not dry. In this way, cracking due to drying shrinkage does not occur because the material is still wet. If cracks are observed in the material, it will be due to physical or chemical processes.

8.7 Why are polymers difficult to image? What can be done to overcome these difficulties?
Answer:
Challenges faced during imaging of polymers include beam damage due to radiation sensitivity, charging, beam contamination, and low contrast. Chemical analysis using EDS/WDS is also difficult since polymers are prone to damage at high probe currents.
Contrast can be enhanced with etching, staining, and replication of polymers. The use of conductive tape and conductive coating allows for electrons dissipation and minimizes beam damage. Replication is also used to completely avoid beam damage especially for highly sensitive samples, whereby a replica is scanned by SEM instead of the sensitive sample.

8.8 What is the difference between microtome and ultramicrotome?
Answer:
Microtome can produce samples of 5 µm in thickness, while ultramicrotome can produce thin slices of 200 nm.

8.9 How is a polymer replica prepared?
Answer:
The specimen is placed onto a thick tape, and a mixture of silicone and a curing agent is poured over the polymer. After curing, the mixture is peeled off and observed in the SEM.

8.10 What is staining of polymers?
Answer:
Staining involves the introduction of heavy atoms into the polymeric sample so as to enhance its density and hence the contrast. Staining can be undertaken by physical or chemical means.

8.11 Why is it difficult to examine biological materials in the SEM?
Answer:
Biological samples are difficult to examine due to their nonuniform structure, organic makeup, and their water content. Radiation damage can occur in the sample. Basic steps involved in the preparation of biological materials are fixation, drying, and coating.

8.12 Explain fixation in biological samples.
Answer:
Fixation is used to stabilize and prevent the structure of the biological sample from decomposition in order to preserve its nature during analysis. This process involves several important stages such as prevention of decomposition of a sample by enzymes activities and removal of the water content of a sample through dehydration processes. Common fixation methods have been divided into two major groups including chemical fixation and physical fixation.

Index

A
Abbe's equation, 51, 129, 133
Accelerating voltage, 2, 5, 8, 9, 11, 17, 18, 20, 23, 27, 30, 33, 37, 39, 45, 46, 50–52, 66, 87, 124, 125, 133, 137, 139, 147, 149, 160, 161, 163, 164, 167–169, 172, 177, 181–183, 185, 186, 189, 192, 194, 206, 208–211, 218, 222, 250, 251, 261, 273, 291, 295, 296, 298, 299, 301, 322
Apertures, 2, 15, 17, 18, 22, 24, 26, 29, 31, 37, 39, 41, 42, 45, 47, 48, 50, 51, 63, 65, 86, 117, 129, 133, 135, 138, 139, 142, 143, 154, 157, 159, 160, 167, 172, 176–179, 182, 190–193, 196, 197, 202, 204, 207, 220, 267, 301
Atom model, 86–87, 233–234
Atomic number (Z) contrast, 9, 95, 97–112

B
Backscattered electron detector
    advantages, 69, 70
    channel plate detector, 71
    collection efficiency, 66, 70, 71, 99, 101, 104, 105, 112
    drawbacks, 69, 70
    scintillator BSE detector, 71
    SSD, 66, 70
    working principle, 66–69
Backscattered electron imaging
    applications, 111–112
    COMPO, 108, 111
    limitations, 112
    SHADOW, 108
    TOPO, 108, 111
Backscattered electrons
    energy distribution, 98
    origin, 68, 95, 190
    spatial distribution, 106
    spatial resolution, 110
    yield, 97
Backscattered electron yield
    directional dependence, 101–103
    effect of atomic number, 98–100
    effect of beam energy, 98
    effect of crystal structure, 101
    effect of tilt, 101–102
Beam damage, 8, 149, 168, 171, 175, 208, 273, 289, 299, 338, 346, 349
Beam energy, 22, 27, 34, 46, 51, 64, 66, 71, 77, 88, 90–91, 94, 95, 97–99, 105, 110, 116, 118–120, 122, 133–136, 139, 140, 147, 164, 169, 171, 181, 182, 186, 194, 198, 208, 238, 239, 246–252, 254, 261, 263, 273, 275, 284, 289, 291–293, 298, 299, 322
Boersch effect, 182, 186
Bright field (BF), 207–209

C
Cathodoluminescence, 13, 61, 123, 175, 181, 230
CeB_6 emitter, 30
Charging
    mechanism, 224
    methods to reduce, 166
Cold field emitter
    advantages, 34
    drawbacks, 34
    service lifetime, 33–34
    working principle, 32–33
Compositional contrast, 7, 9, 77, 95, 97, 106, 109, 111, 182
Computer control, 4, 15, 16, 71, 72, 202, 210, 225
Condenser lens
    beam current, 39, 52, 189

Condenser lens (*cont.*)
  demagnification, 37, 39, 40
  spot size, 36, 39, 40, 52, 138, 175, 189
Contamination, 20, 27, 29, 30, 33, 34, 36, 41, 73, 75, 79, 120, 137, 138, 146, 166, 170, 171, 173, 174, 178, 182, 190, 224, 267, 271, 273, 299
Contrast formation
  COMPO, 108, 111
  compositional contrast, 7, 9, 77, 95, 97, 106, 109, 111, 182
  SHADOW, 108
  TOPO, 108, 111
  Z contrast, 99, 100, 106–110, 112

D
Dark field, 206–209
Depth of field, vii, 6, 8, 31, 43, 129, 136, 146, 152–154, 156, 157, 171, 187, 347, 349
Digital imaging
  advantages, 133, 134
  picture element, 78, 82, 83, 134, 143
  pixel, 80–82
Directly heated cathode, 23, 28
Dwell time, 52, 53, 79, 165, 175, 285, 287

E
Edge effect, 124–125, 146, 147, 169–171
Effective probe diameter, 50–52
Elastic scattering, 86, 87, 89–91, 94, 95, 97, 98, 106, 113, 116, 122, 167, 204, 246–248, 250, 253, 256, 258, 291, 335
Electromagnetic interference (EMI), 76, 138, 169, 182
Electromagnetic lens
  design, 37
  magnetic field, 37, 38, 41, 42
  material, 42
  working principle, 38
Electron backscatter diffraction
  Bragg's equation, 211, 212, 303
  brief history, 211
  crystallography, 210, 215, 217
  crystal orientation, 213, 216
  diffraction, 9, 210, 211, 215, 218
  electron backscatter pattern, 211, 215
  Euler map, 216
  experimental set-up, 213–215
  grain map, 213
  grain shape, 9, 210, 213, 215
  inverse pole figure maps, 216

Kikuchi pattern, 210, 213–215
Kossel diffraction, 211
orientation map, 217, 218
phase distribution, 9, 334
selected area channeling pattern, 211
texture, 210, 216, 227
transmission EBSD, 218
transmission Kikuchi diffraction, 218
working principle, 212
Electron beam-induced deposition (EBID), 181, 225
Electron beam lithography
  computer-aided design, 221
  electron beam resist, 222
  imprint lithography, 219
  lift-off, etching, 222, 223
  optical lithography, 219
  pattern design, 222
  raster scanning, 221
  vector scanning, 221
  working principle, 222
Electron column, 4, 15–20, 24, 34, 37, 39, 52, 56, 59, 72, 74, 75, 159, 190, 192, 193, 301
Electron range, 93–94, 98, 105, 116, 136, 140, 250–252
Electron source/gun
  brightness, 17–18
  energy spread, 20
  size, 19
  stability, 20
Energy dispersive x-ray spectroscopy
  absorption effect, 293
  advantages, 272
  atomic number effect, 291–292
  background correction, 275
  Be window, 259, 267
  Castaing's first approximation, 290–291, BNF–290
  dead time, 276
  drawbacks of, 272
  EDS detector, 11, 58, 60, 238, 259, 265–268, 270–272, 276–278
  EDS resolution, 277
  escape peaks, 281–282
  field-assisted transistor, 34
  first principles standards analysis, 298
  fitted standards standardless analysis, 298
  fluorescence effect, 293–294
  full width half maximum, 276
  full width tenth maximum, 277
  internal fluorescence peak, 283
  line scan, 285–288

Index

low-voltage EDS, 299–300
matrix effects, 291–294
minimum detectability limit, BNF–300
multichannel x-ray analyzer, 271
overlapping peaks, 278
peak acquisition, 273
peak broadening, 278–ENF
peak distortion, 278–279
peak identification, 273–274
Phi-Rho-Zi correction, 295
pulse height analysis, 270
qualitative analysis, 265, 273
quantitative analysis, 261, 265, 290, 294, 296, 299, 300
sensitivity, 265, 272, 275, 299, 300
spot analysis, 53, 284
standardless EDS analysis, 296–299
sum peaks, 282–ENF
working principle, 267–272
x-ray map, 284–285
ZAF correction, 295–296, BNF–295
Everhart-Thornley (E-T) detector
advantages, 64
backscattered electron signal, 59, 60, 63
Faraday, 61
Faraday cage, 60, 61, 63, 122
lateral displacement, 126–127
light guide, 60, 61, 71
photomultiplier tube, 61, 85, 123, 228
scintillator, 61, 65
secondary electron signal, 59, 115, 124, 181
solid angle of collection, 62, 64, 66, 83, 106, 183, 208, 272
working principle, 59–62

F
Field emission electron gun
advantages, 31, 32
drawbacks, 31, 32
quantum tunneling, 31
working principle, 30–31
Focused ion beam
detector, 204
electrospray technique, 202, 203
gas field ion source, 203
instrumentation, 204
introduction, 199
ion imaging, 205
ion-solid interaction, 205
ion sources, 203
lens, 204

liquid metal ion source, 203
stage, 204
Taylor cone, 202, 203
volume plasma source, 203

G
Gas path length, 196
Gaseous secondary electron detector (GSED), 192, 193

H
Heater, 28, 71, 74, 75
High-angle annular dark field (HAADF) detector, 207, 209
High-resolution imaging, 31, 42, 45, 64, 75, 135, 140, 147, 160, 163, 170, 174, 175, 321
High-tension (HT) tank, 74, 170
History of the SEM, 12

I
Indirectly heated cathode, 28
Inelastic scattering, 86, 88–90, 94, 95, 106, 113, 114, 116, 122, 167, 204, 246–248, 250, 256, 258, 291, 335
Infrared (IR) camera, 58
Interaction volume
effect of atomic number, 91–92
effect of beam energy, 90–91
effect of tilt, 93
Ion beam-induced deposition (IBID), 224
Ion imaging, 205

L
LaB_6 emitter
advantages, 28–29
drawbacks, 30
flats, 28, 29
material, BNF–29
tip design, 29
Lens aberrations
astigmatism, 48–50
chromatic aberration, 45–47
diffraction at aperture, 47–48
spherical aberration, BNF–45
stigmator, 48, 160
Light microscopes, 1, 2, 6, 11, 51, 144, 310, 350, 352

Limitations of the SEM, vii, 8
Low vacuum
　gas path length, 196
　history, 191
　introduction, 189
　low vacuum detector, 193
　mean free path, 193
　skirt, 194, 195
　working principle, 192
Low voltage imaging
　beam deceleration, 164
　detectors, 186
　E x B filter, 184
　energy-filtering, 183
　energy selective backscatter (EsB) detector, 183
　low angle backscattered electron (LABe) detector, 186
　r-filter, 185
　upper electron detector (UED), 183

M

Maintenance of the SEM, 168, 170, 178
Mean free path (MFP), 72, 114, 193, 261
Microchannel plate (MCP), 204
Monte Carlo simulation, 90–93, 101, 102, 253, 292

N

Numerical aperture, 129

O

Objective lens
　immersion lens, 42
　pinhole lens, 41–ENF
　real aperture, 43, 160
　snorkel lens, 42
　virtual aperture, 43, 160
Oligo Scattering, 194
Operational parameters
　accelerating voltage, 147
　alignment, 159, 160
　beam energy, 147
　objective aperture, 157
　probe current, 151
　specimen tilt, 158
　spot size, 151
　working distance, 154

P

Pauli exclusion principle, 233, 234
Plural scattering, 194
Pressure-limiting aperture (PLA), 190–193, 196, 197, 344

R

Resolution limit
　Buxton criterion, 132
　edge resolution, 132
　Houston criterion, 132
　maximum spatial frequency, 133
　radial intensity distribution, 132
　Rayleigh criterion, 131
　Schuster's criterion, 132
　Sparrow criterion, 132
Richardson law, 22

S

Safety requirements
　emergency, 179
　handling, 179
　radiation, 179
Sample preparation
　amorphous polymers, 334
　bacteria, 353
　cast, 325, 326
　chemical fixation, 349, 350, 352
　cleaning, 311
　coating, 322
　crystalline polymers, 335, 336, 341
　drying, 174, 325
　dye impregnation, 331
　embedding, 312, 326
　epoxy impregnation, 331
　etching, 315
　evaporation, 321
　fixing, 316
　fracturing, 316
　grinding, 312
　handling, 323
　heart tissue, 352
　impregnation, 315
　insects, 354
　lapping, 312
　mounting, 312
　physical fixation, 349, 351
　polishing, 312
　polymers, 173, 337, 339, 345

replica, 325, 326
replication, 345
sampling, 310
sectioning, 310
sputter coating, 320
staining, 345, 351
stem cells, 353
storage, 323
stub, 165, 173, 174, 311, 316, 322, 323, 326
thermoplastics, 333
thermosets, 334
Wood's metal, 331
Scan coils
  magnification, 53–56
  raster, 4, 52, 53
  working distance, 53–55
Scanning electron microscopy (SEM)
  alignment, 177
  image acquisition, 177
  maintenance, 177
  operation, vii, 31, 41, 56, 71, 72, 129, 138, 160, 178
  sample insertion, 174
  sample preparation, 173
  sample size, 173
Schottky field emitter
  advantages, 36
  design, 35, 36
  drawbacks, 36
  material, 35, 37, 219
  operation, 36
Secondary electron imaging
  examples, 122, 130, 163, 210
  factors that affect, 77, 124, 191
  irregular samples, 126
  particles, 126
  topographic contrast, 113–127
Secondary electrons
  energy distribution, 114
  escape depth, 113, 114
  origin, 95, 96
  SE_1, 115–117
  SE_2, 115–117
  SE_3, 115–117
  SE_4, 115–117
  yield, 112, 113
Secondary electron yield
  directional dependence, 122
  effect of atomic number, 120
  effect of beam energy, 116, 118–120
  effect of tilt, 120, 121
Shot noise, 85

Signal-to-noise ratio (SNR), 41, 52, 65, 82–85, 99, 112, 117, 136–140, 150, 151, 154, 165, 207, 208, 270, 336
Spatial resolution
  airy disc, 129
  effect of accelerating voltage, 137
  effect of beam current, 135
  effect of convergence angle, 135
  effect of probe size, 135
  limiting factors, 140
Specimen chamber, 4, 15, 16, 24, 56, 58–61, 63, 65, 69, 71–73, 76, 77, 101, 117, 166, 174, 182, 189, 190, 193, 202, 204, 207, 211, 213, 214, 267, 301, 302, 309, 320, 324
Specimen stage
  cold stage, 58, 346
  high precision stage, 58
  hot stage, 13, 58
Spot size, 29, 36, 39, 40, 51, 52, 82, 110, 131, 134–136, 138, 139, 151, 158, 171, 175, 176, 182, 189, 192, 203
STEM-in-SEM
  advantages, 208
  applications, 209
  drawbacks, 208
  working principle, 208
Strengths of the SEM, 8, 11, 116

T
Thermionic emission electron gun, 20–22, 24
Through-the-lens detector, 65, 66, 138
Topographic contrast, 7, 77, 96, 97, 100, 108, 109, 111, 113–127
Transmission electron microscopes (TEM), 2–4, 6, 11, 12, 37, 80, 198, 201, 206, 208, 344
Tungsten filament
  advantages, 27
  beam current, 24, 26, 29
  bias resistor, 24–25
  drawbacks, 27
  emission current, 19, 22, 24, 25, 27
  false peak, 27, 58
  filament heating current, 24, 26
  material, 22–23
  saturation, 26
  service lifetime, 27
  work function, 20–22, 27
  working principle, 23–24

U

Useful magnification, 2, 5

V

Vacuum system
  diffusion pump, 73, 74
  ion getter pump, 73
  rotary pump, 73, 74, 318
  turbomolecular pump, 73
Vibrations, 23, 34, 36, 71, 138, 169, 170, 182, 205, 219, 334

W

Water chiller, 15, 74, 75, 178
Wavelength dispersive x-ray spectroscopy
  analyzing crystals, 303
  diffraction, 303, 304, 328
  electron probe micro analyzer, 300
  Johann geometry and Johansson geometry, 303
  Rowland circle, 302, 303
  WDS detector, 301–304
Wobbler, 160, 177
Working distances (WD), 6, 41, 42, 44, 52–55, 58, 65, 71, 99, 112, 135–137, 139, 142, 143, 154, 169, 174, 176, 177, 182, 185, 186, 192, 194, 197, 208, 218, 227, 272

X

X-rays
  characteristics, 237
  characteristic x-rays, 4, 78, 86, 237, 238, 241, 242, 244, 250, 251, 258, 261, 263, 265, 267, 273, 282, 284, 293, 298, 303
  continuous x-rays, 236–238, 240, 241, 250, 251, 276, 278, 284, 285, 301
  critical excitation energy, 244–247
  critical ionization energy, 245, 254, 256–259, 262
  cross section of inner-shell ionization, 246–247
  depth distribution profile, 253–254
  Duane-Hunt Limit, 239–241
  excitation potential, 244–246
  fluorescence yield, 236, 237, 246, 298
  Kramer's law, 239
  mass absorption coefficient, 256–262, 293
  Moseley's law, 244
  orbital transition, 241–249
  production, 234–241
  secondary electron fluorescence, 261–263
  short-wavelength limit, 239
  x-ray absorption, 256–261
  x-ray absorption edge energy, 245
  x-ray absorption edges, 245, 258, 259
  x-ray range, 250–252
  x-ray spatial resolution, 251–252